America's New Vaccine Wars

America's New Vaccine Wars

California and the Politics of Mandates

MARK C. NAVIN AND KATIE ATTWELL

OXFORD
UNIVERSITY PRESS

OXFORD
UNIVERSITY PRESS

Oxford University Press is a department of the University of Oxford. It furthers
the University's objective of excellence in research, scholarship, and education
by publishing worldwide. Oxford is a registered trade mark of Oxford University
Press in the UK and certain other countries.

Published in the United States of America by Oxford University Press
198 Madison Avenue, New York, NY 10016, United States of America.

© Oxford University Press 2023

Library of Congress Cataloging-in-Publication Data
Names: Navin, Mark, author. | Attwell, Katie, 1979- author.
Title: America's new vaccine wars : California and the politics of mandates/
Mark C. Navin, Katie Attwell.
Description: New York, NY : Oxford University Press, 2023. |
Includes bibliographical references and index.
Identifiers: LCCN 2023017316 (print) | LCCN 2023017317 (ebook) |
ISBN 9780197613238 (hardback) | ISBN 9780197613252 (epub) |
ISBN 9780197613269
Subjects: MESH: Vaccination Refusal—ethics | Anti-Vaccination Movement—ethics |
Child | Health Policy—legislation & jurisprudence | California | United States
Classification: LCC RA638 (print) | LCC RA638 (ebook) | NLM WS 135 |
DDC 174.2/944—dc23/eng/20230525
LC record available at https://lccn.loc.gov/2023017316
LC ebook record available at https://lccn.loc.gov/2023017317

DOI: 10.1093/med/9780197613238.001.0001

Printed by Sheridan Books, Inc., United States of America

For Debra (MN)
For Mum and Dad (KA)

Contents

Key Dates

Key Dates in California's Vaccination Policy History

Date	Event
1888	California mandates smallpox vaccine for school entry.
1911	California amends its smallpox vaccine mandate to permit conscientious objection.
1921	California repeals its smallpox vaccine mandate and prohibits local schools and health authorities from introducing vaccine requirements.
1960, 1961	California passes and implements a school entry polio mandate, via Assembly Bill 1940, accompanied by a nonmedical exemption (NME) provision.
1967, 1971	California mandates additional vaccines, accompanied by NMEs.
1977	California consolidates its various vaccine mandates into one law, via Senate Bill 942; NMEs are retained.
2010	Dr. Richard Pan is elected to the California Assembly.
2011	Washington state passes Senate Bill 5005, which requires parents to receive clinician counselling prior to acquiring NMEs.
2012	Pan introduces Assembly Bill 2109, modeled on Washington state's Senate Bill 5005; it passes and is signed into law.
2014, January	Assembly Bill 2109 is implemented.
2014, December	Disneyland measles outbreak occurs.
2015, February–June	Pan and co-sponsors introduce Senate Bill 277 to eliminate NMEs. The bill travels through committees of both houses of California's Legislature, accompanied by frequent protests.
2015, June	The California Legislature passes Senate Bill 277; Governor Brown signs it.

Key Dates in California's Vaccination Policy History

Date	Event
2015, October	California activists attempt to place referendum on the ballot to repeal SB277 but fail to gather sufficient signatures.
2016, April	California activists file legal complaint against Senate Bill 277; Court of Appeals rejects complaint in 2018.
2016, July	Senate Bill 277 is implemented; NMEs are abolished.
2017–2019	California media reports that physicians are providing fraudulent medical exemptions.
2018	Rates of medical exemptions in California increase 250% (from 0.2% in 2015–2016 to 0.7% in 2017–2018).
2019, February–September	Pan and co-sponsors introduce Senate Bill 276 to regulate the provision of medical exemptions. The bill travels through committees, passes the California Legislature, and is signed by Governor Newsom; frequent protests occur.
2019, August–September	Protesters throw menstrual blood onto a table in a Senate committee meeting. An activist opposed to Senate Bill 276 attacks Pan in the streets of Sacramento.
2020, January	Senate Bill 276 is implemented.
2020, March	The United States declares COVID-19 to be a national emergency. California is the first U.S. state to follow suit, imposing lockdowns and restrictions.
2020, April–May	Freedom Angels—a group that protested against Senate Bill 276—organizes "Operation Gridlock" at California's State Capitol to protest COVID-related restrictions. Other alliances form between COVID-19 protestors and activists who opposed Senate Bills 277 and 276.
2020, June	Protesters who object to mask mandates and other pandemic control measures harass California health officials at their homes.
2021, February	Protesters shut down a COVID-19 vaccination center at Dodger Stadium, Los Angeles.

Key Dates in California's Vaccination Policy History

Date	Event
2021, August	California mandates COVID-19 vaccines for health care workers, state employees, and school staff; protestors demonstrate.
2021, October	California becomes the first U.S. state to announce a COVID-19 vaccine for school attendance, beginning in the 2022–2023 school year. Parents protest at the Capitol and around the state.
2022, April	Senator Pan retracts a bill to mandate COVID-19 vaccines for school enrollment.

Preface

We have spent recent years trying to understand parents who refuse to vaccinate their children. Along with many other researchers and activists, we have supported efforts to educate and persuade people to vaccinate. We have written many articles about this topic; Mark even wrote a book.[1] Katie convinced her state government in Australia to fund a vaccination promotion campaign that targeted a community of home-birthing and breastfeeding parents.[2] We hoped that parents would choose to protect public health and their own children if governments could only find the right way to overcome vaccine refusal.

We wish that voluntary immunization policies were sufficient to protect high levels of childhood vaccination in the United States. But many policymakers and public health advocates have decided that immunization policies need to become more stringent, even coercive. In particular, they think that parents who refuse vaccines should no longer be able to send their children to daycare or school if parents object to vaccines for religious or "philosophical" reasons. Across the country, Democratic state legislators and health advocate allies, including the American Medical Association and the American Academy of Pediatrics, have recently worked together to eliminate "nonmedical exemptions" (NMEs) to childhood vaccine mandates. Their goal has been to make it more difficult for parents to opt out of vaccinating their children.

This book is about the origins and the outcomes of America's recent efforts to eliminate NMEs from school and daycare vaccine mandates.

We anchor our discussion in California. It was the first U.S. state to eliminate NMEs to school entry vaccination requirements in response to vaccine refusal and, later, to limit the authority of doctors to provide medical exemptions. These efforts have increased California's immunization rates, but they have also ignited polarizing, nationwide debates about parents' rights, democracy, and the authority of the government to use coercion to promote health. Struggles over childhood vaccination policy have informed fights about COVID-19 pandemic control measures and COVID-19 vaccine mandates. We explore the meaning of these battles for parents, doctors, the politics of public health, and the future of bioethics.

We are writing this book three years into the COVID-19 pandemic. For the first time in over a century, most Americans have had recent, direct experience with the *force* of public health laws. State governments ordered much of the country to stay in their homes or to stay away from workplaces and nonessential businesses. Many citizens refused; some protested at public buildings or outside politicians' homes.[3]

Some people have expressed surprise about mass refusals of COVID-19 vaccines or about resistance against pandemic control measures. But we were not surprised to find vaccine refusers at the forefront of COVID-19 protests, even before vaccines against COVID-19 were available. Some of the battle lines of these COVID-19 fights were drawn during earlier and ongoing struggles over school vaccine mandates. California has been the epicenter of that struggle.

Today's fights about vaccine mandates represent a profound failure of democratic politics. These battles often feature simplistic thinking about public health goals and individual rights, and they invoke black-and-white framing about the supposed goodness or badness of mandates. America's contemporary vaccine wars feature both threats and acts of violence against public health officials, school board members, and representatives of government. These vaccine wars have kept children from attending school, destroyed friendships, and polarized communities. War is always a failure, so our use of the term "vaccine wars" indicates our critical view of current conflicts over immunization policy.[4]

Politics should not be war, but war metaphors are unfortunately apt descriptors for America's contemporary vaccine politics. We have come not to defend America's vaccine wars, but to explain them.

Acknowledgments

Many people supported our work on this book over the past 3 years. While we take full responsibility for the text and any errors in it, we recognize and thank many colleagues, friends, and others for their support and assistance.

We are very grateful to Shevaun Drislane for her capable and methodical research assistance. Shevaun not only assisted us with referencing, source-checking, and fact-checking but also provided intellectual guidance from her own public policy expertise.

We appreciate the time, energy, and input of the California experts and activists who agreed to be interviewed, including those who asked not to be named in the book. Some of these individuals—including Leah Russin and Hannah Henry—also read chapter drafts, and they provided many helpful suggestions. Katie appreciates that Dorit Reiss and her family hosted her in California during field-work for this book. Dorit had a significant impact on this book—as a research participant, a fellow researcher, and a facilitator of research.

We are grateful for the work that all the members of Vaccinate California did to promote vaccination in their state. We recognize that they may not always agree with the arguments we make in this book, but we hope that our admiration for them shines through nonetheless.

In the background of our work as researchers are many others who have shaped and supported our contributions to the field. There are too many to mention, but Katie wants to give a special shout out to Saad Omer, who connected us to some of our research participants, as well as to Julie Leask, David T. Smith, and Adam Hannah. Katie also thanks the VaxPolLab and Coronavax teams and her research manager, Sian Tomkinson. Mark especially wishes to thank Mark Largent, Alberto Giubilini, and Bernice Hausman for their early encouragement of this project.

Katie has benefited from the support of student researchers in the University of Western Australia's Bachelor of Philosophy program. She and Mark thank Maddison Ayton for her research assistance in Chapter 7, and acknowledge Breanna Fernandes for her work on recent legislative attempts to restrict or expand mandates during COVID-19. Mark thanks his student research assistants, Ethan Bradley and Ariel Pierce.

Some of our friends and colleagues read earlier drafts of this book and provided comments. We are especially grateful to participants in Mark's Michigan manuscript workshop: Ethan Bradley, Abram Brummett, Russ Faust, Nick Gilpin,

Kate Guzman, Anna Kirkland, Naomi Laventhal, Karen Smith, Kayte Spector-Bagdady, Daniel Thiel, Sean Valles, Abram Wagner, and Jason Wasserman. We also extend our heartfelt thanks to others who read and commented on the manuscript, including Hillel Levin, Heidi Larson, James Colgrove, Brendan Nyhan, Michael Deml, Benjamin Reilly, Doug Diekema, and Denise Lillvis.

Katie acknowledges research support from the Australian Government's Australia Research Council, which funded her Discovery Early Career Researcher Award (DECRA) fellowship under grant number DE190100158. She also acknowledges the ongoing support of past and present management of the University of Western Australia, including Graham Brown, Amanda Davies, Romola Bucks, and Anna Nowak, as well as Research Office staff Peter Elford, Leylani Taylor, and Melina Wood.

Katie and Mark are grateful for the anonymous reviewers who gave many helpful comments on two drafts of the manuscript. We also thank Lucy Randall, Brent Matheny, and everyone else at Oxford University Press who helped this project come to fruition.

Finally, Katie and Mark thank each other's families for hosting them at the beginning and end of the writing period, so that at least some of the work on this book could be completed in person. Katie thanks her family (Ian, Albi, and Chas) for supporting her work and her travel. Mark likewise thanks his family (Deb, Lillian, Phoebe, and Ezra) for their consistent support, and he thanks Deb, Lillian, and Phoebe for reading an early draft and providing comments and suggestions. (Maybe next time, Ezra!)

1

Introduction

Sacramento, April 8, 2015

The air was electric at California's Capitol. At a rally on the building steps, one speaker after another railed against a new bill to regulate parents' vaccination choices. If it passed, parents could no longer skirt California's daycare and school vaccine requirements by claiming religious or philosophical objections to vaccines. In response to attempts to eliminate these nonmedical exemptions (NMEs), Robert F. Kennedy Jr. shouted to the crowd that "parents know best" when it comes to their children's health. Bob Sears, the pediatrician author of bestseller *The Vaccine Book*, called on parents to "get out there and fight for your rights!" Protestors, many of them dressed in red shirts, chanted, "My Child, My Choice." Signs amplified their message: "Force My Veggies, Not Vaccines" and "Protect the Children, Not Big Pharma."

After the rally ended, protesters lined up to enter the Capitol building. They populated most of the queue, but Hannah Henry lined up, too. Holding her toddler in her tattooed arms and wearing a smarter version of her usual hipster clothes, Henry shared a "crunchy granola" aesthetic with many of the protestors. Her blue outfit contrasted with the protestors' sea of red, but despite the different hues of their clothing, Henry shared many of their values. She followed the "attachment" parenting practices championed by Bob Sears—and popularized by his parents, William Sears and Martha Sears—including birth bonding, breastfeeding, baby wearing, and co-sleeping.[1] Also, her children were enrolled in a Waldorf school in Napa, just north of San Francisco, which was part of an international network of Steiner schools that have notoriously low vaccination rates.[2]

Even though the Sacramento protestors were Hannah Henry's "people" in many ways, she was there to support a different community. Like Henry, they were wearing blue, and many sported T-shirts or buttons with the "I ♥ Immunity" logo that Henry had designed. Others were adorned with the blue "Protect Me" stickers that Henry had created for vulnerable people to wear. Henry and her allies were at the Capitol to demand stricter immunization laws.

The Senate Health Committee was convening to discuss Senate Bill 277 (SB277). If it passed, SB277 would dramatically change California's vaccine laws. For decades, parents in California had been able to send their unvaccinated children to school if they objected to vaccines for religious or personal reasons. SB277 would eliminate those NMEs. If it became law, parents would be able to send unvaccinated children to school only if they received a medical exemption, meaning that they would need to convince a physician that vaccines were bad for their children.

The Committee began its work by introducing the text of SB277. It later called invited witnesses to speak. Then came the public comment period.

Several members of the public spoke in favor of SB277. Palo Alto mother Leah Russin argued that vaccination was essential to protect the community, including her infant son. Along with Hannah Henry and some other California parents, Russin had recently co-founded the parent-led advocacy group, Vaccinate California. Dorit Reiss also spoke. Reiss, a law professor at the University of California Hastings, had recently become one of the nation's preeminent vaccination policy experts. She was raising a young family in the Bay Area and was at the Capitol to speak about the importance of immunization for her family and community.

Joining the pro-vaccine parents were representatives from many civil society associations, including the California chapters of the American Medical Association (AMA) and the American Academy of Pediatrics (AAP). Kat DeBurgh spoke for the California's County Health Officers Association. Like other medical experts in California and across the country, California's physicians were concerned that too few children were being vaccinated and they believed that eliminating NMEs would increase immunization rates.[3] Catherine Flores Martin spoke for California's Immunization Coalition. She had been working closely with other civil society organizations and with parent activists to build support among California's legislators for eliminating NMEs.

Advocates of SB277 had persuaded many legislators, and their bill would eventually pass. But the presence of so many protestors at the Capitol illustrated that SB277 was divisive. Public comments were overwhelmingly negative. Of the 497 members of the public who spoke, 438—or 88.1%—opposed the bill.[4] Most of the bill's opponents claimed that their children were medically fragile or had been injured by vaccines. They demanded the right to make health care decisions for their families, and they attacked SB277 as an unethical and illegal violation of informed consent. They claimed to speak on behalf of many other parents and children and to represent a long list of organizations, including Awake California, Million Mamas Movement, Vaccine Injury Awareness League, Canary Party California, California Coalition for Health Choice, Our Kids Our Choice, Educate Advocate, and Californians for Medical Freedom.

Most speakers directed their words toward a smooth-faced man who stood at the podium. A researcher, a physician, a former assembly member, and now a state senator, Dr. Richard Pan was the author of California's SB277. He was admired by public health advocates, but his leadership of the effort to eliminate NMEs made him a target of criticism by vaccine-refusing parents. Pan was there to bear respectful witness to the opposition. "You don't see it all on camera," Kat DeBurgh explained later, "but Dr. Pan stood there and took it all. Every single person, he looked them in the face and took whatever they threw at him."[5]

Two months later, after it passed the Legislature, California's Governor, Jerry Brown, signed SB277 into law. It was a tremendous victory for Senator Pan, for the bill's other legislative sponsors, and for the alliance of other legislators, civil society organizations, and parents who worked to create and pass SB277. Their network had changed California's vaccination landscape, and they hoped the new law would deliver higher coverage rates and stronger protection against deadly diseases for their communities.

In 2015, the future of immunization in America appeared bright. California's success in eliminating opt-outs for vaccine mandates seemed to show that U.S. states could increase immunization rates through small and inexpensive policy changes. The AMA and the AAP soon called for other states to follow California's lead.[6] But SB277 was not a small policy tweak, and it had a potentially large price. This law replaced a policy that had nudged parents to vaccinate with a policy that sought to coerce them. SB277 had good outcomes, including moderate increases in immunization rates and new public attention to the importance of immunization for public health. However, among its bad outcomes was a new kind of political polarization about immunization policy, and perhaps even about vaccines themselves.

It can be tempting to contrast the optimism with which public health advocates greeted California's SB277 with the clear failures of America's rollout of COVID-19 vaccines. (As of April 2023, only 69.4% of Americans older than age 12 years had completed the primary series for COVID-19 vaccines, compared to rates higher than 90% in many other industrialized societies.)[7] However, the status of California's—and America's—immunization governance in 2015 was less rosy than it may have appeared to be. The vaccine opponents who turned out in force to speak against SB277 did not return home quietly. Many did not comply with the new legislation when it was implemented, and some of them violently resisted further attempts to regulate vaccination. Across the country, the California experience seeded further legislative attempts to crack down on vaccine refusal, and in Texas this prompted vaccine opponents to establish the country's first political action committee to protect parents' rights to send unvaccinated children to school. Activists in Ohio, Michigan, Oklahoma, and Oregon did likewise.[8]

It may seem as if most of the current dysfunction in America's immunization social order is an artifact of the COVID-19 pandemic. We have all witnessed dramatic political conflicts about COVID-19 disease control measures. But fissures were already appearing in American's routine childhood immunization programs in the 2010s. California's SB277 was both a symptom and a cause of significant problems with America's immunization social order.

How Did We Get Here?

Before 2015, immunization policy was not a salient political issue for many Americans. Every state had mandated vaccines for school enrollment for decades, and all but two (Mississippi and West Virginia) permitted NMEs. Accordingly, vaccine mandates operated as nudges, prompting parents to vaccinate their children. Parents who chose not to vaccinate for religious or, in some states, personal reasons could usually still send unvaccinated children to school with an NME.

Under these policies, childhood vaccination coverage rates were sufficient to protect most communities against most vaccine-preventable diseases. Indeed, at the beginning of the 2010s, national childhood immunization rates were at or near record highs. For example, in 2010, among children aged 19–35 months, 91.5% had completed one or more doses of measles, mumps, and rubella (MMR) vaccine, 95% had completed three or more doses of diphtheria, tetanus toxoids, and pertussis (DTP/DT/DTaP) vaccine, and 90.4% has completed three or more doses of *Haemophilus influenzae* type b (Hib) vaccine.[9] The Patient Protection and Affordable Care Act (Obamacare) had recently been implemented; its requirement that health insurance companies offer no-cost vaccines provided reason to hope that immunization rates could continue to increase.[10] Other than researchers and technical experts who worked in public health and epidemiology, or clinicians who were experiencing vaccine refusal in their clinics, few people seemed to pay attention to immunization policy.

Nevertheless, there were warning signs about the instability of America's immunization social order.[11] Some parents were refusing to vaccinate their children, whereas others were asking their pediatricians to follow selective or delayed schedules, including one developed by Dr. Bob Sears in *The Vaccine Book*. A 1998 *Lancet* publication by Andrew Wakefield—which drew spurious links between the MMR vaccine and autism—is widely credited with driving reduced uptake of that vaccine.[12] Denise Lillvis and colleagues point to concerns about the use of a mercury-based preservative, thimerosal, as another key driver of parents refusing to vaccinate their children at the beginning of the new millennium.[13] Arizona, Texas, and Arkansas introduced new philosophical exemptions for their school-entry mandates in the early 2000s in response to these new vaccine scares.[14] By the early 2010s, states with easier-to-acquire NMEs had lower immunization rates and more outbreaks of disease compared to states that made it more difficult for parents to opt out of vaccinating.[15]

There seemed to be more vaccine-refusing parents every year. Between 2012 and 2018, the percentage of NMEs to school enrollment vaccine requirements for increased nationally from 1.2% to 2%.[16] In California, the numbers grew from 1.24% to 2.54% in the 10-year period from 2004 to 2014.[17] These numbers appear small, but the rate of increase was high, and communities need vaccination coverage of up to 95% to protect against diseases such as measles. Moreover, parents who eschewed vaccination often congregated in the same residential or social communities, and this "geographic clustering" generated higher risks of local outbreaks.[18]

When outbreaks did occur, people got sick and sometimes died. Yet even as all these problems were building, the mechanisms for governing vaccination in American states seemed stable; policymakers and the public appeared to have little incentive to change.

Then Disneyland happened.

Disneyland is supposed to be "The Happiest Place on Earth," but between December 17 and 20 of 2014, guests to the theme park in Anaheim, California, encountered an insidious and invisible threat. Some developed fevers, sore throats, dry coughs, conjunctivitis, Koplick's spots, and blotchy skin rashes. A few got ear infections, bronchitis, and pneumonia. These fun-seekers contracted measles at Disneyland. When they arrived home, they exposed their loved ones to one of the most contagious diseases on earth.

In 2013, the United States recorded 187 cases of measles. This was somewhat higher than previous years, indicating that America's local elimination of measles (accomplished in 2000) was at risk.[19] In the first quarter of 2014, there were further outbreaks resulting in 288 confirmed cases.[20] However, these pre-Disneyland outbreaks received minimal national news coverage and prompted little public consternation about America's vaccination governance.

In 2015, the United States reported 188 cases of measles,[21] but that year, by contrast, American media were overwhelmed with coverage of the fallout from the Disneyland outbreak.[22] Eventually, only 125 cases of measles would come to be linked to Disneyland,[23] yet this event marked a turning point in popular and political thinking about vaccine policy, especially in California. Disneyland is what policy scholars call a "focusing event" because it prompted a pivot to a radically novel approach to governing vaccine acceptance.[24]

We do not know who started the Disneyland outbreak. Patient zero likely traveled from a country where measles outbreaks are common.[25] But once it started to spread in the United States—mostly among unvaccinated American children—the outbreak became an American problem.

State health officials used social media to track the spread of disease and raise awareness of exposures. News stories warned about the risks of infection, and people began speaking up about their communities' risks of outbreaks. They were worried about vulnerable Californians, including babies too young to be vaccinated, and sick or old people with fragile immune systems. Even healthy people were at risk: In rare cases, vaccines can be ineffective, and the protection they offer decreases over time. More Californians began paying attention to how individual vaccination choices affect the community's protection against outbreaks.

Most Californians had long supported vaccination for themselves and their families, but they had not otherwise advocated for it. For over a decade, they had watched other people invoke "parents' rights" to support vaccine refusal. They had been quiet when vaccine skeptics insultingly referred to pro-vaccine parents as sheep-like followers of a herd instinct.

The Disneyland outbreak mobilized some Californians to fight for their community's protection from disease. Members of this pro-vaccine majority started using new terms to express the moral ideal of a community that cares for its members through vaccination: Rather than "herd immunity," they talked about cultivating "community immunity" as a symbol of love and protection, epitomized by Hannah Henry's "I ♥ Immunity" button. And they started to pursue legislative change.

Why California?

We focus our attention on immunization policy changes in California to illustrate the emergence of national and international trends in immunization governance that precede and will likely extend past the COVID-19 pandemic.

California is America's most populous state, the world's fifth-largest economy, and has been characterized as "America, only sooner."[26] For a sense of California's long and ongoing tradition of cultural and policy innovation, consider that this state witnessed the invention of blue jeans (1873), the motion picture studio (1902), McDonalds (1948), and the internet (1969).[27] It is the home to Hollywood, countless record labels, and many of the world's largest and most influential technology companies, including Apple, Alphabet (Google), Meta (Facebook), Intel, HP, Cisco, and Twitter.

California's state political conflicts have often predicted and reshaped national debates. California's Republican party in the 1960s and 1970s married libertarian small-government activism with White racial resentment, remaking the national Republican Party and propelling Californians, Richard Nixon and Ronald Reagan, to the White House.[28] In the early 1990s, California's politics were dominated by White racial anxiety about the supposed "demographic re-placement" of White Californians by immigrants from Mexico. This xenophobic panic led to draconian laws such as Proposition 187, which prevented undocu-mented immigrants from accessing public services, including education and health care.[29] Nationally, American politics is just now starting to reckon with racist mass paranoia about "replacement theory," a conflict California has faced for decades.[30]

California also has a long history of innovating in environmental and public health. Its 1959 State Air Quality Program and 1969 Porter–Cologne Water Quality Act were the first state efforts to regulate air and water pollution, and these laws were model legislation for later federal bills. In 1995, California was the first U.S. state to ban smoking in restaurants, and in 2007, San Francisco became the first U.S. municipality to ban baby products containing bisphenol A and to regulate the use of phthalates in consumer products.[31] California was the first U.S. state to order a lockdown during the 2020 COVID-19 pandemic, and it was the first state to announce that COVID-19 vaccines would be added to the list of vaccines required for school entry (although it did not follow through on that intention).

We focus on California because it was the first state to eliminate NMEs to school and daycare vaccine mandates in response to vaccine refusal. The passage of SB277 kicked off similar efforts in other U.S. states; NME elimination bills subsequently passed in New York (2019), Maine (2019), and Connecticut (2021). Washington state eliminated its personal belief exemption for MMR vaccinations in 2019 but kept its religious exemption. Other states have attempted or are cur-rently attempting to eliminate or restrict NMEs.[32] California's Immunization Coalition has provided informal advice and strategy to its counterparts in other states.[33] And legislators and pro-mandate activists across the country have learned from California's success. For example, Maine Families for Vaccines formed in 2019 to overturn that state's NMEs, and it seeded a national organiza-tion, the Safe Communities Coalition, now with chapters in seven states.[34]

California was also a leader in immunization policy reforms from an inter-national perspective, as its efforts anticipated similar policy changes in sev-eral other high-income jurisdictions, including Australia, Italy, France, and Germany. These countries all had their own local reasons to revise or implement vaccine mandates, but California's successes played some role. For example, Italian policymakers explicitly invoked California's recent policy changes in their legislative debates about whether to mandate additional vaccines and impose new consequences for vaccine refusal.[35]

Perhaps most significantly, California's passage of SB277 illustrates how America's two major political parties oriented themselves on opposite sides of a new policy debate. Democrats, who dominate the California legislature, embraced state intervention to protect individuals and communities from other people's vaccination choices. In California and other blue states, Democrats led the fight to eliminate NMEs to vaccine mandates. In contrast, Republicans had long sought to protect parents from state encroachment into family decisions, and their party became associated with the defense of NMEs.

COVID-19

We are finishing this book in 2023. The past 3 years have been marked by tremendous uncertainty and conflict about how governments should prevent infection and promote vaccination. In the context of the COVID-19 pandemic, battles over immunization policy have become increasingly salient in many countries, and especially in the United States. It is horrifying to us that so many people have suffered and died from COVID-19 and that political conflict has stymied efforts to protect individuals and our communities.

This is not a book about COVID-19 vaccination conflicts, but we sometimes use examples from the pandemic as lenses through which to view the conflicts this book focuses on. We trace connections between political polarization surrounding COVID-19 pandemic governance and the polarization that emerged in the context of efforts to eliminate NMEs to school vaccine requirements. We also attend to how difficult it can be to enforce coercive public health measures in the context of political polarization, distributed responsibilities, and countervailing incentives. For example, the inability of some schools to enforce mask mandates during the COVID-19 pandemic echoes the widespread failure of many California schools to prevent unvaccinated children from enrolling in classes, even after SB277 was enacted. Finally, we consider how structural failures of the American public health system can explain both some of the disappointing results of COVID-19 vaccination efforts and the unwinding of the "mandates & exemptions" regimes that were the norm in U.S. states from the 1970s to the 2010s.

Even though we draw many connections between the COVID-19 case and the case of childhood school vaccine mandates, we are wary of trying to explain or evaluate aspects of COVID-19 pandemic governance. There are significant differences between the diseases, vaccines, government policies, and political contexts for routine childhood immunizations and COVID-19. COVID-19 is a novel disease, and we know little about its long-term effects, but we have decades of experience with—and mountains of scientific knowledge about— other vaccine-preventable diseases. Similarly, we have ample safety and efficacy data for routine childhood vaccines, whereas COVID-19 vaccines were initially authorized under emergency protocols. COVID-19 vaccines decrease risks of hospitalization and death, but they are not very effective at preventing infection, especially from new variants. In contrast, many routine childhood vaccines offer nearly 100% protection against infection. Accordingly, there is often more room for reasonable disagreement about using or mandating COVID-19 vaccines than there is about using or mandating routine childhood vaccines.[36] Also, COVID-19 vaccines were rolled out to (nearly) the entire U.S. population, beginning with older adults, whereas most routine childhood vaccines are delivered in the first 4 years of life. It follows that ethical and political conflicts about COVID-19 immunization policies may be very different from conflicts about mandates for routine childhood vaccines, in light of how differently we think about the governance of adults as opposed to children.

Finally, conflict about COVID-19 vaccines has taken place against a backdrop of massive disruptions caused by both the COVID-19 pandemic and pandemic control measures. Even before COVID-19 vaccines were developed, America experienced divisive political conflicts about mask mandates, school closures, and stay-at-home orders. We acknowledge that debates about routine childhood immunization policies are sometimes informed by outbreaks of diseases such as measles or mumps—as in the case of the Disneyland measles outbreak—but these outbreaks have not been nearly as disruptive or divisive. Accordingly, conflicts about routine childhood immunization programs cannot entirely explain or predict conflicts about COVID-19 immunization programs, although there are significant similarities between these two conflicts.

Some Reflexivity

Before we outline our thesis, we want to say something about our social identities, the facts and values we take for granted, our motivations for framing this book, the expertise and experience we bring, and how we hope to engage our readers. We want to be as explicit as possible about the "common ground" we presume, with the hope that the book's structural monologue can invite as much dialogue as possible.[37] We acknowledge Olúfémi O. Táíwò's observation that "when we act in social contexts, we treat the information in the common ground as if it were true," and we would like to our readers to know what we think is true.[38] Many kinds of books could and should be written about America's childhood immunization programs. This is the book *we* had to write.

We are White, middle-class, progressive academics in our mid-forties. Although we live on separate continents, we have markedly similar lives. Our class socialization and university job security incline us to treat institutions as tools that can be useful for us and those we care about. We have no qualms about speaking up when those institutions fail to operate in ways we think they should.

Our lives are not perfect; for example, we both experience chronic illnesses. However, we move more-or-less freely through a world that sometimes seems to be made for people like us.

Katie has two children, and Mark has three. At the time of writing, our children are in middle school and high school. Both of us raised our families following "natural" or "attachment" parenting practices, including midwife births (Katie birthed her son at home), extended breastfeeding (Mark's wife was a La Leche League leader), babywearing, and cloth diapering. Our children are fully vaccinated—including against COVID-19—but our other parenting practices give us much in common with some of the more visible groups of America's vaccine refusers.

We also resemble some of the actors who worked to eliminate NMEs in California. Like Leah Russin, we were both galvanized to work on vaccine issues after we confronted vaccine refusal in our communities. We wanted to understand why parents in our neighborhoods, mother-and-baby classes, and playdate groups were not vaccinating their children, and why they were placing our young unvaccinated infants at risk.

The work in this book combines our different disciplinary perspectives and our diverse research interests. Mark is a bioethicist and philosopher whose previous book examines the ethics and rationality of vaccine refusal. Mark has led vaccination social science projects in his home state of Michigan. He also works with patients, families, and health care teams in his consulting work as a hospital clinical ethicist. Katie is an Australian political scientist and public policy scholar whose empirical and conceptual work has focused on vaccine refusal, uptake, and policy and practice responses. She has special expertise in international comparative work on vaccine mandates, including the U.S. context.

In our previous work together, we have analyzed new vaccine mandates across a set of countries, developed a taxonomy for different types of vaccine mandates, considered how mandates mobilize or serve ethical values, and examined empirically how policymakers mobilize values in developing mandates.[39] This book draws on our diverse experiences and different kinds of expertise, but it also pushes us into new areas of inquiry, including the historical and legal dimensions of America's vaccine policies. We have also conducted new interviews with key players in California's efforts to revise and eliminate NME laws.

We take for granted that routine childhood vaccines are very safe and highly effective at protecting against diseases that can cause serious harms. We credit mass immunization programs with the low incidence of those diseases in our communities. Indeed, we regard mass vaccination programs as perhaps the most significant government intervention in population health. We therefore expect our governments to promote the accessibility, affordability, and desirability of vaccines through a range of policy instruments. We focus on *mandates* as one component of mass vaccination programs because they are the primary means of governing vaccine uptake in America, but we acknowledge the importance of noncoercive efforts to promote vaccination.

We believe that governments should protect individual liberties and provide robust opportunities for health, education, and access to privileged social positions. Governments should also regulate inequalities of wealth and income in ways that benefit all people, especially those who are worst off. Although parents should have discretion in raising their children, we think that the state should intervene to promote children's well-being and to prevent abuse and neglect. We believe that social justice—including immunization justice—requires attention to ongoing and historical oppression of people of color, women, immigrants, and the poor. We recognize that social institutions powerfully shape people's beliefs, attitudes, and behaviors, and that adequate explanations of social phenomena require attention to gender, race, class, and other social categories.

Although we embrace vaccination, we recognize that people who refuse vaccines love their children as much as we love ours, and we acknowledge that they have often made sincere efforts to become informed about vaccines.[40] America's vaccine refusers are often doing the best they can—both epistemically and morally—in a country whose major institutions are often untrustworthy and unjust.[41] When we discuss vaccine refusers, we often focus on mothers because they usually make decisions for children's health and because almost all of the leaders of recent anti-mandate movements identify as mothers.[42] Of course, fathers also often refuse vaccines for their children (and, in the case of COVID-19, for themselves).

Our book focuses on institutional explanations for shifts in America's childhood immunization governance. When we zoom in on individual actors, we look mostly to California's pro-vaccine parents, lobbyists, and politicians—rather than to vaccine refusers—because these are the individuals who drove the recent policy changes our book focuses on. Many books and research articles have centered the voices of vaccine refusers or have focused on them as a problem to solve (we offer a list in Chapter 4's notes). Our book instead focuses on the new immunization social order that pro-vaccine activists, lobbyists, and politicians have created.

Many everyday efforts to promote vaccination take place in families, clinics, schools, and other social spaces. Many articles and books have highlighted these contexts of immunization contestation and governance, often with a focus on transforming these encounters to promote vaccine acceptance.[43] Our focus is instead on policy responses by the state and the role of policy actors in bringing them about.

We suspect that many of our readers will, like us, identify with the actors who abolished California's NMEs. The purpose of this book is to contribute to critical conversation among *people like us*, which includes *people like them*. That is, we imagine that our readers support vaccines, believe the government has an obligation to protect public health, and are sympathetic with developments in California's vaccine laws. We hope that our readers will join us in evaluating and reimagining America's immunization social order.

Our Thesis

California's abolition of NMEs to childhood vaccine mandates is a watershed moment in the history of American public health.

The decision to eliminate NMEs in California and other states came at the end of a series of experiments in policy design whereby public officials made exemptions more difficult to access without removing them altogether. But eliminating NMEs is not a mere policy tweak. It is not just one more small step toward making it difficult for families to refuse vaccines. Instead, SB277 and other attempts to abolish NMEs radically reshape immunization governance and the immunization social order. They replace efforts to persuade and nudge people to vaccinate with an ultimatum: Either vaccinate your children or they cannot attend school.

For many decades, West Virginia and Mississippi were the only U.S. states that did not permit NMEs to their vaccine mandates. But those states' policies do not provide a model for what California has done. This is because West Virginia and Mississippi's policies were not designed to respond to vaccine refusal, and their passage was not politically contentious. By contrast, California altered a long-standing policy that had normalized vaccine refusal for nearly 60 years, with the goal of making vaccine refusal socially deviant. And California did so in the context of emerging political polarization about immunization policy. Furthermore, California's elimination of NMEs is not analogous to the creation of smallpox vaccine mandates in the 19th and early 20th centuries. Those early coercive immunization policies responded to different political and epidemiological conditions among populations whose interests and expectations are radically dissimilar from those of people today.

We also argue that eliminating NMEs is more problematic than people may realize. Few American states will be able to eliminate NMEs now that such efforts are tightly associated with the Democratic Party, while efforts to preserve NMEs are now associated with the Republican Party. In addition, in the few American states in which it may be possible to eliminate NMEs, doing so is likely to be less effective than advocates suppose. We offer data from California to show that its elimination of NMEs did not convert many vaccine-refusing parents into vaccinators. Instead, SB277 motivated parents and schools to find other ways to enroll unvaccinated children, or prompted parents to choose the penalty of school exclusion rather than vaccinate their children.

This book cheers the victories of the individuals and institutions that fought for community protection by trying to eliminate NMEs. We laud their good intentions, and we admire their skills and political acumen. But we simultaneously despair of a world in which NME abolition is the only available means by which such actors can try to address America's immunization policy problems. We worry that their efforts may have counterproductive effects and that they will fail to achieve their goals.

Thinking About Policy Change

Vaccine refusal can be surprising if you think that essential public policies have a natural stability or that history moves always toward the creation of better conditions for human flourishing. It can be natural to think that we would soon get vaccine-preventable diseases under control after vaccines were created and distributed, and that once we got vaccine-preventable diseases under control—or even eliminated or eradicated—that they would stay that way.

Likewise, it can be reassuring to assume that political fights can be definitively resolved and that politics lays foundation stones for a stable social life. But that is the wrong metaphor. Public policies—even ones that last decades—are not stone foundations on which we build the houses of our shared civic life. The U.S. Supreme Court's decision in *Dobbs v. Jackson Women's Health Organization* (2022) illustrated the precarity of a half-century of abortion rights.[44] Justice Clarence Thomas' concurring opinion in that case indicates that many other rights—for example, to contraception and same-sex marriage—may be less secure than they appear to be.

Rather than stable stone foundations, perhaps it makes more sense to think of public policies as machines. A successful policy can seem like a well-oiled appliance, whose components include "devices"—that is, people and technologies that perform their part for the policy to work well.[45] If we visualize public policies as machines, then, even if we recognize their fundamental fragility (machines are not stones, they have moving parts that can break down or fail), we might be tempted into thinking that we can keep them running forever. Perhaps all we need to do is pass a new law, or add a new recommendation or regulation. In *Drift into Failure*, Sidney Dekker argues that this is a common strategy: Locate the broken part—the particular area of failure—and try to fix it.[46] In the immunization space, this would involve meeting vaccine refusal with more education, better targeted recommendations, and maybe even harsher requirements.[47]

But Dekker argues that true systems failure happens when background structures no longer support existing strategies. For immunization governance, those background structures include the identities, values, and practices of each generation of Americans. They include what John Rawls called the "basic structure" of society, which includes the political constitution, the legal system, the economy, the family, and formal and informal civic associations.[48] In contemporary America, these background structures are changing (and often failing) in ways that make it impossible to fix immunization policies with simple tweaks. Inasmuch as these institutions explain the breakdown of America's immunization social order, it makes little sense to conceive of vaccination governance as a discrete machine that can be tinkered with to keep it working.[49]

At the crux of our thesis about institutional failure, then, is an image of complex public policies as neither stable foundation stones nor discrete machines that can be repaired. Instead, immunization governance—and the immunization social order it makes possible—is an expression of deep historical, sociological, political, and legal dynamics.

We are also skeptical of simplistic accounts of the kind of policy change that SB277 represents. One insufficient narrative focuses narrowly on the Disneyland measles outbreak as an exogenous cause of California's efforts to eliminate NMEs. We agree that the "punctuated equilibrium model" of institutional change helps explain how some policies can remain relatively stable for a long time and then change in response to dramatic external events, such as wars, financial collapse, or *disease outbreaks*.[50] And it seems likely that the Disneyland outbreak, along with other outbreaks of previously well-controlled vaccine-preventable diseases, contributed to efforts to eliminate NMEs, as we noted above, and as we discuss in Chapters 3 and 4. However, recent changes to California's and other states' routine childhood immunization policies also seem to have powerful *endogenous* causes.

From one point of view, the elimination of NMEs for childhood vaccine mandates may appear to be an instance of what has been called policy "layering," according to which institutions transform slowly as new elements become attached to existing institutions.[51] In a layering process, new developments do not replace or otherwise fundamentally alter the previous policies but, rather, introduce new rules that coexist with old rules. But the elimination of NMEs is more like what has been called institutional "displacement," which occurs when one rule or set of rules replaces another, in ways that often radically reshape the nature and goals of a policy or institution.[52] In particular, SB277's elimination of NMEs transforms California's governance of childhood immunization from a set of nudges to something that seeks to be coercive.

Key Concepts

We use several key terms in the book, and in this section we define them and explain how they fit with the arguments we advance.

A person possesses **immunity** against a disease when their body is able to avoid infection. A person who is **immunized** has had their immunity deliberately cultivated. One form of immunization, called **variolation**, was practiced in India at least 1,000 years ago. This process involved infecting people with a mild form of smallpox to cultivate immunity against more serious strains of that disease. In contrast, **vaccination** history begins with Edward Jenner who, in 1796, immunized James Phipps against smallpox by using pus from cowpox sores.[53] There are now vaccines that can prevent or control 29 infections. We use the terms immunization and vaccination interchangeably.

When the vast majority of a population is vaccinated against a disease, or has otherwise become individually immune (e.g., through recovery from disease), it is highly unlikely that individuals who lack immunity will become infected or that outbreaks will occur. This phenomenon is often called **herd immunity**, as the high immunization level achieved by the population functions as a firewall to protect vulnerable members of "the herd" from infection. We use a newer term to refer to this phenomenon—**community protection**—because mass vaccination safeguards a *community* of persons rather than a herd, and it provides *protection* to vulnerable persons rather than immunity.[54] The California activists and civil society actors we interviewed for this book often used the phrase **community immunity** to try to communicate the pro-social norms of vaccination.[55]

People are **vaccine hesitant** when they have concerns or doubts about the safety or efficacy of vaccines.[56] Many vaccine hesitant people nonetheless vaccinate. **Vaccine refusers** deliberately refuse vaccines.[57] Both of these terms also have more technical meanings—and there is some contestation about them—but we use them in the straightforward way we have listed here.[58] Both hesitancy and refusal are problems of **vaccine acceptance** because they are about the attitudes and actions of individuals.[59] We generally avoid using the term **anti-vaccine** to describe vaccine refusers, although we use it to refer to some organizations and campaigns.[60] Applied to individuals, this term is unnecessarily pejorative and may be inaccurate in ascribing someone's behavior (refusal) to a belief (vaccine opposition). However, our decision to abstain from pejorative descriptions of vaccine refusers does not denote an acceptance of their laudatory alternative self-descriptions—for example, "pro-science" or "health literate."

We can contrast vaccine acceptance problems with **vaccine access** problems, which occur when people remain unvaccinated or undervaccinated because immunization programs fail to adequately reach them.[61] We describe these people as experiencing vaccine access problems or vaccination system issues, rather than being vaccine hesitant. However, complex relationships may exist between vaccine access problems and vaccine acceptance problems, especially for members of disadvantaged communities.[62] People in subaltern positions may doubt or refuse vaccines because of their complex relations with state power or following histories of abuse. In such cases, governments and other social institutions have not cultivated sufficient trust or provided appropriate outreach.

We use the term **government** to include legislative, executive, or judicial branches, but we aspire to specificity where possible. We use government and **the state** interchangeably. **Vaccination governance** or **immunization governance** refer to the efforts that governments undertake to ensure that their populations are vaccinated. Immunization governance often differs dramatically between states or countries. A core component of immunization governance in the American context—and in many other countries—is a **vaccine mandate**. A vaccine mandate is a policy instrument that creates a requirement for individuals to vaccinate, either directly or as a condition of accessing other goods. Mandates vary according to their *scope* (how many vaccines are required), the *sanctions* they impose, how *severe* those sanctions are, and how *selectively* the sanctions are imposed.[63]

This book focuses on **exemptions** to vaccine mandates, which allow people to avoid vaccination without experiencing the sanctions that mandates would otherwise impose. Exemptions can be medical or nonmedical. A **medical exemption** (ME) excludes people from sanctions based on either their heightened vulnerability to vaccine side effects or their inability to cultivate individual immunity from a vaccine.[64] All American states provide MEs.[65] **Nonmedical exemptions** (NMEs) allow unvaccinated persons to avoid sanctions when they do not comply with the mandate for nonmedical reasons, usually their religious or moral objections to vaccination. Public policy sometimes distinguishes between religious NMEs and secular NMEs, and the latter are sometimes called **personal belief exemptions** (PBEs). For example, 29 American states allow only religious NMEs, whereas 15 states allow both PBEs and religious NMEs, meaning that 44 states offer at least one kind of NME.[66] Yet there is equivocation about PBEs in the literature, as some states that provide exemptions for both religious and secular objections call their NME policies "personal belief exemptions." (For example, this is how California's policymakers, experts, and activists spoke about their policies.) To preserve terminological consistency, we use the term NME to refer to exemptions that are offered for either religious or secular reasons.

Immunization governance forms one part of the **immunization social order**. Following Anna Kirkland, we use this term to refer to the "institutions, laws, pharmaceutical biotechnologies, and social practices that work together to produce high levels of vaccine coverage to prevent a wide range of diseases"

(p. 2).[67] So, in addition to vaccine mandates and education and outreach efforts by governments, the immunization social order includes complementary efforts by physicians and other actors, including from civil society. It includes scientists who create vaccines, federal agencies that approve and recommend them, and the clinicians who promote them. The major accomplishment of America's immunization social order has been the widespread freedom from illness that most Americans experience. It is a further accomplishment that this high degree of public health has been achieved through policy instruments and practices that promote voluntary vaccination.

The elimination of NMEs can make immunization policies much more coercive. By coercion, we mean threats that, if not complied with, will leave a person much worse off, so much so that a coerced individual will have little reasonable choice but to comply. There are many theories of coercion,[68] and this book's arguments are consistent with many of them. Whether the chief attribute of a coercive act is the magnitude of its threats[69] or the impact it has on the psychological states of the coerced person,[70] an effective threat to prevent unvaccinated children from enrolling in school can surely be coercive for many families. Importantly, eliminating NMEs will make vaccine mandates coercive *only if* the mandates are enforced. But California's schools sometimes have not enforced vaccine mandates after SB277 was implemented, and we argue that similar difficulties with enforcement are likely to plague immunization governance in other communities that eliminate NMEs.

A voluntary system of immunization governance does not rely on coercion. Accordingly, a system of mandates & exemptions could be considered voluntary if NMEs were easy to obtain. Other voluntary forms of immunization governance eschew mandates entirely, in favor of persuasion. One way to persuade people to vaccinate is to activate their social identities and align them with values that support vaccination. Governments can also orient people toward vaccination in more manipulative but still voluntaristic ways. In particular, the behavioral sciences have identified many ways to change people's behaviors by bypassing their active reasoning processes and instead using environmental cues, affect, or other methods to influence them.[71]

Finally, we name the pieces of legislation we discuss in this book, and we usually refer to those names rather than to bill numbers. For example, most people refer to California's Nonmedical Exemption Bill by its bill number—Senate Bill 277 or SB277—as California's bills do not come with simple descriptive titles such as the "Patient Protection and Affordable Care Act" (itself commonly called "Obamacare"). But in addition to not being descriptive, California's bill numbers are reused in each legislative session, making it difficult to ascertain their chronological order. Accordingly, we have appended our own titles to California's legislation. We call 2012's Assembly Bill 2109 (AB2109) the Clinician Counselling

Bill, 2015's Senate Bill 277 (SB277) the **Nonmedical Exemption Bill**, and 2019's Senate Bill 276 (SB276) the **Medical Exemption Bill**.

Chapter Summaries

This book consists of nine chapters in addition to this one.

Chapter 2 explains the emergence of comprehensive state-based school vaccine mandates across the United States in the 1960s and 1970s. Many of these policies included NMEs from their origin. Accordingly, recent efforts to eliminate NMEs represent a dramatic transformation of a long-standing form of immunization governance. This chapter explains why U.S. state governments focused on schools as the location for mandates, and it highlights the history, law, and ethics of religious and philosophical exemptions.

Chapter 3 explores how state policymakers responded to rising rates of vaccine refusal in the 2000s and early 2010s by employing clever tweaks to their existing "mandates & exemptions" policies. In particular, legislators in some states embraced mandatory immunization education as a new requirement for parents who wanted NMEs. These simple and inexpensive policy changes were backed by social science, but they did not reduce NME rates as much as some policymakers had hoped.

In Chapter 4, we explain how a network of parents and civil society organizations supported elected officials' efforts to eliminate NMEs in California. Parent activists responded to outbreaks in their communities by mobilizing against the permissive attitude that California took toward vaccine refusal. After the Disneyland measles outbreak of 2014, pro-vaccine parents contacted Senator Richard Pan, who had led successful efforts to tighten California's NMEs policy in 2012. Pan encouraged parents to be the public face of his new campaign to eliminate NMEs. Their organization, Vaccinate California, lobbied for Pan's new bill and activated other parents to join their movement. Along with other health and medical civil society organizations, Vaccinate California shepherded the Nonmedical Exemptions Bill through California's legislature.

Chapter 5 examines how decades of NME availability normalized vaccine refusal in California and across the United States. California's Nonmedical Exemptions Bill therefore attacked a way of life that had long been tolerated and even cultivated by laws and social institutions. This chapter frames conflicts over the three major California vaccine mandate bills—Clinician Counselling Bill, Nonmedical Exemptions Bill, and Medical Exemptions Bill—as disagreements about whether to protect or stigmatize ways of life that NMEs had fostered. The chapter concludes by considering how the politics of these three pieces of legislation have informed conflicts around COVID-19 mitigation measures.

In Chapter 6, we criticize appeals to earlier eras of immunization in attempts to justify contemporary efforts to make mandates more coercive. Previous eras' more coercive vaccine mandates were enforced against populations whose ethical and political sensibilities were very different from those of today. In particular, recent decades have witnessed the creation of conscientious objector protections, the institutionalization of restrictions on public health police powers, and the Bioethics Revolution. Furthermore, previous instances of coercive vaccine mandates sometimes caused popular backlashes against immunization governance, as in the case of late-19th-century England. Such backlash remains a risk for today's efforts to introduce greater coercion into vaccine policy.

Chapter 7 addresses the historical and contemporary roles played by physicians and their professional societies in advocating *for* vaccine mandates and *against* investments in public health infrastructure that could have made voluntary immunization more feasible. In particular, we highlight how the AMA and, to a lesser degree, the AAP have been among the most effective independent voices for health-related political advocacy but have also been among the loudest voices in calling for more coercive immunization policies. America's immunization governance has an unfortunate history of seeking to punish rather than provide, and the political activities of physicians have been significant drivers of that dynamic.

Chapter 8 evaluates efforts to eliminate NMEs from ethical and pragmatic points of view. It first examines arguments that critics often raise against coercive daycare or school vaccine mandates: parents' rights, informed consent, and children's rights to care and education. Then it evaluates arguments that proponents offer in defense of such policies: children's right to receive vaccines, the importance of preventing harms associated with disease transmission, the duty to contribute to herd immunity, and the overall social benefits associated with high rates of vaccination. Finally, it expresses skepticism about the potential for today's coercive vaccine mandates to eradicate, eliminate, or even control diseases. Accordingly, today's vaccine mandates need to be justified by their short-term and often more moderate benefits, and the potential harms of such policies need to be given greater comparative weight.

In Chapter 9, we argue that American efforts to eliminate NMEs for school and daycare mandates are likely to produce underwhelming benefits and higher-than-expected costs. In states whose legislatures can pass NME elimination laws, many of their local communities and schools will not fully enforce those laws. More important, only a handful of U.S. states are likely to eliminate NMEs, in light of new political polarization about immunization policy: Democratic state legislators favor NME elimination and Republicans oppose it. Finally, we argue that we cannot blame partisan political fights about vaccine mandates for routine childhood vaccines on the COVID-19 pandemic or on President Trump's response to it. Efforts to tighten and eliminate NMEs—led by Democrats—have driven partisan conflict about immunization policy from the 2010s.

Chapter 10 suggests that America's immunization social order may be breaking down. Decreased public trust prevents vaccine mandates from being consensus public projects, and the decline of liberal democratic government undermines the legitimacy of coercive immunization policies. Even as voluntary vaccination programs are unlikely to achieve sufficiently high vaccination rates to prevent outbreaks, efforts to implement coercive measures may fail because of political polarization. Accordingly, America's public health institutions— and individual Americans—should prepare for a world in which outbreaks of vaccine-preventable diseases are more common. Drawing parallels with the current climate crisis, this chapter identifies mitigation and adaptation strategies for dealing with a resurgence of vaccine-preventable diseases.

2

The Mandates & Exemptions Regime

Introduction

School enrollment vaccine mandates have been the lynchpin of America's governance of immunization since the second half of the 20th century. But one might wonder why we had to mandate vaccines at all, given that vaccines have helped American children to be so much healthier. This chapter locates the origin of late-20th-century school vaccine mandates in underfunded and underwhelming efforts to promote new vaccines against polio and measles in the 1950s and 1960s. Schools became natural places for immunization governance because almost all American children pass through them, state and federal governments possess authority over them, and school-enrollment mandates make it possible to create stable and wide-reaching vaccination programs without creating new institutions or spending additional money.

The school vaccine mandates that American states created in the 1960s and 1970s frequently incorporated nonmedical exemptions (NMEs) from the beginning. These "mandates & exemptions" regimes aimed to orient parents toward vaccination and to create pro-vaccination social norms, but they did not coerce or punish committed vaccine refusers. Furthermore, the widespread availability of NMEs helped state vaccine mandates to be objects of political consensus. Recent efforts to eliminate NMEs therefore represent a dramatic change in the goals and consequences of America's school vaccine mandates.

Epochs of Disease and Prevention

As late as the 1890s, 20% of Americans did not survive childhood, and most children died from diseases we can now prevent.[1] Before the introduction of vaccines, clean water, and sewer systems, people living in the United States were regularly exposed to pneumonia, tuberculosis, diarrhea, and enteritis. Outbreaks of smallpox, yellow fever, and diphtheria were less common but were often deadlier. For example, in 1735, a diphtheria epidemic hit the town of Hampton Falls, New Hampshire, and killed 210 of the 1,200 residents; 95% of the deaths were among children.[2] In wartime, most deaths were from disease rather than from battlefield injuries.[3] Oppressed social groups—including Native Americans, enslaved persons, and poor immigrants—often faced heightened risks of infection, disease, and death, due to their immunological status, their poor working and living conditions, and the few resources they possessed to prevent and treat illness.[4] We can therefore describe most of human history as an "epoch of disease," marked by regular exposure to dangerous and deadly diseases, especially during childhood. But the past 100 years or so have introduced a radically new era for public health: an "epoch of prevention," marked by increasing expectations of a healthy childhood.

The Sanitarian reformers of the late 19th and early 20th centuries brought clean water, sewers, trash collection, and other disease-reducing technologies and institutions to America's cities. Then the Progressive Era reformers of the early to middle 20th century drafted armies of educated health workers to promote the well-being of Americans through school health inspections, community clinics, and visiting nursing services. However, it was vaccination—first for smallpox and then for cholera, rabies, tetanus, typhoid, diphtheria, polio, and many more diseases—that may have done the most to help Americans reconceive of childhood as a presumptively healthy period.

Until 2020, many people living in wealthier societies may have had little to worry about from infectious disease. American parents likely assumed that their children lived in an epoch of prevention. But COVID-19 provided a dramatic reminder that we are unlikely to ever completely escape the epoch of disease. More than 1 million Americans have now died from COVID-19, with millions more enduring possibly lifelong conditions as "long-haulers."[5] Globally, more than 6 million people have perished. The pandemic has further illustrated pervasive health injustices that make some Americans—for example, Black persons and immigrants—disproportionality vulnerable to novel infections. It has reminded us of the importance of vaccines, the personal and social vulnerability we face without them, and the importance of institutional justice to ensure the health of all members of the community.

Unjustified and Underfunded Optimism

In the 19th and early 20th centuries, American cities and states experimented with coercive immunization policies for smallpox vaccines. But by the end of World War II, America's public health officials expected to promote vaccination—including many new vaccines—through voluntary means. They believed that people would choose vaccines for themselves. This faith was part of a broader optimism about science and medicine that Americans possessed in this period and which helped achieve the mid- to late 20th century's epoch of prevention. The number of vaccines exploded in the middle of the 20th century.

These great scientific and medical successes led America's public health experts to expect a rapid end to vaccine-preventable disease outbreaks.[6]

The mandates employed during smallpox outbreaks of the 19th century and the first years of the 20th century had gone out of fashion by the post-World War II period. On this point, historian James Colgrove notes that "through the middle decades of the century, a voluntaristic ethos prevailed with respect to vaccination. When the Salk [polio] vaccine was licensed [in 1955], mandates were felt to be superfluous at best and philosophically objectionable at worst."[7] Even medical societies—which have often led today's calls for vaccine mandates—were skeptical about coercive measures to promote vaccines in the post-war period.[8] The few state-level vaccine mandate bills that passed in the 1950s responded to arguments about saving the government money (a vaccinated child costs less to care for than a child with polio) rather than seeking to save lives.[9]

In this spirit of optimism and voluntarism, U.S. health officials in the 1960s made disease eradication a top public health goal. This new era for vaccination was marked by federal leadership (both in policy development and in funding) and a much longer list of diseases to be vaccinated against.[10]

However, the measles eradication campaigns that began with full force in 1967 stalled shortly thereafter. In 1969 and 1970, there were massive outbreaks across every U.S. state.[11] Outbreaks occurred primarily among Black and Latino communities and in poor neighborhoods, whereas uptake of the vaccine was highest among Whites and suburbanites. Although the rhetoric of the time blamed access issues for low vaccine uptake—Black and Latino families were characterized as "hard to reach"—it seems likely that ongoing racist oppression also contributed to acceptance issues in these populations. Hence, even though the vaccine mandates that emerged from this period were race-neutral on their face, they promised to provide disproportionate benefits to children in Black and Latino communities, in light of those communities' heightened vulnerability to infection. However, it likely would have been better to invest in efforts to improve vaccine uptake, for example, through outreach campaigns and targeted engagement with the concerns of undervaccinated populations. The consistent underfunding of these kinds of public health efforts is an unfortunate theme that runs through this book.

The instability of vaccine funding appears to have contributed to low immunization rates in the late 1960s and early 1970s. The 1962 Vaccination Assistance Act (VAA), signed by President Kennedy, had provided funding and support for four vaccines: diphtheria, polio, tetanus, and pertussis. The VAA also gave the Communicable Disease Center (currently called the Centers for Disease Control and Prevention [CDC]) a leadership role in national immunization governance. This law provided funding for immunization education and persuasion efforts, doing for a larger set of vaccines what March of Dimes had previously done for polio vaccine.[12] But President Nixon later refused to reauthorize the VAA; federal funding for vaccinations evaporated. We do not have reliable or complete data on vaccination uptake rates across the United States before the 1970s. Accordingly, we cannot be sure how much immunization rates declined after the end of VAA funding, although Richard Altenbaugh and Elena Conis have argued that they dropped.[13] Regardless, the retreat of the federal government from funding immunization programs likely left the campaign for the new measles vaccine with less support than was required for success.[14] In 1977, President Carter responded to underwhelming uptake of the measles vaccine by signing the Childhood Immunization Initiative. However, by the time this law went into effect, most states had already introduced new school vaccine mandates.

Turning to School Mandates

When the United States started ramping up its school-based vaccine mandates in the 1960s and 1970s, it drew in part on a patchwork of prior policies, many of which were rarely enforced or applied only during outbreaks. Some preexisting mandates harked back to much earlier periods. For example, in the 19th century, Boston had been the first U.S. city to require proof of smallpox immunization for enrollment in school.[15] Other cities had followed, and by the mid- to late 19th century, some state governments had imposed similar requirements. These include Massachusetts in 1855, New York in 1862, and Connecticut in 1872. But few of these mandates survived to the middle of the 20th century—having expired or been repealed—and even fewer were actively enforced. As a result, the mid-20th-century push for state vaccine mandates represented a novel intervention in the context of faltering campaigns to promote the uptake of new vaccines.

The failure of the 1960s measles eradication campaign—and inconsistent funding for vaccination from the federal government—sparked renewed interest in using schools to govern vaccination.[16] A major driver was the Joseph P. Kennedy Foundation, which focused on preventing "mental retardation," one of the main causes of which was measles-caused encephalitis. The Kennedy Foundation lobbied for states to impose vaccine mandates, including a letter-writing campaign from prominent members of the Kennedy family, which, at the time, played the role of American royalty.[17] Subsequent federal anti-measles campaigns included advocacy for states to pass laws requiring vaccines for school enrollment. The idea was that parents were "basically willing" to vaccinate and that mandates would help push them over the line.[18]

Efforts to promote measles vaccination during the 1960s and 1970s paid little attention to the complex reasons why people might not vaccinate their children. Advocates of new mandates were especially disinterested in those who might deliberately refuse. At the time, vaccine rejection was thought to be a small problem, easily overcome by better institutions. In the early 20th century, it had been common for public health leaders to attribute a person's unvaccinated status to their supposed ignorance and deliberate refusal. But by the middle of the 20th century, it was more common for vaccine advocates to blame nonvaccination on failures of the public health system and to relabel the unvaccinated as "hard to reach."[19] The idea was that underfunded public health systems left many people either unable to access vaccines or else difficult to target with education and persuasion efforts that could motivate them.[20] Hence, the modern school vaccine mandates of the 1960s and 1970s were motivated by a recognition that the public health system had failed to provide vaccines or to persuade people to accept them, rather than by a desire to coerce or punish noncompliant parents. Governments realized that in the absence of consistent funding to support vaccine advocacy work, voluntary immunization policies would be unlikely to sustain high immunization rates or to ensure sufficient uptake of all the new vaccines that were becoming available. They needed an inexpensive way to promote vaccine uptake in light of institutional and economic limits.

During outbreaks, widespread fear of infection often cultivates support for emergency vaccination measures. For example, America's smallpox vaccination campaigns and mandates were often activated only during outbreaks. But, after outbreaks receded from the public imagination, communities were often less willing to fund vaccination campaigns or to tolerate mandates. As another example, consider how willing the U.S. federal government was to purchase and distribute COVID-19 vaccines at the height of the pandemic but how quickly federal funding became contentious once infection rates declined.[21] In early 2022, the U.S. Congress slashed funding for COVID-19 testing, treatment, and vaccines from the federal budget, even though those cuts would undermine America's ability to respond to the ongoing pandemic.[22]

The historical instability of funding and support for immunization programs creates a problem for immunization governance. How can communities maintain high levels of vaccination if funding for programs declines as emergencies abate? How can public health leaders cultivate stable pro-vaccination social norms in communities that revert to indifference after outbreaks have diminished? School-entry vaccine mandates present a potentially attractive solution to these problems.

Why Schools?

Schools are vulnerable to outbreaks because they gather children and adults together in enclosed rooms for extended periods. Accordingly, one reason to focus on school vaccine mandates is because they protect *people in schools*. School vaccine mandates also help maintain the functioning of schools, which is important for the education of students and for the ability of parents to participate in the formal workforce. The COVID-19 pandemic demonstrated the necessity of school and daycare for working parents.

Perhaps more important, school vaccine mandates can effectively govern immunization for the *broader community*. Nearly everyone attends school when they are children, so school mandates offer an opportunity to govern the immunization status of everyone in society, given enough time. Furthermore, state governments have constitutional authority to govern their communities' schools, such that it is legally possible to use school mandates for immunization governance.

Schools are also a central location where American governments have tried to solve social problems. For example, activists and policymakers in the Progressive Era focused on schools as a way to improve children's overall well-being, particularly their physical health. This included school medical exams and hearing and vision tests.[23] School meal programs have aimed to address youth hunger. School desegregation and integration aimed to address racial inequalities. And, even as America retreated from its commitment to the welfare state—for example, with 1996's Personal Responsibility and Work Opportunity Reconciliation Act—American politicians of both parties heralded the potential for education to generate economic security and class mobility.

Because schools already did so much to govern the lives of children and to influence their broader communities, school vaccine mandates appeared to be an especially efficient way to promote immunization. Instead of creating and funding new institutions to deliver or promote vaccines, states across the country needed only to add a new requirement to existing school enrollment processes. Walter Orenstein and Alan Hinman, former Directors of the National Immunization Program (CDC), explain this pragmatic case for school vaccine mandates: "School laws establish a system for immunization, a system that works year in and year out, regardless of political interest, media coverage, changing budget situations, and the absence of vaccine-preventable disease outbreaks to spur interest" (p. S23).[24] Orenstein and Hinman's quote illustrates why large public institutions make good governance machines. Schools already initiate parents and children into new patterns of state-supervised behaviors. Every year, parents fill out forms, administrators review rosters, and teachers keep roll. Vaccine mandates become another part of the "back to school" agenda.

Finally, running the community's immunization governance through schools can help avoid some of the controversy that might arise if the state tried to mandate vaccines through other means, such as in the workplace. There is a long American tradition of conducting public health policy on "captive" populations, including children, to avoid political conflict. Indeed, America's public health infrastructure came of age using coercive measures against subordinate populations in imperial and domestic contexts, as we discuss in Chapter 6. Yet, U.S. public health officials have experienced stiff resistance whenever they tried to directly govern the White working or middle classes. On this point, the historian James Colgrove argues that immunization advocates in the post-World War I period began to focus more on children because they, like Black and immigrant communities, were a population over which the state already had institutionalized coercive control.[25] Indeed, children and schools had often been governed by other kinds of politicomedical power, including the Polio Pioneer trials for the Salk vaccine.[26] Accordingly, if the government wanted to govern White people's vaccination choices, then it needed to focus on a population of White people over whom it could claim institutional control: children.

Mandates and the Expansion of State and National Power over Schools

The introduction of modern state-level school mandates—and the national pressure to develop and enforce them—illustrated the increased role that state and national governments were playing in American primary and secondary education by the 1960s and 1970s. For most of American history up to that point, schools had been under local control or under the control of private associations. In 1835's *Democracy in America*, Alexis de Tocqueville celebrated the private associations and civic institutions of America's diverse local communities. He was inspired by their tendency to cultivate democratic ideals and to resist the totalitarian tendencies of larger governments. He especially commended local communities for maintaining independent schools and for resisting central political control. By comparison, de Tocqueville took a dim view of what was happening on his own continent. Forces in Europe were pushing for "the prerogatives of the central power," which would keep individual citizens "more tenuous, more subordinate, and more precarious" in their relationships to the national state.[27] From de Tocqueville's point of view, the local control of U.S. schools (among other associations) was a key expression of this new country's commitment to democracy.

American local communities still exercise substantial authority over schools, but state and national governments have claimed governance of key components of education. In the years since de Tocqueville's American tour, most U.S. states radically expanded their oversight of public schools, establishing credentials for teachers, developing standard curricula, and seizing control through funding. Even before de Tocqueville's visit, some U.S. states created state systems of education. For example, in 1827, Massachusetts became the first state to offer tuition-free high school to all students throughout the state.[28] The decades after de Tocqueville left America saw a dramatic expansion of state involvement in education. By the late 1800s, almost every state provided free primary and secondary schooling.[29] The Progressive Era witnessed increased professionalization, standardization, and centralization.[30]

Individual state governments expanded their influence over local schools through their authority to directly govern education. Some also centralized school funding, which further cemented their control. Accordingly, the tension between state and local power over schools has been about state authority versus local operations, and about the power of the purse. In contrast, the federal government's involvement in American education policy has occurred mostly through the Spending Clause of the Constitution—which authorizes Congress to "provide for . . . the general welfare of the United States"—along with "conditional grants," which require states to comply with federal dictates if they wish to receive grant monies.[31] The federal government has also interjected itself into school governance through its role in guaranteeing the protection of constitutional rights, including equality, liberty, and due process.

The federal government's entrance into the governance of primary and secondary schools occurred mostly in the second half of the 20th century. It primarily manifested in efforts to combat discrimination and poverty, in light of 1954's *Brown v. Board of Education*, the Civil Rights Act of 1964, the Elementary and Secondary Act of 1965, the Education Amendments of 1972, the Rehabilitation Act of 1973, and the Education for All Handicapped Children Act of 1975.[32] Recent decades have seen a further expansion of federal funding of (and, thereby, indirect control over) schools throughout the country. Thus, the local democratic control over schools that de Tocqueville so admired has been mitigated by greater state and federal involvement seeking to better meet students' needs and to protect them from local forms of discrimination and disadvantage.

Following broader patterns of school governance, school vaccine mandates were largely a matter for local governments in the 19th and early 20th centuries, but then state governments took them over between the 1960s and 1970s. The federal government intervened indirectly to encourage states to ensure that schoolchildren in their communities were vaccinated. In the 1970s, Congress provided states with grants to keep better immunization records (including for students), and it promoted school immunization requirements. The CDC pushed for states to adopt school entry mandates, creating and distributing sample legislation. The CDC director also contacted each state's governor individually to press the point.[33]

In 1968, only half of U.S. states required at least one vaccine for school entry, but by 1974, 40 states had vaccine mandates, and by 1981 every state did.[34] Over the following decades, states expanded their school immunization requirements beyond initial enrollment (usually first grade or kindergarten) to include Head Start early childhood education, health and nutrition programs, preschools, childcare facilities, and secondary schools. While the initial impetus for vaccine mandates in this period was to increase uptake of measles vaccine, the new laws often mandated many other vaccines.[35]

Local schools frequently resisted attempts to compel them to require vaccines for school entry, and they sometimes refused to enforce the new vaccine mandates. State health departments often had to pressure schools—sometimes threatening lawsuits against recalcitrant school officials—before schools excluded unvaccinated children.[36] When these laws were finally enforced, they often had dramatic effects. For example, in 1977, schools in Los Angeles County finally excluded children who were not vaccinated against measles, turning away 23,000 in one day.[37] California was taking a leading role in (re)establishing vaccine mandates; other major U.S. cities followed Los Angeles' lead, leading to further mass exclusions.[38]

America's public health leaders turned to school-based vaccine mandates because school mandates were effective and politically feasible. But we should not understate the social significance of using schools to enforce immunization policies. Schools are potentially some of the most powerful and radical political institutions that states have at their disposal. Populations do not always accept government intrusion into their local schools, which have been sites of dramatic political and legal conflict in American life. Indeed, a striking number of prominent U.S. Supreme Court cases—addressing *general* issues of constitutional law—have been fought on school turf. These include cases about free speech, religious liberty, religious establishment, due process, and privacy, among others.

We have been fortunate that even though school-based vaccine mandates have sometimes been politically contentious, they have generally enjoyed bipartisan support among state and national political leaders. But as that consensus has eroded, we should not be surprised to witness greater conflict surrounding school vaccine mandates. Indeed, we should expect efforts to eliminate NMEs to heighten political conflict around vaccine mandates, given that a primary goal of NMEs is to make mandates less contentious.

Nonmedical Exemptions

In the 1960s, the main advocates of school mandates did not envision these policies to be coercive, but instead thought mandates should prompt parents to seek out vaccinations for their children. For example, CDC leader Alan Hinman wrote in 1979 that "some additional stimulus is often needed to provoke action on the part of a basically interested person who has many other concerns competing for attention."[39] Similarly, Colgrove notes an apparent irony in the Great Society's efforts to empower and mobilize communities for *self-improvement* while also *mandating* vaccinations. But Colgrove suggests that this apparent contradiction dissolves once we realize that mandates were intended as activation devices. That is, mandates were supposed to be what we now call "nudges," not forceful interventions to compel vaccination.[40]

Advocates for school immunization mandates in the 1960s and 1970s were well aware of ongoing battles over the racial integration of schools and the political power of private school advocates. Suburban White parents from across the country resisted the racial integration of their children's public schools.[41] From the 1970s to early 1980s, the most salient political issue for America's political right—especially in the South—was resistance to public school integration and the right of private schools to racially discriminate.[42] Accordingly, advocates of school vaccine mandates were operating against the background of fierce political battles about America's schools, and they could not afford to awaken passionate groups to oppose their policies.

Contemporary discussions about vaccine mandates sometimes treat exemptions as corruptions of originally pure policies, as if exemptions were destructive afterthoughts. "In the good old days," such storytelling may go, "everyone had to vaccinate." "But then," so goes the narrative, "exemptions undermined the whole policy."[43] It follows that the current wave of states rolling back or eliminating their NMEs aims to restore a "golden age" of good vaccination governance. But there was no such "golden age." America's modern system of school-based immunization governance came preloaded with NMEs,[44] and advocates of these policies were usually explicit about their commitment to voluntary vaccination.[45]

Traditional and Modern Religious Exemptions

Modern vaccine mandates emerged during a period when American government was changing how it thought about religious liberty and, in particular, the rights of religious conscience.[46] The traditional view of religious exemptions—operative from the 18th century to the middle of the 20th century—was that they resolved a tension between church and state.[47] On this view, religious exemptions were not about protecting *private* conscience but were instead about trying to solve a pressing problem of *political* life in the post-Reformation West: how to achieve stable political union among people from different faiths. For example, Mennonites will not take oaths, Quakers will not fight in wars, and Catholic priests will not testify about what they learn in the confessional. How could these potential lawbreakers remain citizens in good standing? From its early days, the United States often answered that question by exempting members of individual religious communities from laws they found to be objectionable, based on explicit prohibitions from their churches.[48]

On this traditional account, it is a person's membership in communities of conflicting authority that grounds a claim to a religious exemption, rather than the inconsistency of their private beliefs with their political obligations. State representatives can therefore assess whether people qualify in a straightforward way (e.g., "Are you Quaker?"). They can design exemption policies that are easy to enforce (e.g., only exempt Quakers) and for which the number of exempted people is easy to predict (e.g. the number of Quakers will equal the number of people who receive religious exemptions).[49]

To be clear, what we are calling the "traditional" account of religious exemptions is an idealized reconstruction of the logic behind 18th- and 19th-century policies. In some cases, U.S. states created religious exemptions only for one or a handful of minority denominations and not for others. Indeed, some states went out of their way to create new laws that, although general on their face, clearly aimed to persecute religious minorities. We should not understate the role of religious bigotry in the creation of the 18th- and 19th-century religious exemptions laws, even as we reconstruct a logic behind these laws that prioritizes the *institutional* church's requirements, rather than the value of *individual* conscience.

The traditional conception of religious exemptions faces substantial problems. From a practical point of view, it is often not obvious what a particular religion requires of its members. Churches sometimes provide only general moral guidance, and the principles of moral theology can appear to pull in different directions. The possibility of this sort of intradenominational ambiguity and disagreement led the U.S. Supreme Court—in *Thomas v. Review Board* 405 U.S. 707 (1981)—to focus on *private* religious convictions, rather than denominational membership, when determining whether someone's objections qualified as religious. Furthermore, if the government were to recognize the institutional teaching authority of a church as having higher status than an individual's religious conscience, then the government would thereby "establish" some religions while "disestablishing" others, which the Constitution's 1st Amendment prohibits. Also, Americans' religious beliefs and practices have become less tied to institutional churches than they were in the past, such that institution-based conceptions of religious belief increasingly fail to capture the reality of religion in America.[50]

There is ongoing debate in the academic literature and in the courts about whether *secular conscience* also qualifies as a religious belief from the point of view of the U.S. Constitution. For example, Micah Schwartzman argues that "religion is not special," but because the U.S. constitution insists that religion must be special, the way to marry the demands of ethics with the constraints of law is to embrace an expansive conception of "religion."[51] The Supreme Court had an opportunity to do just this in *Seeger v. United States* 380 U.S. 163 (1965) and *Welsh v. United States* 398 US 333 (1970), which addressed whether wartime conscientious objectors who objected to war for secular reasons were covered by protections for religious conscientious objectors. However, the Supreme Court chose to treat these as cases matters of statutory interpretation, and it avoided addressing the constitutional questions about the relationship between secular and religious conscience.[52]

Although the Supreme Court has not embraced an expansive definition of religion in its interpretation of the 1st Amendment, the Congress and many states have embraced expansive conceptions of conscience in various conscientious objection policies.[53] This includes religious exemption policies for vaccine mandates. Twenty-nine states offer only religious exemptions to vaccine mandates, but many of those states will offer "religious" exemptions to persons who object to vaccines for secular reasons. For example, Douglas Diekema observes that Oregon's religious exemptions law defines "religion" as "any system of beliefs, practices, or ethical values."[54]

We can now see how contemporary U.S. jurisprudence about religion can complicate exemptions policies: If any deeply held value can underpin a "religious" objection to a law, then attempts to determine whether someone objects to a law for a religious reason must focus on the role that the reason plays in the life of the objector, rather than on its substantively "religious" content. (For example: Is the objection sincere? Is it based on a deeply held belief? Is the objection consistent with other choices the objector has made?) It can be difficult and expensive for the government to evaluate the sincerity, importance, or consistency of a person's private beliefs.[55] Furthermore, in light of the wide array of values and commitments that can give rise to vaccine refusal, it can be very difficult to predict which people—or how many people—will object to vaccines before they are mandated.[56]

Nonmedical Exemptions to Modern Vaccine Mandates

State and federal courts have so far not required religious exemptions to vaccine mandates. Even in the context of today's heated debates about religious liberty, the Supreme Court has consistently held that general laws need not exempt people who have religious objections. For example, on December 13, 2021, the Court turned down two requests by petitioners who claimed religious objections to COVID-19 vaccine mandates.[57] Accordingly, U.S. states offer NMEs not because the Constitution requires it but, rather, for pragmatic or other principled reasons.

One reason why late 20th-century school vaccine mandates often came preloaded with NMEs was because religious exemptions were sometimes offered for other coercive state policies. In particular, the U.S. government had strategically used conscientious objector rights for military conscription to dampen popular resistance to the Vietnam War, a conflict in which 2.2 million American youth were conscripted to fight. Anyone who had a deep moral conviction against serving in combat was allowed to avoid doing so, although they were sometimes assigned alternative service. Such policies were largely supported by military leaders and war advocates because peaceniks often make bad soldiers and they undermine morale in the ranks.[58] Allowing conscientious objectors to avoid combat service also undercut public resistance to the war. If objectors would not have to fight in the war, they might be less likely to mobilize politically against it. Indeed, this is one reason why military leaders—and warmongering civilian politicians—are often committed to an "all-volunteer" military. It makes it politically easier to fight wars when no one is being "forced" to participate.[59] American states likely had similar hopes for NMEs to vaccine mandates. By leaving refusers free to opt out, they could avoid fueling resistance to the new mandates and to vaccination more generally. Policymakers likely also anticipated that very few people would request NMEs and that overall compliance would remain high.

Even though almost all states included NMEs in their modern school vaccine mandates, those NMEs have not been static artifacts. State-based contestation over NMEs has been common for at least the last couple of decades.[60] In particular, opponents of vaccine mandates have often sought to widen the scope of exemptions or to make them easier to receive; they have sometimes succeeded.[61]

Attention has frequently focused on two states, West Virginia and Mississippi, neither of which has had NMEs in recent decades. West Virginia is the only state whose vaccination mandate has never included NMEs. In contrast, Mississippi had a religious exemption from 1960 until 1979, but it was struck down by the state's Supreme Court after a parent objected that his religious denomination was not covered by the exemption law. The Mississippi Supreme Court did not accede to this parent's request to expand its NMEs to cover additional religions or secular objections. Instead, it eliminated all NMEs. As Colgrove and Lowin note, "The court's argument was not that the [NME] law preferred one religion over another . . . but that it elevated one person's beliefs above the general public welfare."[62]

Colgrove and Lowin suggest that there are three reasons why West Virginia and Mississippi have not implemented NMEs. First, these states likely had few committed vaccine refusers. West Virginia and Mississippi are two of the poorest and least educated states, and vaccine refusers tend to have higher incomes and greater levels of educational attainment.[63] Another factor is that both states have strong alliances between health officials, legislators, and professional medical organizations. In particular, Colgrove and Lowin note that chairs of legislative health committees in those states have worked with public health leaders to prevent NME bills from advancing past their committees. Finally, given that both West Virginia and Mississippi have relatively poor outcomes in many other areas of population health, Colgrove and Lowin suggest that health officials have rallied around their states' high immunization coverage rates as "a rare positive achievement that is worth maintaining."[64]

It is noteworthy that advocates for eliminating California's NMEs did not look to West Virginia and Mississippi for guidance. Catherine Flores Martin of California's Immunization Coalition noted that "[p]eople didn't want to use West Virginia and Mississippi as a model, because they viewed them as lesser states, maybe." She continued in a tone of mock snobbery, "We don't want to be compared to Mississippi, they're the worst in education, they're the worst in healthcare." Flores-Martin attributed this sentiment to a California arrogance.[65]

The outliers of Mississippi and West Virginia notwithstanding, the ubiquity and spread of NMEs across American states has long constituted "the great compromise" of U.S. immunization policy. This mandates & exemptions compromise would remain stable from the 1980s into the first years of the new millennium.[66]

Nonmedical Exemptions in California

California led the country in many aspects of its mandatory vaccination policies, including its use of exemptions. For example, in 1911, it amended its 1888 smallpox vaccine mandate to permit conscientious objection.[67] This made California one of the few states to attach NMEs to their 19th-century mandates.[68] Formal protections for people who conscientiously objected to vaccines were very new; they were first implemented in England only in 1898 (see Chapter 6). And California was eager to dispense with vaccine mandates altogether when risks of smallpox outbreaks diminished. In 1921, it repealed all mandates and prohibited local schools and health authorities from adopting rules or regulations "on the subject of vaccination."[69] Jonathon Kuo and Elena Conis report that "over the next thirty years, as smallpox dissipated, state authority to control the disease was rarely invoked, and talk of mandatory vaccination was scarce in the legislature."[70]

California was once again an early adopter when, in 1961, it passed Assembly Bill 1940 (AB 1940), which required polio vaccination for school entry. The passage of this act placed California near the beginning of the history of the mandates & exemptions regimes. (As we discussed above, most of the push for state-based school mandates occurred in the late 1960s and 1970s, after the disappointing result of the measles vaccine rollout.) AB 1940 was the brainchild of two Democratic politicians, William Byron Rumford and Umbert J. DeLotto. The pair were motivated by their professional and social experiences with polio, and they wanted Californians to be safe from this disease. There had not been any large polio outbreaks in California for many years, and there did not appear to be a groundswell of parent support for mandates. Nevertheless, there had been some outbreaks in undervaccinated populations, and some other states were also considering polio mandates.[71]

Rumford and DeLotto faced opposition to AB 1940 from non-mainstream health actors, including chiropractors, homeopaths, and natural healers. Don Matchan, from the local *Herald of Health* publication, and his close colleague Fred Hart, from the National Health Federation, represented organizations that resisted government regulation of alternative medical therapies. This duo and their supporters persuaded Rumford and DeLotto to include a "belief" (later "personal belief") exemption in their new mandate bill. By incorporating the exemption into the original mandate bill, Rumford and DeLotto followed the example of a similar bill that had recently passed in Ohio.[72]

California's permissive polio mandate formed the blueprint for the state's subsequent vaccine mandate policies. Additional vaccines were later added to the list of required immunizations according to the same mandates & exemptions policy logic, including measles in 1967 and diphtheria, pertussis, and tetanus in 1971. In 1977, California passed Senate Bill 942, which combined all of the state's school vaccine mandates into one law.[73] By that time, the mandates & exemptions model had become ubiquitous throughout the country.

Rumford and DeLotto were not physicians, unlike Richard Pan, who later led California's Clinician Counselling Bill, Nonmedical Exemption Bill, and Medical Exemption Bill. Indeed, Rumford and DeLott's efforts did not receive much support from California's physicians or from their professional societies, which were "ambivalent towards compulsion" and did not resist the addition of the "belief" exemption.[74] In the 1960s, California's mainstream medical professionals had not called for new mandates, nor did they mobilize support for the bill.

This is in stark contrast to the present day. One of the notable features of contemporary battles over school vaccine mandates is the prominent role played by physicians and their professional societies. (We dig deeper into this theme in Chapter 7.) The leadership role played by physicians in contemporary efforts to eliminate NMEs diverges from physicians' indifference toward the creation of the mandates & exemptions regime in the 1960s and 1970s. Accordingly, today's attempts to eliminate NMEs are not only efforts to radically remake school-entry vaccine requirements but also novel forms of physician engagement in immunization governance.

The American regime of state-based school vaccine mandates with NMEs was remarkably stable from the 1970s to the 2000s. But the early years of the new millennium witnessed a resurgence in vaccine hesitancy and refusal and also new struggles to revise immunization policies. Chapter 3 tells the story of the "last tweaks" that state governments made to their mandates before they turned to eliminating NMEs.

3
Last Tweaks

Introduction

There is a compromise at the core of the "mandates & exemptions" model of immunization governance that we described in Chapter 2: The state requires vaccination, most people comply, a few request exemptions, and the exempted people stay quiet. All parties must play their roles for this kind of policy to work. Otherwise, mandates risk becoming controversial or ineffective.

The mandates & exemptions compromise started becoming unstable in the 2000s. Nonmedical exemption (NME) rates began rising in many U.S. states. Some parents were becoming increasingly vocal in their objections to vaccines, and they occasionally found sympathetic partners in state legislators—usually Republicans—who attempted to eliminate or curtail mandates on their behalf. Those efforts were usually ineffective, but they succeeded in Arizona, Arkansas, and Texas.[1] At the same time, some state legislators—almost all of them Democrats—began tinkering with their states' mandate policies to try to increase immunization rates without having to eliminate NMEs. These policy developments made use of insights from social psychology and cognitive science about how to shape people's behavior. Policymakers who advocated for "tweaks" to their states' mandates & exemptions regimes also often assumed that distributing accurate information about vaccines would be enough to increase vaccination rates.

This chapter focuses on various American states' legislative interventions that aimed at reducing NME rates during the 2000s and 2010s. We pay particular attention to Dr. Richard Pan's Clinician Counselling Bill, Assembly Bill 2109 (AB2109). Pan is a central figure in California's recent immunization policymaking, and both his personal history and political activity illustrate some of the broader themes this book addresses. Accordingly, we provide some background about Pan, which we return to in later chapters. We conclude this chapter by discussing the failures of Pan's Clinician Counselling Law, which set the stage for Pan's later fight to eliminate NMEs in California via the 2015 Nonmedical Exemptions Bill, Senate Bill 277 (SB277).

The New Vaccine Refusal

The early years of the 21st century witnessed increased skepticism about vaccine safety and efficacy in the United States, United Kingdom, Europe, Australia, and beyond.[2] Some refusers were motivated by potential links between autism and the measles, mumps, and rubella (MMR) vaccine. Other concerns focused on ingredients in vaccines, including aluminum and mercury. Parents also questioned the safety, efficacy, or necessity of vaccines for reasons that connected to their values and identity.

Parents in communities such as Hannah Henry's often believe that organic foods, long-term breastfeeding, and responsive parenting will raise up healthy, well-adjusted children.[3] There is evidence to support these claims.[4] But some members of these communities see vaccines as unwelcome, chemical intrusions into their forms of "natural" living,[5] whereas others believe vaccines are unnecessary for their healthy children.[6] Many such parents think their children's immune systems are better able to fight off disease because of their "good parenting" choices.[7] Some even believe that vaccine-preventable diseases are beneficial for children. Another subset of vaccine refusers believe their children have suffered from vaccine complications or are likely to experience them.[8] Other parents may reject vaccines because of their deep distrust of government. Also, parents of color may be hesitant about vaccines because of histories of racist medical abuses, ongoing forms of racialized health injustices, and recent experiences of racism in clinical or government service encounters.[9] Regardless of how they come to that decision, all vaccine-refusing parents believe that not vaccinating is a way to protect or promote their children's health. As Elisa Sobo notes, sometimes rejecting vaccines is not a negation but, rather, an affirmation: A parent who refuses vaccines may be declaring that they have found a different way of being a good parent, and that they identify with other parents and communities that eschew vaccines.[10]

Our focus in this book is not the *origins* of vaccine refusal but, rather, the political and legal responses to some of its *consequences*. For those who want to learn more about the diverse origins and manifestations of vaccine refusal, we have listed some excellent recent books and journal articles in the notes.[11]

In the 2000s, vaccine-refusing parents and their political allies in state governments often tried to make it easier to send unvaccinated children to school.[12] For example, in 2003, Texas Republicans—who controlled all branches of state government—introduced a new philosophical exemption to their state's vaccine mandate. Texas had previously permitted only medical and religious exemptions. Rekha Lakshmanan and Jason Sabo observed that "[Texas] state politicians were eager to reward favored constituencies, such as home school factions and other groups advocating for strict parental rights and less government oversight."[13] One of these groups was Parents Requesting Open Vaccine Education (PROVE), a local organization of 3,500 families who claimed the new philosophical exemption as a policy victory.[14] After Texas introduced philosophical exemptions, demand grew dramatically, from 2,314 in 2003 to 64,176 in 2018 (a 26-fold increase).[15]

At the same time that vaccine-refusing parents were advocating for easier ways to send unvaccinated children to school, a spate of disease outbreaks in the 2000s led public health leaders and their political allies to push for policy changes in the opposite direction. Outbreaks of pertussis, mumps, and measles illustrated that some communities in America had insufficient vaccine coverage. Public health policy researcher Denise Lillvis and colleagues suggest that early 21st-century outbreaks of pertussis in states such as Washington, California, and Oregon were a source of shame for policymakers in those states and that this shame motivated them to act.[16] In particular, outbreaks of previously well-controlled vaccine-preventable diseases were causing state policymakers to reassess their long-standing tolerance of vaccine refusers. By 2010, many studies had debunked the MMR–autism hypothesis,[17] worries about other vaccine "injuries,"[18] or concerns about supposedly dangerous vaccine ingredients.[19] When this scientific consensus was not enough to stem the rising tide of vaccine refusal, policymakers looked to legal and policy interventions.

Clever Hacks for Failing Mandates

Public health policymakers in the 2000s and early 2010s often remained optimistic about their ability to change refusers' minds with increased education and information provision despite limited evidence of the effectiveness of such efforts. A core component of early 21st-century efforts to shore up existing mandate & exemption policies involved "increasing information rather than persecuting parents or restricting choices."[20] Reformers in this period sometimes focused on mandatory vaccine education as a potential tool for reducing NME rates. Also, policymakers and researchers worried that some parents who were not particularly opposed to vaccines were obtaining NMEs because this was easier than getting their children vaccinated.[21] So, reformers identified policy tweaks to disincentivize the pursuit of NMEs, without eliminating access to NMEs entirely.

Lillvis and colleagues name this period of reform the "West Coast Period of Exemption Contraction" (although it was not limited to states on the West Coast). Policies in this period made nonmedical exemptions more difficult to access, often by requiring parents to first receive counselling about vaccines and vaccine-preventable diseases. Underpinning the West Coast Period were several insights: In states where it was easier to get NMEs, there were higher rates of exemptions.[22] These states were also more likely to have disease outbreaks. Furthermore, as the West Coast Period unfolded, states that introduced additional burdens to their application processes for NMEs began to see reduced NME rates and increased vaccination rates. Making NMEs more difficult to access seemed to work, and education requirements were a useful "barrier."

One version of these events played out in Michigan. In December 2014, the state government announced that parents and guardians who wanted NMEs would have to attend in-person education sessions at their local health department offices.[23] When that policy began, health departments in some counties required parents to bring their children to the education sessions so that nurses could vaccinate them immediately afterwards.[24] But these officials were overly optimistic. There is little reason to think that Michigan's individual education sessions changed many people's minds. Mark's study in Oakland County tracked the vaccination status of the 4,098 children who received NMEs in that county during the 2017–2018 school year. Children of religious objectors subsequently received at least one previously refused vaccine only 4.4% of the time, and children of parents who voiced safety concerns did so only 8.1% of the time.[25]

Despite these low post-education-session vaccination rates, Michigan's statewide NME rate nevertheless declined by 35% in its first year, from 5.2% to 3.4%.[26] The *burden* of an education session seems to have done more to reduce exemption rates than did the *education* itself.[27] In other words, only a few people who attended the sessions changed their minds, but fewer people applied for NMEs when they became more difficult to obtain. Unfortunately, Michigan's new policy achieved only a one-time reduction in NME rates; they started slowly increasing in subsequent years.[28]

A similar strategy had previously been implemented in Washington state. However, in that state, parents who wished to acquire NMEs had to be counselled by a vaccine provider (e.g., their family physician), who would then sign the exemption form. Unlike in Michigan, the state did not provide the vaccine education sessions. Washington state's Clinician Counselling Bill, Senate Bill 5005 (SB5005), was passed in 2009 and implemented in July 2011.[29] By the following year, the number of NMEs had declined substantially,[30] which appeared to be a tremendous public health success. However, as in the case of Michigan, Washington's education sessions created only a one-time reduction.

Nudging, Information, and the Quick-Fix Policy Hack

The West Coast Period of Exemption Contraction was characterized by small policy tweaks that promised big rewards. The idea was to provide parents with accurate information about vaccines and also to modify the environments in which parents made vaccination decisions to orient them toward vaccination. In this way, the West Coast Period employed a popular trend in public policy: "nudging."

When people make decisions, they can be unconsciously affected by behavioral factors that are not relevant to the facts they are considering. Social psychologists and cognitive scientists have shown that our decisions can be influenced by the formats in which people present us with information and by the beliefs we have about other members of our social groups.[31] Someone who is able to direct these (and other) unconscious cognitive processes can sway our decisions. To use language that Thaler and Sunstein introduced, someone who (re)structures our "choice architecture" can incline us toward a particular choice; this is a nudge.[32]

Nudges can be effective in public policy and in other institutional contexts. For example, they have been used to successfully increase enrollment rates in workplace retirement plans.[33] More relevantly, clinicians have used nudges to enhance parental uptake of vaccinations in health contexts.[34] Douglas Opel and his collaborators demonstrated that physicians who conduct vaccine encounters on the premise that vaccination will occur (the "default option," using language such as "Well, we have to do some shots") face only one-third as much vaccine refusal compared to physicians who approach the encounter in a more open way, such as by inviting parents to first ask questions.[35] This pro-vaccination framing nudge is one of the only evidence-backed discrete interventions for increasing vaccine uptake in a clinical setting.[36] The American Academy of Pediatrics has recommended it since 2016.[37] It was predictable that governments would try to use similar measures to incline parents away from NMEs.[38]

One of the chief attractions of nudges is that they can be used in settings in which the population might resist a more coercive approach.[39] That makes them particularly useful in societies that prioritize liberty, like America, and in which efforts to limit existing liberties may be met with fierce opposition. A further reason to use nudges to promote compliance with existing voluntary policies— rather than creating new coercive policies—is that legislative inertia usually makes it easier to work within current policies than to make new ones.[40] The literature on "path dependence" provides ample evidence of the inertial force of existing policies and practices.[41]

In addition, small tweaks to vaccination laws can be accomplished without a large investment of state resources. Effective public health institutions are expensive, and America's public health institutions are notoriously underfunded, as we elaborate in Chapter 7. It takes time, effort, and a large amount of resources for public health institutions to create relationships and to cultivate public trust. Community health clinics, "mother and baby" visiting nurses, school health outreach, and public-facing communication and advertising cost money. By contrast, smart hacks promise big rewards for minimal investments.

From this perspective, it is worth comparing how burdensome it was for states—and their employees—to enforce different kinds of mandatory education tweaks to their vaccine mandates.

Washington's clinician counselling requirement was less expensive for government than was Michigan's. Whereas Michigan's governments paid for public health workers to speak with parents, Washington placed the financial burden on parents (and their insurance companies) to compensate clinicians for their time providing vaccine education. However, Michigan retained control over the content of mandatory education sessions because they were provided by local public health workers. By contrast, Washington state relinquished control to clinicians in the community. Although many clinicians likely did a good job, a minority almost certainly "sold" their NME signatures to vaccine refusers without providing any pro-vaccine counselling.[42] We examine Californian physicians' practice of selling of *medical exemptions* to childhood vaccine mandates in Chapter 5; similar kinds of fraud occurred across the country in response to COVID-19 mandates.

We noted at the beginning of this chapter that some parents are firmly committed to vaccine refusal for reasons that are related to their communities, identities, and their ways of life. However, until recently, health professionals and policymakers often regarded vaccine refusal as a result of insufficient information or understanding. According to this "information deficit model," people are vessels for information, and their decisions are a consequence of the information they possess. If you want people to vaccinate, then you need only to provide them with accurate information to counter their false beliefs. Scholars and public health communicators in previous decades often tried such strategies but usually failed.[43] We human beings do not easily dismiss our beliefs, and being presented with new information—even evocative information—can sometimes have unpredictable results. For example, Brendan Nyhan and colleagues found that parents who were presented with scary images of children suffering from vaccine-preventable diseases became *more* likely to believe that vaccines cause autism, and that parents who were presented with a narrative about a vulnerable infant *increased* their worries about vaccine side effects.[44]

Social scientists have now debunked the information deficit model, and their work has paved the way for forms of engagement that draw on refusers' values, social identities, and internal motivations.[45] But these *values-first* interventions can be costly to implement and difficult to oversee. For example, even if public health officials wanted to pay every physician in Washington state to learn how to use evidence-based persuasion techniques when counselling vaccine-resisting parents, they could not ensure that physicians would use those methods with their patients. As we noted above, physicians who are sympathetic with vaccine refusers may be willing to sign the paperwork without counselling parents to vaccinate. And at the other end of the spectrum, the fact that nearly half of American pediatricians dismiss vaccine-refusing parents or do not accept them in their practices indicates that many physicians have given up on persuading these families to vaccinate.[46]

Physician Counselling Requirements and the Vaccine Policy Debut of Dr. Richard Pan

After Washington state introduced a clinician counselling requirement for parents to obtain NMEs for their children, legislators in other states pushed for similar policies.[47] These policies cost governments almost nothing, they created little political controversy, and they promised near immediate declines in NME rates. California's legislators watched for evidence that Washington's new physician counselling requirement was successful. Among them was newly elected assemblyman and pediatrician, Dr. Richard Pan.

Pan had attended medical school at the University of Pittsburgh, where he was taught by Professor Julius Youngner. (Decades earlier, Younger had developed the polio vaccine with Jonas Salk.) Pan's professors would often tell him and his fellow students that unless they practiced medicine overseas, they would be unlikely to see vaccine-preventable diseases such as measles and polio. But Pan did not have to move so far away. One of his fourth-year clinical placements was with a Public Health Service clinic in the Germantown neighborhood of Philadelphia during the 1991 measles outbreak.[48] "I got to see measles live there," Pan explained, "Nine children died. Over nine hundred people got measles."[49] Pan cited this experience as a guide star for his later immunization policy work.

After graduation, Pan's experience as a resident physician further cultivated his admiration for mass vaccination programs. In particular, he was struck by the dramatic impact of *Haemophilus influenzae* type B (Hib) vaccine, which first went into wide use in 1988.[50] Pan's attending physicians told him that "prior to the vaccination, there would always be at least one patient who was in the ICU who had a form of invasive Hib infection." By the time Pan entered clinic, just a few years later, the Hib vaccine "had already diminished the infection of this very serious disease, down to the point where I only saw one case in my entire period of training."[51]

Pan's commitment to public health continued as he completed his Master of Public Health at Harvard University and began a fellowship at Boston Children's Hospital and Massachusetts General Hospital. After completing his training, Pan moved to the University of California (UC), Davis, to work at its children's hospital and to serve as a clinical professor in the medical school. Pan wanted to "make communities healthier," and the Medical Board of California awarded him the Physician Humanitarian Award in 2010 for his efforts to expand children's access to health care and to improve community health. When Katie interviewed Pan, he was eager to emphasize that his work in medicine and public health was part of a very successful "first career." The seeds of Pan's "second career" in politics were planted during his experiences at UC Davis, which is located just outside of Sacramento. Pan's physical proximity to the corridors of state power created opportunities for him to lean into politics. He became Chair of the Council on Legislation for the California Medical Association and President of the Sierra Sacramento Valley Medical Society.[52]

Vaccine-preventable diseases became an important policy issue just as Pan's political star was rising. In 2010, California experienced an outbreak of pertussis among babies and children. Hundreds of people were hospitalized and 10 infants died. Pan knew that "the pertussis vaccine is probably one of our less effective vaccines; it wears off after a time."[53] Still, vaccination was the best defense available. The fact that the 2010 outbreak had been concentrated in areas with low coverage rates was further confirmation.

In the year of the pertussis outbreak, Pan ran for office as a California Assembly member and won. His immediate priority was reinstating funding for vital social services that had been cut following the global financial crisis of 2008. But Pan's attention soon returned to vaccination. Rates of NMEs had been trending upward in California for over a decade, and he wanted to find a way to "bend the curve." In 2012, preliminary data showed that NME rates had decreased in Washington state after it implemented its new clinician counselling requirement.[54] Pan soon thereafter introduced a copy-cat bill to the California Assembly: AB2109.[55]

Pan was motivated to write his Clinician Counselling Bill because he thought it was too easy for parents in his state to get NMEs:

> Essentially, all you had to say is: "I don't want to get my kids vaccinated," and your kids could go to school. . . . We were one of the most lax states in the entire country. . . . And our personal belief exemption rates were climbing up and up and up.[56]

Kris Calvin, the CEO of California's chapter of the American Academy of Pediatrics (AAP-C), echoed Pan's remarks. California had high rates of NMEs because there were "very low barriers to remain[ing] unvaccinated," and school staff were handing out exemption forms to busy parents who forgot to bring their children's vaccination records.[57]

Pan and his allies hoped that California's Clinician Counselling Bill would replicate Washington state's success. If it succeeded, the policy would increase

immunization rates without requiring additional resources for vaccine promotion or the political will to address the structural problems stymying America's
immunization governance. "There are higher barriers to being vaccinated in our
country because we don't have universal health insurance," Calvin explained.[58]
Instead of addressing these barriers by investmenting in public health
infrastructure—community clinics, visiting nurses, and targeted persuasion
campaigns for the hesitant—the Clinician Counselling Bill shifted the burden
onto people who wanted to opt out of vaccines.

The architects of the Clinician Counselling Bill originally wanted applicants
for NMEs to satisfy the "counselling" requirement only by discussing vaccines
with physicians. But in the face of political pressure, they expanded the list of
eligible health care professionals to include nurse practitioners (NPs), physician assistants (PAs), credentialed school nurses, and—most controversially—
naturopaths.[59] Almost all physicians, PAs, NPs, and school nurses are
pro-vaccine,[60] but many naturopaths reject the germ theory of disease and ad-
vise their patients against vaccines.[61]

The Clinician Counselling Bill would come to include an even more significant
political compromise. California's existing "personal belief" exemption covered
religious as well as nonreligious beliefs, and the state had always treated them
similarly. Governor Jerry Brown, however, was uniquely concerned about the
imposition of the counselling requirement on people with religious objections.
"Governor Brown went to seminary school," Pan explained, "He was very concerned about religion."[62] Accordingly, Governor Brown's Executive Signing
Statement created greater protections for people who objected to vaccines
for religious reasons. On September 29, 2012, Brown wrote, "I will direct the
[California Health Department] to allow for a separate religious exemption
on the form. In this way, people whose religious beliefs preclude vaccinations
will not be required to seek a health care practitioner's signature."[63] The effect
of the signing statement was to block the physician counselling requirement
from applying to any parents who claimed religious objections to vaccines. Pan
could have tried to pass a new law to override Brown's executive order. But he
concluded that the new protection for religious exemptions "was a reasonable
compromise."[64] He took comfort from the fact that Brown used an executive
order to impose his will. "At least he wasn't asking me to put it in the law."[65]

The Limits of Physician Counselling

Pan and his fellow advocates of California's Clinician Counselling Bill initially thought they were importing a successful policy from Washington state.

But by the time the California version came into effect in 2014, its supporters had already soured on it, in light of diminishing returns from Washington state's policy.[66] In the first years of that state's mandatory counselling law, there had been increases in vaccination coverage and decreases in NME rates, but there was little further improvement in subsequent years. Also, it was worrisome that rates of *medical* exemptions had begun to creep up even before the bill's implementation. Between the 2010–2011 school year (during which the bill was passed) and the 2013–2014 school year, the rates of medical exemptions among children in Washington state increased from 0.4% to 0.7%.[67] But since many more parents were consulting with their physicians, at least some of the new medical exemptions would have previously been documented as NMEs.

Interviews for this book reinforced this "failure narrative" of the Washington Clinician Counselling Bill and invoked that narrative as a reason why California's vaccine policymakers quickly gave up on AB2109.[68] However, Washington state's own coverage data indicate that passing the Bill dramatically reduced NMEs between the 2010–2011 school year (covering 5.5% of children) and the 2013–2014 school year (down to 4.4%). Even with the replacement effect of some families migrating to medical exemptions, the overall rate for exemptions of any kind dropped from a high of 5.9% (2009–2010) down to 5% (2013–2014).[69] Washington's SB5005 delivered some positive outcomes, despite the prevalence of the "failure narrative" in some circles.

Pan was committed to letting California's Clinician Counselling Law be implemented before assessing whether additional laws were needed to reduce his state's NME rates. The Clinician Counselling Law went into effect in 2014, and at the time, Pan had no plans to intervene further in exemptions.[70] On this point, Catherine Flores Martin of the Immunization Coalition observed, "If there was a master plan [for additional vaccine mandates bills], I was not aware of it."[71] Kris Calvin from the AAP-C concurred, stating, "We believed in [the Clinician Counselling Bill]."[72] Indeed, even as Pan and others later identified shortcomings in Washington state's new law, there was little support for further modifying California's vaccination policies. Flores Martin, whose Immunization Coalition co-sponsored the Clinician Counselling Bill, noted that it had been exhausting just to achieve that modest change.[73]

Following its 2014 implementation, California's NME rates initially decreased to 2.5%, from 3.1% in 2013. This was promising. But in 2015, NME rates decreased only 0.2% more, to 2.3%.[74] This pattern mirrored what scholars would subsequently conclude about Washington state's experience: Counselling requirements generated large one-time gains, with marginal improvement (and even some backsliding) in subsequent years.[75] More worrisome for California was the fact that substantial geographic clustering remained, even after the clinician counselling requirement had been in effect for 2 years.[76] The law had brought down the overall NME rate, but it did little to decrease NME rates among clusters of committed vaccine refusers.[77] And those clusters remained prime targets for outbreaks.

The clinician counselling requirement had been somewhat successful in the aggregate, but it seemed to be doing little to influence what Kris Calvin called the "I don't want to" people.[78] These committed vaccine refusers were getting "counselled," as the law required, but were not changing their minds. If they fulfilled the counselling requirement with sympathetic clinicians—for example, naturopaths—they may not even have received accurate information about vaccine safety and efficacy.

Even though the limitations of the Clinician Counselling Bill were well known, all of the policy actors Katie interviewed insisted that it could have remained their last effort to modify California's vaccine mandates. For example, Pan was planning to pivot from mandates to a new bill that would require schools to publicize their vaccine coverage data.[79] His goal was to let parents see that data and then exert social pressure on each other and on school institutions through their school enrollment choices. This was another way that Pan could help California promote public health without spending much money—by shedding light and bringing data transparency so that parents could exercise "choice and voice."[80]

If Pan and his allies had remained committed to making only narrow policy tweaks, they may also have been able to optimize the encounters that the Clinician Counselling Bill required. Indeed, there has been little empirical investigation into what happened in these sessions, or what continues to happen in the states that require clinician counselling. It is possible that California's requirement could have done more to decrease NME rates if clinicians had been trained and resourced to communicate better with vaccine-hesitant parents. We will never know. The policy's advocates would soon come to embrace a far more radical solution: eliminating NMEs.

On February 19, 2015—only a little more than a year after California's Clinician Counselling Law was implemented—Senator Ben Allen and Pan (now also a state senator) introduced a bill to eliminate NMEs in California. SB277 was passed by the California legislature later that year, subsequently signed by the governor, and went into effect on July 1, 2016. That day marked the end of the clinician counselling requirement. No one would need to be counselled about exemptions they could no longer receive.

In the post-Disneyland political landscape, it was predictable that California's activists and legislators would push to eliminate NMEs. This change may have seemed like yet another low-cost and small-scale policy hack in the spirit of those that had already been enacted. After all, the only thing required was to remove an exemption from an existing practice. But regardless of what its advocates believed, the Nonmedical Exemptions Bill would radically reshape immunization governance in California.

4

Mobilizing for the Nonmedical
Exemptions Bill

Introduction

The passage of California's Nonmedical Exemption Bill was a new development in American immunization policy, both because it made vaccination policy more coercive and because that change was influenced by the emergence of a new political constituency: pro-vaccine parents. These ordinary citizens were concerned about their communities' vulnerability to disease, and they were angry that other people were putting their children at risk. The new political mobilization of pro-vaccine parents heralded the end of the old "mandates & exemptions" order, in which managing vaccine mandates was the business of experts engaged in technocratic tinkering, while vaccinating parents remained on the sidelines.[1] After the passage of California's Nonmedical Exemption Bill, more jurisdictions across the United States began considering coercive measures to promote vaccine compliance, often in partnership with newly mobilized parent groups.

Origin Stories for Mobilized Pro-Vaccine Parents

Social science research about parents' vaccination decision-making explains vaccine refusal in terms of a now-standard set of "origin stories." These stories often focus on how parents developed fears or doubts about vaccines, how they responded to their children's unexplained health conditions, or how they had bad clinical encounters with supposedly judgmental doctors.[2] In Chapter 3, we directed readers to some of the research about the origins of vaccine refusal. But this chapter tells a different kind of origin story. It is about how some pro-vaccine parents mobilized to help eliminate nonmedical exemptions (NMEs) in California.

The parents discussed in this chapter were worried that California's government was enabling vaccine refusal. Pan's Clinician Counselling Law made it more difficult to refuse vaccines, but they thought that law did not go far enough.

Vaccine refusers could still send their children to school, and the counselling requirement had done little to increase immunization rates among geographically clustered communities of the most committed refusers. The parents who mobilized to eliminate NMEs were done tinkering with mandates & exemptions policies. They wanted radical change.

Leah Russin has degrees in biology and law, and she worked both as an environmental lawyer and as a staffer for a U.S. senator before she took time away from paid work to care for her baby boy. Russin relished the friends she made and the networks she developed through her mother & baby group during the raw months of early motherhood. But she soon discovered that some of her fellow moms had decided not to vaccinate their babies. A few of these mothers were also bringing unvaccinated older children to the group meetings. One of those children had whooping cough.

Russin did not want to share space with people who were so cavalier about communicable diseases. She asked leadership to exclude unvaccinated families from the mother & baby group. They refused, Russin explained, "because the law said that personal belief exemptions are just fine." They "felt it was important to their mission to say: we support all parents' journeys within the law." Russin decided to leave the mother & baby group to avoid exposing her baby to vaccine-preventable diseases. She was angry that California's lax attitude toward vaccine refusal made him unsafe in a community space that she valued. Although Russin's group was not governed directly by school and daycare vaccine mandates (or by NME policies), she believed that California's NME policies contributed to a broader cultural permissiveness about vaccine refusal. She wanted to eliminate NMEs so that vaccine refusal would be less tolerated in her community.[3]

Another California parent, Renée DiResta, developed a similarly powerful aversion to the presence of vaccine refusers in public spaces. DiResta, now a Stanford-based expert in social media algorithms and data,[4] used to ride Northern California's Bay Area Rapid Transit (BART) to work when her son was a baby. In early 2014, well before the Disneyland measles outbreak, there was a measles exposure on BART. The news left DiResta concerned that she had brought the disease home to her infant son, who was too young to be vaccinated. Distressed by this possibility, she contacted both her state assembly member and senator to talk about how to better protect her community from outbreaks:

> I'd never done anything in local politics. I didn't know who you were supposed to call; I'd only been in California for two years. So I called my state representative and I said, "What can I do? It seems ridiculous that you have these personal belief exemptions out here. Can San Francisco be somehow exempt by population, or something along those lines? Like, what can we do in San Francisco to avoid these exposures on BART and elsewhere?"[5]

Her political representatives told DiResta that there was "not a lot of momentum" to do anything about unvaccinated people spreading disease. But that news motivated DiResta to get to work. She discovered the vaccine coverage rates for the schools her son might attend. She downloaded 10 years' worth of California Department of Public Health data to evaluate trends in vaccination rates and to identify geographic clusters of under-immunized children. She was further galvanized when she encountered anti-vaccine sentiment in online parenting forums:

> The idea that somebody's scientific illiteracy was a justification for them to put other children, especially mine, at risk, that was my biggest motivator. I felt like this was one of the areas where the collective good was more salient, particularly as the trend of non-vaccination had continued to exacerbate over time.[6]

These experiences showed DiResta how big a problem vaccine refusal had become and also how California's vaccination policies were making things worse. DiResta was a recent arrival from New York, and California felt like a different country when it came to vaccines. "The idea that you could just voluntarily wave your hands and opt out" of vaccination did not exist in New York, as "there's never been a personal belief exemption [there]." She recalled that religious exemptions were available, and she believed that the state interrogated the veracity of applicants' beliefs. That is not to say that New York was perfect. "People were lying with the religious exemption, I came to find out," DiResta explained. However, the fact that vaccine refusers in New York had to be dishonest meant that they were violating social norms. By contrast, "there was a much more visible anti-vaccine movement in California than there was in New York" with more "hippie granola types," and DiResta thought that California's permissive NMEs were a big part of the problem. "California let you write on a piece of paper that you don't want to vaccinate, and then you just weren't part of the social contract." Even as vaccine refusers violated the "social contract," California's NME laws allowed them to remain full members of society. DiResta wanted to fight back.[7]

In Chapter 1, we introduced Hannah Henry, who had much in common with the vaccine refusers who were placing DiResta and Russin's communities at risk. Henry practiced attachment parenting with her four children and resembled the "granola mom" stereotype to which DiResta referred. (We remind our readers that these were also *our* communities and that *we* followed similar parenting practices with our young children.) Henry also sent her children to a Waldorf school. Waldorf schools implement the "anthroposophic" teachings of their founder, Rudolph Steiner, who thought that children thrive best with play-based learning, natural materials, limited technology, and regular physical movement. Steiner also believed that young people experience spiritual growth through illness and that vaccines deprive children of the benefits of being sick.[8] We do not know how many Waldorf parents refuse vaccines because they embrace Steiner's ideas about the benefits of disease, but Waldorf parents refuse vaccines at far higher rates than do parents at public or other private schools. This is true of Waldorf schools around the world, from Perth, Australia, to Denver, Colorado, and of the California Waldorf school that Henry's children attended.[9] However, unlike many of her Waldorf family peers, Henry passionately supported vaccination. Like Russin and DiResta, she wanted to do something about the fact that not enough children in her community were getting vaccinated.

The Clinician Counselling Law appeared to be modestly reducing NMEs in California, as we discussed in Chapter 3. But that new policy did not seem to make a difference in Henry's community. Parents at her children's school were committed refusers. It was a core part of their identities that they rejected childhood vaccines, and the fact that they were clustered together meant that they often reflected and reinforced this practice with each other. A single conversation with a physician was highly unlikely to change their minds. The "geographic clustering" of these vaccine-refusing families made their community especially vulnerable to outbreaks.[10] But they seemed to think that intensive parenting practices would protect their children from harm.

Henry's commitment to vaccination—and her worries about outbreaks— made her an outlier among her neighbors and the parents at local schools. She responded to the weakening of her community's protection against outbreaks like the child who saw through the naked emperor's nonexistent "new clothes." Everyone kept pretending that their community was safe, but Henry was willing to speak the embarrassing truth. Furthermore, Henry's family was especially vulnerable to vaccine-preventable diseases. Her mother was seriously ill and had a compromised immune system. Her youngest child was not yet fully vaccinated. The people who surrounded Henry—*her people*—were both the source of the threat and its potential victims.[11]

Henry spent months sitting with her frustrations, trying to figure out what to do. She spoke to people in her community, including her pediatrician, who was also worried. Henry explained, "[The pediatrician] said she's really frustrated because she couldn't get to some of these parents who were refusing vaccines [and] 'did I have any ideas as a parent?' . . . These were casual conversations in the examination room with my infant."[12] Henry regarded herself as "very naïve" about vaccine refusal. At first, she thought it was "an education issue" that could be corrected with fact-based outreach. (Henry was not alone in this assumption, as we noted in Chapter 3's discussion of the information deficit model.) Early in her public advocacy, Henry reached out to an administrator at one of her children's schools and suggested they hold a roundtable discussion. Could she invite her child's pediatrician to speak to the school's parents about vaccines? "[The school administrator] said, 'Oh my God. No, no, no!' I mean, not in so many words, he didn't outright, but it was very clear to me that this was gonna be a very contentious issue."[13] After this rebuff, Henry said she continued to obsess over vaccine refusal in her community. She spoke to other parents who shared her concerns. Like DiResta, Henry started tracking data about vaccination rates in schools. She was similarly unhappy with what she saw, and she began to worry that it was only a matter of time before something very bad happened.[14]

From the Disneyland Measles Outbreak to Eliminating NMEs

When measles broke out at Disneyland in late December 2014, the problem could have been contained to the theme park. California's vaccination policies may have had little to do with the *origin* of the Disneyland measles outbreak. The park's visitors came from all over the world, and the United States did not require international tourists to be vaccinated. But what happened in Disneyland spread, at least in part, through unvaccinated Californians and into their own communities. Although the Disneyland outbreak may have had international origins—specimens from several cases matched samples from a recent outbreak in the Philippines—it soon became a California problem.[15]

Renée DiResta was not surprised about the Disneyland measles outbreak, especially after her possible exposure to measles on the BART during a previous outbreak.[16] But Leah Russin found it "eye-opening" to see "measles spread so fast." Russin had been vaccinating her son according to the schedule, and it sometimes seemed like she was barely managing to keep him protected against outbreaks: first pertussis in 2010 and then measles in 2014. When Russin looked at vaccination coverage rates, she saw that there were many children with personal belief exemptions enrolled at her local schools. "It was a moment where I realized the extent to which public health and government were failing us."[17]

The fact that California's vaccine governance focused on mandates for school enrollment shaped pro-vaccine advocacy by priming pro-vaccine parents and communities to see their local schools as a locus of risk. Data about rates of NMEs oriented current school parents (like Henry) and future school parents (like DiResta and Russin) to consider how a protected minority—vaccine refusers—made schools and broader communities unsafe for everybody. In Henry's community, and especially in the Waldorf school, vaccine refusers sometimes seemed like they were in the majority.

Like Russin and DiResta, Henry also decided that her state needed to eliminate NMEs. Henry drafted a petition for the MoveOn online grassroots policy
advocacy group. The petition language was succinct and specific: She called on
California to "require that students be vaccinated and not have a personal belief
exemption." Henry did not ask California's lawmakers to raise vaccine coverage
rates. She did not ask them to invest in better vaccine promotion. (By then, she
believed that pro-vaccination initiatives would likely fail in her community.) She
"just needed [legislators] to restrict that personal belief exemption, which had
been abused."[18]

Henry took her petition to the community and the media, while other pro-
vaccine parents went directly to their state legislators. And anyone who knew
anything about California's vaccination policy history understood that there was
one legislator they needed to reach: Dr. Richard Pan. Now a state senator, Pan
became a rallying point for politicians, activists, and community members who
were calling for abolishing NMEs.

Russin knew that Pan was the person to call:

> I knew he had done AB2109 [the Clinician Counselling Bill], and I knew that
> it had been what I considered a baby step, albeit a successful baby step. But it
> clearly had not gone, in my opinion, far enough, and I wanted the law to be
> much more strict. So I reached out to his office: "Why didn't you go for gold
> with AB2109? What can we do to get rid of non-medical exemptions to vaccine
> requirements for school?"[19]

Russin's call was answered by Pan's Chief of Staff. Unlike congressmen, state
legislators usually do not have large office staffs. After the Chief of Staff realized
that Russin was a vaccine supporter—and not a critic calling to attack Pan—they
had a long and constructive conversation. "I'm thinking about doing a voter initiative to eliminate NMEs," Russin offered. Pan's Chief of Staff explained that their
team was considering its own legislation, if enough people like Russin called for
it. "Oh, I'll call for it," Russin replied. "Here, I'm calling for it."[20]

The next day—February 4, 2015—Russin drove 2 hours to Sacramento for a
hastily called press conference with Senator Pan and other legislators who were
writing a bill to eliminate NMEs. These included Pan's co-sponsor Senator Ben
Allen—a former lawyer, human rights advocate, and education leader—and
Assembly Member Lorena Gonzales, among others. All were Democrats.[21] In
front of California's media, Russin repeated her call for the state to abolish NMEs
to daycare and school vaccine mandates. "The Bill hadn't yet been introduced,
the language had not yet been crafted, but we were announcing our intention as
a group to do it."[22]

Vaccinate California

In joining forces with Pan and his legislative allies, Leah Russin was now a member of a team:

> I had this energy, I had this knowledge base and skill set. How should I best use it? And Senator Pan basically said: I need you to show up as a mom. And so he definitely crystallized the most useful way for me to be effective.[23]

Russin described her first role as "nursing [her] baby in a T-shirt in the front row, looking like a mom."[24] Behind this cheeky remark was Russin's choice to enact a key messaging strategy for Senator Pan's new Nonmedical Exemption Bill, one that would deviate from the approach he took with the Clinician Counselling Bill. During the fight for that previous legislation, Pan explained, "We had physicians and public health experts come and testify how important this bill is." In contrast, the Bill's opponents were parents who claimed that their children had been vaccine-injured. "Often that's a code word for autism, which has been debunked," he contended, "But it's not a good thing to have experts . . . usually male experts . . . telling (mostly) mothers that they're wrong. So even though we were successful [in passing the Clinician Counselling Bill], we knew that [this] dynamic didn't play out well."[25] Pan was right. It is not a good look for male technical experts to challenge the fierce and heartfelt claims of mothers who assert that their children have been injured by vaccines. Legislators and members of the community empathize with mothers—with whom they often have more in common—rather than with the experts. Pan wanted to approach things differently in his efforts to eliminate NMEs. "We cannot have doctor or expert against parent. . . . Parents have to lead the bill; they have to be the sponsor of the bill. . . . They're the face of the bill." Pan did not have to look far to find his parent leaders: They had already found him. Like Russin, many concerned parents were calling his office and were demanding action. Pan and his staff asked individual parents to combine their efforts. "You guys start talking to each other, right. You know, get them some resources to mobilize."[26]

The parents swiftly cohered into a new organization: Vaccinate California. Vaccinate California's other founding members included Tisha Terrasini-Banker (an actor), Rachel Deutsch (a lawyer), and Jennifer Wonnacott (a public relations and communication expert with a history of consulting on political campaigns).[27] Wonnacott was on maternity leave from her job running communications for another state senator when Russin met her at the February 4th press conference.[28]

Hannah Henry was also a founder of Vaccinate California. Her MoveOn petition was getting plenty of attention, and she also applied her visual merchandising skills to Vaccinate California's efforts. When MoveOn suggested that she brand her campaign with a picture of measles-infected child, Henry objected. She knew that you do not sell products by upsetting people. You have to give them something to believe in. Henry's "I ♥ Immunity" logo, which would come to represent Vaccinate California and their broader alliance, would be far more effective.[29]

Law professor Dorit Reiss soon became one of Vaccinate California's most important members. Reiss had arrived in California from Israel a few years earlier to complete her PhD in administrative law. After meeting her future husband, she stayed and built her life and family in the Bay Area. Reiss first encountered vaccine opponents in 2012 while reading parenting blogs. Her first response was to provide accurate information about vaccines and vaccine-preventable diseases in the comments sections of online news stories. "The way I saw it was: I'm going to make some comments online about this because people should speak up. And then I'll go back to my life." But in 2014, Reiss joined Vaccinate California, became an expert about California's vaccination laws, and shifted her research to focus on vaccination law, policy, and jurisprudence. She supported Senators Pan and Allen's legislative staffs, identified how other American states were limiting NMEs, and made key additions to the Nonmedical Exemptions Bill.[30]

Russin reflected on the remarkable skillset of the Vaccinate California parents by noting that "we each had our own expertise."[31] Many members of Vaccinate California, like Reiss and DiResta, were employed in high-profile professional positions. But other members of the group—like Russin—had taken breaks from their careers to focus on motherhood, and they wanted to apply their workplace skills in the home and for the community. Collectively, these highly educated, proficient communicators were experts in politics, law, public relations, and visual merchandising.

Individual members of Vaccinate California, like Reiss, applied their specialized skills to support the Nonmedical Exemptions Bill. Collectively, the organization also worked to encourage other parents to demand that their legislators vote for the bill. For years, the *critics* of vaccines and mandates had been the only visible parent activists, but Vaccinate California wanted to change that. "Our role was not to convince people about vaccine science, that's what the doctors are for," explained Henry. Similarly, "Our role was not to strategize deeply about how to move the legislation, that's what the lobbyists were for." Instead, "Our role was really to help shape the media narrative, and make it clear both to legislators and to the general public that parents were demanding this legislation."[32] The politics of vaccine refusal used to be defined by vaccine refusers. It was now going to be shaped by pro-vaccine parents.

Mobilizing the Coalition

Vaccinate California's efforts to promote the Nonmedical Exemptions Bill involved the cultivation of partnerships with other civil society organizations, including the American Academy of Pediatrics California, the California Medical Association, the Health Officers Association of California, the California Immunization Coalition, the California State Parent Teacher Association, and the California School Nurses Organization. Many of these professional groups had worked together previously, including to pass the Clinician Counselling Bill in 2012. The addition of Vaccinate California prioritized pro-vaccine parent voices in a new way.

One of the wider alliance's first tasks was to write the bill. Here, the group relied on Reiss's expertise. In reflecting on how an outsider like her could help draft legislation, Reiss explained that the U.S. legislative process "is less expert-based and less professional than in many other places," which provide better support for legislative staffs.[33] The fact that many American state legislatures have strict term limits—for example, California imposes a lifetime maximum of 12 years of service in its legislature—means that few politicians can become expert policymakers. Or, as Reiss put the point, "Most of our [American] legislation is amateur hour."[34] The political success of California's Nonmedical Exemption Bill substantially depended on the fact that it was championed by a pediatrician and seasoned political operative (Pan) and that it was written by a law professor (Reiss).[35]

Vaccinate California's primary role was to increase public support for the Nonmedical Exemption Bill and to insulate lawmakers against opposition. After the Disneyland measles outbreak, legislators in many states were inspired to crack down on NMEs. But in 2015, only California succeeded in eliminating them. Vaccinate California played an important part in that success.

In March 2015, PBS reported that at least 19 states were considering legislation to try to reduce NME rates. The bills included all of the policy strategies we have discussed in this book and some other ones, too. For example, bills in Arizona and Oregon sought to publicize rates of NMEs in schools via the "sunlight principle" that also attracted Senator Pan after the passage of his Clinician Counselling Bill. Meanwhile, a Missouri bill sought to require schools to notify parents directly about unvaccinated students at their children's schools. Bills in Maine and Washington would abolish personal belief exemptions, while bills in Minnesota, Oregon, and Maine sought to introduce clinician counselling requirements. Bills in Illinois, Connecticut, Rhode Island, Maryland, New Jersey, North Carolina, and Vermont sought either to tighten eligibility for religious exemptions or to eliminate them altogether.[36]

In April 2015, *The Guardian* reported that some of these bills—in Oregon, Washington, and North Carolina—were withdrawn due to powerful opposition from anti-vaccine activists, who "bombarded [state legislators] with emails and phone calls, heckled them at public meetings, harassed their staff, organized noisy marches and vilified them on social media."[37] The North Carolina bill, which was sponsored by two Republicans and a Democrat, sought to remove religious exemptions. Senator Terry Van Duyn, the Democratic sponsor of the bill, told local radio in April 2015 that she was "not prepared" for the "public reaction to the bill," which she described as "very swift and very furious," creating "an environment that made it difficult to just even talk about it."[38]

In 2019, Dorit Reiss explained that legislators in other states usually failed to pass new vaccine mandate bills because they lacked adequate coordination between legislators and civil society groups and because they had not mobilized pro-vaccine parents:

> So a politician thinks it's a good idea to put in a bill that removes nonmedical exemptions by themselves, without organizing [their state's] Immunization Coalition, without talking to anyone, without talking to the Department of Health to see if they'll be on board. That's a good way to get legislation that's not going to go very far.[39]

That is not to say that adequate coordination and mobilization was *sufficient* to revise state vaccine laws, but only that these factors were *necessary*. For example, in Oklahoma and Oregon, legislators coordinated with activists, but other barriers nevertheless prevented their bills from passing.[40]

Vaccinate California helped their state succeed where other states failed. They met with representatives and senators, and they focused their social media outreach on legislators whose support appeared to be wavering. Vaccinate California also shared information and strategized with representatives of the civil society organizations who were knocking on doors in the Capitol. Catherine Flores Martin explained the work of one of those groups, the Immunization Coalition, a national pro-vaccine nonprofit whose California chapter she led: "We'll go to the legislative offices and meet with the staff to let them know we're in support and see if they have any questions. I don't consider myself a lobbyist, but you are lobbying when you're doing that."[41]

County health officers and their professional organization were also working to cultivate legislative support for the Nonmedical Exemptions Bill. In America's largely privatized health care system, few physicians work directly for the government. County health officers are a major exception. These physicians are appointed as the public health authorities in their counties or cities, and they are usually employed by their local health departments, which makes them account-able to local elected politicians. One of a county health officer's roles is to ensure that daycare and school vaccine mandates are implemented at the local level. They also direct community responses to outbreaks.[42]

Kat deBurgh, the Executive Director of the Health Officers Association of California (HOAC), noted that county health officers generally had to be careful to avoid political controversy around immunization policy. If a health officer became too critical of vaccine refusal, they could lose their job:"We have seen the anti-vaccine advocates could very easily overwhelm a board of supervisors meeting and demand that the health officer be fired and replaced with a more anti-vaccine health officer."[43] However, although individual health officers might be vulnerable, they could speak with a powerful voice by coordinating their efforts. It was this kind of reasoning that motivated HOAC to support Pan's Nonmedical Exemption Bill. "We don't have a lot of money to throw around," deBurgh explained, "but we do have expertise, and we do take positions on bills."

When Senator Ben Allen agreed to work with Senator Pan to abolish NMEs, he first reached out to deBurgh. She "warn[ed] him what a big deal it is to do a vaccine bill." Ultimately, she said, Allen insisted on proceeding because it was "the right thing to do for public health." She offered HOAC's support. HOAC was already sponsoring another bill, SB792, which would require daycare workers to be vaccinated against measles, pertussis, and flu. "Our bill was sort of seen as the *other* vaccine bill. It got a lot less attention," she explained. That bill passed, too.[44]

HOAC's strong public health orientation was echoed by the California Medical Association, which co-sponsored the Nonmedical Exemptions Bill, and by the American Academy of Pediatrics California (AAP California). In her March 30, 2015, submission to the Senate Health Committee, AAP California's Kris Calvin stated,

> Practicing pediatricians . . . have tremendous respect for the parent's right to make choices that affect their child's individual health and well-being. But vaccines are not about a single child—they are about the public health, and what we as a society agree is evidence-based to protect all of our children. To that end, California pediatricians strongly support SB277 and urge its enactment now.[45]

Calvin was echoing the views of her organization's members, but she was also heralding changes in the national AAP's orientation to NMEs. In 2016, the national AAP followed its California chapter by coming out strongly against NMEs. It called on state chapters across the country to help their state legislatures eliminate NMEs.[46]

From a Bill to a Law

The Nonmedical Exemption Bill had a long journey through both chambers of the California Legislature. It was referred to multiple committees and had to be passed twice by the Senate due to changes the Assembly made. Testimony from expert witnesses and community members was often heated, as we discussed in Chapter 1. For example, Mary Holland, a New York–based attorney and critic of vaccine mandates, was brought in as an expert witness by legislators who opposed the bill. She argued that eliminating NMEs would undermine parents' informed consent to vaccination, and she compared the elimination of consent in medicine to other troubling forms of coercion, including rape.[47] (We dig into the details of Holland's claims about informed consent in Chapter 8.)

In navigating the passage of the Nonmedical Exemptions Bill, Pan applied the lessons he learned from passing the Clinician Counselling Bill:

1. *Parents are the public face*: "[E]very time we went for, whether it's hearings or media stuff, we always tried to get a parent there, not just a doctor," Pan explained. "So therefore, when the [opposing] parent says . . . [their] child was damaged by vaccines," the pro-vaccination parent can say, "My child was actually infected or put at risk. My child would die if they were exposed. . . . My child needs to be protected."

2. *Parents argue with parents*: "[W]e wanted to be sure that [parent advocacy] was the focus of the bill," Pan explained. This would not be sustainable with "doctor[s] saying, 'Well, you know, you were wrong, and we need to have vaccines to keep us safe.' It was *parents* saying: 'I'm trying to protect my child.'"

3. *Doctors validate parents*: Doctors would offer their expert opinion only after pro-vaccine parents staked their political claim. "[T]hen the doctor says: . . . 'She's the one who's correct, right, the evidence supports [this parent's] worries [about the spread and impact of vaccine-preventable disease]."[48]

Pan's strategic and tactical insights were central to the success of the Nonmedical Exemptions Bill. Kris Calvin from AAP California also noted his "expertise" and "integrity" and his status as a pediatrician; she did not believe "we would have gotten it otherwise."[49]

The Nonmedical Exemption Bill's journey through the State Legislature was not always easy. For example, some lawmakers wanted to ensure that children with medical reasons for not being vaccinated could still be exempted from school vaccination requirements. In response to this concern, Pan and his team agreed to insert an additional provision in their bill. This new text clarified that

California's doctors could continue to consider "family history" when they determined if children were at heightened risk of vaccine complications, and hence eligible for medical exemptions.

Hannah Henry summarized the significance of the "family history" legislative compromise:

> There was some pretty powerful testimony—emotional testimony—against the bill, that was very focused on the need for medical exemption. Like, the need for broad exemptions that a typical doctor might not identify.... We had to add something to the bill that allowed you to say: If a family member was affected [by a condition or a vaccination], you might also be medically exempt.... And that was definitely a big concern, but we had to proceed.[50]

After NMEs were eliminated, medical exemptions increased dramatically. The "family history" provision likely played a role in that development, as we discuss in Chapter 5. For now, we continue our focus on the passage of the Nonmedical Exemptions Bill.

Scholars and commentators have credited the successful passage of the Nonmedical Exemption Bill to the skillful communication strategy its authors and supporters deployed.[51] One sign of their success was that almost every state newspaper and many national news outlets endorsed the bill and echoed the arguments of its advocates.[52] This was no small achievement, since "Telling the story of Vaccinate California is not as interesting as an anti-vaxx one," as Leah Russin told the scholar Samantha Vanderslott in 2017.[53] Media coverage focused on two key narratives that the Nonmedical Exemption Bill's advocates cultivated and disseminated. The first narrative was that the community lacked sufficient immunity—the Disneyland measles outbreak was often offered as the core piece of evidence—and that NMEs were a major cause of that problem.[54] The second narrative was that curtailing NMEs would protect vulnerable members of school communities, including students such as 5-year-old Rhett Krawitt (who lived with leukemia), baby siblings, and adults with compromised immune systems.[55] The coalition behind the Nonmedical Exemptions Bill framed their bill as a way to protect schools as the networked hearts of broader communities.

Hannah Henry's brainchild—the "I ♥ Immunity" campaign—expressed the core value that motivated the Nonmedical Exemptions Bill. When the bill's advocates spoke in public settings and when they spoke to Katie for this book, they frequently highlighted the relationships of immunological interdependence and responsibility that exist within communities. This kind of messaging highlighted a deep contrast between the groups of parents who were advocating on opposite sides of the debate. Parents who opposed the bill tended to focus on how it would affect their own families, but advocates appealed to broader ethical ideas about public health and well-functioning communities.

Community and Privilege

The parents of Vaccinate California had good reason to believe that their state's laws could be changed to better serve them and their communities. Privileged and progressive—like the leaders of their state's Democrat-dominated government—these talented parents had generally been treated well by America's social and political institutions. They had attended good schools, had worked in high-status jobs, and had enjoyed generally comfortable lives. Their privilege does not make their goals or achievements any less significant. However, the means by which they were successful tells us much about the privileges that certain Americans can expect, as well as how their shared identities would help them to succeed.

In 1965's *The Logic of Collective Action*, Mancur Olson argued that members of large groups have strong incentives not to voluntarily contribute to shared projects if they have reason to believe that other people can be found to do the work instead. In contrast, Olson thought that members of small groups tended to be better motivated to contribute to social projects. In such groups, members often have more in common, and there are fewer people available to do your share of the work.[56]

Most California parents vaccinate their children. They constitute a large and diverse group, with differences marked by race, ethnicity, education, economic status, and religion. Attempts to coordinate the members of this large group for collective action would likely be prone to misunderstandings, missteps, and perhaps even internal conflict. By contrast, the mobilization of vaccine refusers is more straightforward. This smaller group shares not just opposition to vaccination and mandates but also often a commitment to ideas about birthing practices, infant feeding, parenting styles, the environment, nature, and Western medicine. Scholars recognize non-vaccinators as a health social movement.[57] Vaccine refusers' shared values and identities facilitate group organization, common appeals, and collaborative resistance. Indeed, as we discuss in Chapter 5, California's vaccine refusers responded in exactly this way to the set of bills that California passed to make vaccine refusal more burdensome.

The parents of Vaccinate California were members of the *large* group of California parents who vaccinate their children, but they were also members of their own *small* group. Like their opponents, they were mostly mothers. They were college educated, and many had previously engaged with politics, whether professionally or through previous citizen activism. They were a relatively homogeneous small group that aimed to represent the larger population of California parents who supported vaccination and vaccine mandates.

The Vaccinate California parents did not require large financial investments for their advocacy work. They needed far more time than money. But time is it-self a valuable and scarce resource. Having time to participate in the legislative process often means having a partner who earns enough money to help support the family. Having time to draft legislation, meet with legislators, or promote bills on social media may mean having paid parental leave or a job that paid well enough so you can stay at home with the kids for a while.[58] It may also mean that one possesses an "amazing ability to work obscene hours," as Russin described DiResta, who was "working at a startup she'd co-founded" during Vaccinate California's activism.[59]

But Vaccinate California did cost money, in addition to investments of time. For example, Russin explained that she and DiResta reached into their own pockets to pay for targeted Facebook advertisements:

> I spent twenty dollars here and there. If our callers [to Vaccinate California] told us that a specific Senator was on the fence, I would run an ad in that Senator's district, targeting its zip codes. And for a few days, people in that district who identified as interested in science and parenting or pediatrics would see an ad in their Facebook feed giving them their senator's name, phone number, and talking points. Between Renée and me, we spent less than a thousand dollars combined during the whole effort.[60]

For middle-class professional families, a half-share of a thousand dollars in twenty-dollar increments may be a small investment, but it is not insubstan-tial, and it likely contributed to the success of the movement for the Nonmedical Exemptions Bill.

Sociologist Jennifer Reich identified the networks of privilege and power that facilitate the vaccine refusal of White, middle-class American mothers in her 2014 groundbreaking article.[61] Reich's more recent work with Courtney Thornton demonstrates that similar avenues are often not open to Black mothers, who instead worry about navigating and falling afoul of authorities, face eco-nomic penalties, or lament that moving to states (like Texas) with more lenient NME laws would expose them to higher risks of racist harms.[62] If we flip the lens to consider the impediments to participation in *pro-vaccine* activism, we should be unsurprised to again find White, middle-class educated mothers leading the charge.

Leah Russin clarified that one part of her social privilege—that she was par-enting healthy children—informed her obligation to get involved. "The people in most need of strong vaccination policies are families with children who are immunocompromised," she explained, "And those families are often very busy caring for their child, or advocating for resources for their child's condition. It isn't fair that they should also have to step up to change [vaccination] laws."[63]

The participants in the coalition that mobilized to eliminate NMEs in California gave their time and their money, but they also often paid more significant costs for their advocacy work. Online trolls catfished Dorit Reiss using images of her father, and they manipulated an image of her son to suggest he suffered from a set of ailments that her supposedly neglectful parenting had caused.[64] Hannah Henry's activism pulled her away from her small business, which she eventually folded. Henry also lost friends in her school community after her activism became visible, and members of her community subjected her to threats and verbal attacks. Her family eventually left the Waldorf system and Henry stepped back from social media. When she spoke to Katie for this book in 2019, she was not active in ongoing immunization advocacy.[65]

5

Social Meaning and Political Conflict

Introduction

The pro-vaccine parents described in Chapter 4 believed nonmedical exemptions (NMEs) normalized vaccine refusal. They thought that California's decision to offer NMEs treated vaccination as another activity about which reasonable parents might disagree, rather than an enforceable moral obligation to promote community health. Therefore, their efforts to eliminate NMEs were focused not only on a *policy* intervention but also on the *social identities* of vaccine refusers. California's Nonmedical Exemptions Bill sought to construct vaccine refusal as a form of social deviancy.

This chapter focuses on the social and political conflicts surrounding California's Clinician Counselling Bill, Nonmedical Exemption Bill, and Medical Exemption Bill, and it explores the meaning of these conflicts for the social positions occupied by vaccine refusers. The chapter concludes by considering connections between the politics of these three pieces of legislation and more recent conflicts about COVID-19 mitigation measures.

Nonmedical Exemptions Normalize Non-Vaccination

America's social and political institutions protect a wide scope for parental discretion, including the right to subject children to risky and even dangerous activities. Many American parents enroll their children in full-contact sports—for example, football and hockey—that can cause brain damage and other serious injuries.[1] Some families embrace conservative sexual moralities that dramatically increase risks of suicide and mental illness for LGBTQI children.[2] Parents often also allow their children hours of screen time each day, which can be bad regardless of what children view; the fact that children sometimes interact with destructive social media apps or view pornography is especially distressing.[3] The state generally tolerates wide discretion for such suboptimal parental choices, although governments are often more eager to regulate the choices of parents of color or parents who are poorer or less educated.[4]

In states that offer NMEs for vaccine mandates, vaccine refusal becomes just another way to be a parent. The existence of a mandate signals that the state believes vaccination is *optimal*, but the availability of NMEs denotes that vaccine refusal is an acceptable kind of *suboptimality*.

From the 2000s to the 2010s, California's vaccine refusers occupied prominent spaces in mainstream society.[5] Celebrities such as Alicia Silverstone, Jenny McCarthy, and Mayim Bialik publicly shared why they had decided to refuse vaccines or to follow alternative schedules.[6] Many online and in-person parenting communities embraced non-vaccination as a part of a broader set of "alternative" practices.[7] Members of some vaccine refusers' social circles surely disapproved of their decisions, but these disagreements were often presented as reasonable differences of opinion, much as parents sometimes disagree about how long to breastfeed or whether to circumcise infant boys.

However, NMEs did more than accommodate already committed vaccine refusers. They also fostered a culture that tolerated and contributed to vaccine refusal, and which made disease outbreaks more likely. For example, a 2004 study in *Pediatrics* reported that California—whose NMEs were at that time easy to acquire—had one of the highest percentages of unvaccinated children in the country.[8] And a 2013 study in the same journal reported that clustering of NMEs was a relevant factor in California's 2010 whooping cough outbreak.[9] Also, consider Elisa Sobo's anthropological study of a California Waldorf school, which found that 71% of seventh-graders in the school had an NME in the 2011–2013 period, while 57% of students in such schools across the state held an NME.[10] Sobo observed,

> Joining the school's community . . . often intensified vaccine avoidance and even propagated it among previously vaccinating parents. The social fabric of the school served as an incubator, fostering the extraordinarily high PBE rates . . . and encouraging the noted drop in vaccinations for a family's younger children.[11]

California seemed to be undermining public health through its permissive NME policies.

The mere fact that California offered NMEs was unlikely to have made someone become a vaccine refuser. But Sobo's work shows how local cultural values can intersect with statewide policies to create unvaccinated clusters in schools and communities. Indeed, there is good evidence that vaccine attitudes spread socially and that the vaccine attitudes of others in our social networks are more predictive of our vaccination status than are our own beliefs about the safety and efficacy of vaccination.[12] Furthermore, mere proximity to refusers can lead parents who previously have vaccinated their children to "find exemptions both more acceptable and more desirable."[13]

Vaccine refusal is heterogeneous and it has many drivers, including social media, contemporary norms of parenting, risk framing, economic privilege, and alternative lifestyles.[14] Accordingly, the easy availability of NMEs in California was not the only explanation for the growth in vaccine refusal in that state. But California's long-standing NME policies surely played a prominent role in *normalizing* vaccine refusal.

Making Vaccine Refusal Deviant

California's initial 21st-century battle over NMEs was fought over Dr. Richard Pan's Clinician Counselling Bill. This bill, which passed in 2012 and was implemented in 2014, required a clinician to attest that they had counselled parents about their decision to refuse vaccines. As we discussed in Chapter 3, the Clinician Counselling Bill was a small policy tweak, especially compared to the Nonmedical Exemptions Bill that followed it. But in 2012, California's vaccine refusers responded to the prospect of mandatory counselling as if it violated their fundamental rights.

Dr. Bob Sears criticized the counselling requirement for imposing a time-consuming burden on parents.[15] Barbara Loe Fisher, from the National Vaccine Information Center—an organization that spreads falsehoods about the relationship between vaccination and autism—concurred that the Clinician Counselling Bill forced parents to pay for an "expensive appointment at a medical doctor's office."[16] California's vaccine-refusing parents worried that they would not be able to find clinicians who would sign NMEs, since providers had the right to refuse. Critics also claimed that the policy would financially burden the state when families on public assistance or state-provided health insurance requested additional physician appointments to discuss vaccines. Some argued that the law discriminated against complementary and alternative medicine users because it allowed parents to meet the counselling requirement only by meeting with mainstream clinicians. This objection was somewhat addressed by an amendment that included naturopaths among the clinicians who could provide required vaccine counselling.

This brief outline of objections demonstrates that vaccine refusers experienced the imposition of a small additional burden as a profound encumbrance. Perhaps vaccine refusers sensed that the cultural tides were turning against them, that their good social standing was precarious, and that additional attacks were forthcoming. Advocates of the Clinician Counselling Bill argued that their goal was to ensure that vaccine-refusing parents were making "informed" choices, but critics claimed that the true goal of the legislation was to make it much more difficult for California parents to refuse vaccines. Sears noted,

> The sponsors of this bill may have some good intentions, as their primary "public" reason for the bill is to make sure that parents who don't vaccinate their

children are making an informed medical decision under the guidance of their
doctor. But it isn't difficult to see the REAL reason for the bill: to increase vaccination rates in our state by making it more difficult for parents to claim the
exemption.[17]

Sears was correct. The Clinician Counselling Bill's proponents wanted to make
vaccine refusal more difficult. Parents who valued vaccine refusal were right to
feel under siege.

Leaders of the movement against the Clinician Counselling Bill called on
their supporters to contact their political representatives and to lobby against the
bill.[18] For example, California parent Holly Blumhardt led a Change.org petition
to oppose the new legislation, and she sent it to 20 senators with 3,317 signatures
of support.[19] Parent activists created organizations to mobilize resistance.[20]
Celebrities took their star power inside the Capitol buildings, where actor Rob
Schneider testified against the bill in front of the Senate Committee on Health.
Schneider also networked with Republican politicians at rallies against the bill.
He and Republicans jointly denounced the Clinician Counselling Bill as a kind
of "tyranny."[21]

The Nonmedical Exemption Bill's Direct Attack

Chapter 4 told the story of the press conference on February 4, 2015, during
which an alliance of pro-vaccine parents and politicians announced their intention to abolish NMEs.[22] Two weeks later, on February 19, Senators Richard
Pan and Ben Allen introduced the Nonmedical Exemptions Bill to the Senate.[23]
Vaccine refusers saw Pan and Allen's bill as a declaration of war, and they soon
rallied to defend themselves.

Celebrities again featured prominently in public-facing messaging against
the new bill. Champion mixed martial arts fighter Uriah Faber spoke out against
it,[24] and celebrity vaccine-skeptical physician Andrew Wakefield came to town,
too.[25] Many other well-known anti-vaccine activists and celebrities—such as Jim
Carrey, Jenny McCarthy, and Rob Schneider—spoke at rallies and mobilized online. They argued that vaccination was unsafe and that eliminating NMEs was
going to hurt children. Some celebrities resisted the Nonmedical Exemptions Bill
by appealing to the importance of parental choice rather than to the supposed
evils of vaccines. For example, actress Jenna Elfman tweeted, "Parents should
vaccinate their children as much as they wish to in accordance with current law.
It's THEIR RIGHT."[26] Elfman depicted the mandate & exemptions model as a
reified artifact ("current law"). This positioned the proposed changes as an ille-
gitimate attack on the status quo.

Vaccine-refusing parents set up an organization called Our Kids Our Choice, and they urged their supporters to contact elected officials, spread the word, finance opposition, and attend hearings. When the Nonmedical Exemptions Bill was discussed in the Senate Judiciary Committee, protesters packed the room with signs such as "SB277 makes my child a truant."[27] This message focused on the new kinds of social deviance that the Nonmedical Exemptions Bill would create for vaccine refusers and their families.

Hundreds of opponents rallied on the Capitol steps on the day the Assembly passed the Nonmedical Exemptions Bill. Many wore red shirts and carried placards with slogans. Among them were Assembly representatives who opposed the bill, including Republicans Shannon Grove, Devon Mathis, and Jim Patterson. Grove declared that the politicians behind the bill "don't trust you as parents" and "assume you're ignorant or stupid." She also told the crowd to keep fighting because if they "cowered down" on this bill, it would not be the last time the government attacked them. Patterson claimed that letting the government remove unvaccinated children from school was like placing them in internment camps, although he later retracted this statement.[28]

When Governor Brown signed the Nonmedical Exemptions Bill on June 30, 2015,[29] opponents claimed that it unfairly restricted their choices, deprived unvaccinated children of their constitutional right to an education, and that Senator Pan was "facilitating hate" against them. They also signaled plans to challenge the legislation. The Nonmedical Exemptions Bill had succeeded in placing vaccine refusers outside the confines of polite society, and they were not content to stay there.

Vaccine refusers continued to fight the Nonmedical Exemptions Law by pursuing a referendum to overturn it and by initiating other legal challenges. They clothed these (ultimately doomed) efforts in the rhetoric of parental rights, equal treatment, and fundamental freedoms,[30] and their struggle breathed life into existing and new resistance organizations.[31] Circle of Mamas (an anti-vaccination awareness-raising online initiative) and Learn the Risk (founded by former pharmaceutical employee Brandy Vaughan to resist SB277) spoke out against vaccines through rallies, online organizing, billboards, and truckside ads.[32] Voice for Choice targeted the Nonmedical Exemptions Bill and mobilized the language of "informed choice" to criticize the new law.[33]

Medical Exemptions

Another way that California's vaccine refusers responded to the elimination of NMEs was to seek out medical exemptions. After the Nonmedical Exemptions Bill passed, parents were soon circulating contact information for doctors who were willing to write medical exemptions without examining children or their records. These efforts had dramatic effects. Health researchers soon noted a "replacement effect" in exemption rates at California schools as many families who would have previously received NMEs now submitted medical exemptions.[34] Between the 2015–2016 and 2017–2018 school years, rates of medical exemptions among kindergartners in California more than tripled, from 0.2% to 0.7%. Mohanty and colleagues also noted that these new medical exemptions were often geographically clustered: "Counties that had high PBE [personal belief exemption] rates before SB277 also had the largest increases in medical exemptions during the first year of SB277 implementation."[35]

California parents may have underutilized medical exemptions when NMEs had been available. It is possible that some parents applied for NMEs even though their children were eligible for medical exemptions, perhaps because the NME application appeared to be easier to complete. But this kind of case likely explains only a small part of the post-SB777 increase in medical exemptions rates, and there is good reason to think that many of the new medical exemptions were fraudulent. Recall that one compromise that facilitated passage of the Nonmedical Exemption Bill was that doctors could continue to take "family history" into account when determining whether to offer medical exemptions. This seemingly small measure cut a deep cleft through the law, and vaccine-refusing families were clamoring to pass through it. A small cadre of pediatricians, some of whom had previously been sources for easy NMEs (under the requirements of the Clinician Counselling Bill), now invited parents to apply for no-questions-asked medical exemptions.

After the Nonmedical Exemptions Law was implemented, California journalists soon began reporting about the ready availability of medical exemptions and the dramatic increase in their numbers.[36] A *Los Angeles Times* report noted a proliferation of websites that coached parents about how to get a fraudulent exemption.[37] The California Medical Board investigated complaints and, in 2016, sanctioned Dr. Bob Sears for granting a medical exemption to a 2-year-old without obtaining a medical history.[38] By 2017, the *Los Angeles Times* identified 51 complaints about improper exemptions.[39] "Is it an abuse? Of course it's an abuse," Dorit Reiss told the newspaper, "The law left discretion to the doctors and of course that means doctors can abuse that discretion."

By 2019, the *Voice of San Diego* was on the case. It lodged a public records request to determine how many medical exemptions had been validated by clinicians in their area. Journalists found that one San Diego physician, Tara Zandvliet, had authorized one-third of her school district's medical exemptions in 2015. Fortunately, few physicians appeared to be abusing the system as badly as Dr. Zandvliet, since the *Voice*'s database identified only a handful of doctors who awarded more than 10 medical exemptions in 2015.[40] But even if few physicians were writing fraudulent medical exemptions, they were causing a significant problem, and there was little that California could do to stop them.

Before 2019, there was little that anyone could do to rein in fraudulent medical exemptions. The Medical Board of California could intervene—as it did in the case of Dr. Bob Sears—but only after it received complaints about individual physicians. Families who received fraudulent medical exemptions were not going to complain, and vaccine-supporting parents were unlikely to be aware of which physicians were granting fraudulent medical exemptions. (The example of Sears' censure is an exception that proves the rule: He was the most prominent physician critic of California's new vaccine laws and had been public about his willingness to offer medical exemptions to vaccine-refusing families.) Furthermore, California's state and local governments lacked the ability to verify whether individual exemptions were legitimate or to investigate physicians who seemed to be granting a suspiciously high number of exemptions.[41]

These barriers did not prevent public health leaders from trying to address the issue. For example, the Santa Barbara County Health Officer, Dr. Charity Dean, wrote to her county's school and child care administrators in June 2016, directing them to submit copies of all medical exemptions to her Immunization Program staff. She indicated that the exemptions would be reviewed individually and would contribute to an anonymized data set so that her request would not breach federal privacy legislation.[42] Under Dean's leadership, Santa Barbara County intended to apply oversight to medical exemptions via what it called a "Medical Exemption Pilot Program," which Dean hoped other counties would emulate.

Vaccine refusers resisted Dean's initiative. A Voice for Choice member called Dean "power hungry" and accused her of making herself the "czar" of medical exemptions.[43] When Katie interviewed Health Officers Association of California Director Kat DeBurgh in 2019, DeBurgh described the "chilling effect" of Dean's harassment on other California health officers:

> The names of her kids got published, it was a really brutal thing. And the health officers as a whole are brave and good individuals, but you think twice when you see a colleague go through what she went through, before you take action. And it had nothing to do with how legal something was. It was just purely harassment.

Other county health officers faced similar challenges when they tried to restrain the avalanche of medical exemptions their counties' daycares and schools received in the period following the implementation of the Nonmedical Exemptions Law.[44] In response, policymakers, activists, and researchers started to call for greater state-level governance of medical exemptions.[45]

Fixing the Loophole: The Medical Exemptions Bill

After the passage of the Nonmedical Exemptions Bill, pro-vaccine parent activists could have returned to their lives, careers, and families. After all, they had won, and vaccine refusers were now on the outside looking in. Civil society organizations could likewise have shifted their focus to other pressing health or social problems. But increased medical exemption rates indicated that the work of their alliance was not yet done. Vaccine refusers were still finding ways to enroll unvaccinated children in daycare and school. The Nonmedical Exemptions Bill had tried to make vaccine refusal socially deviant, but easy-to-receive medical exemptions helped vaccine refusers continue with their normal lives.

Senator Richard Pan saw that he needed to mobilize California's civil society organizations and Vaccinate California to help reform medical exemptions. Dorit Reiss agreed to help lead the charge. Pan and Reiss argued in *Pediatrics*— the official journal of the American Academy of Pediatrics—that the government should regulate medical exemptions because those exemptions were an exercise of state power rather than personal medical decisions: "Policymakers should recognize that granting MEs . . . is not the practice of medicine but a delegation of state authority to licensed physicians to protect public health and individuals. Essentially, physicians are fulfilling an administrative role."[46] Pan and Reiss's publication in *Pediatrics* placed their advocacy for regulating medical exemptions in the context of efforts by elite medical institutions (the American Medical Association and the American Academy of Pediatrics), Democratic politicians, and progressive activists to further marginalize vaccine refusal.

State oversight of medical exemptions was already an established practice in West Virginia, which provided a useful policy template for California's proposed policy change. When they pushed to eliminate NMEs, Pan and his team had shied away from parallels with West Virginia, but they did not hesitate now.

Pan's Medical Exemptions Bill, Senate Bill 276 (SB276), would permit physicians to authorize medical exemptions only on standardized forms, and it would require those forms to be submitted to a statewide database. The bill would also authorize California's Department of Public Health to review exemptions and to identify and revoke those that were not consistent with national guidelines. Physicians who were found to have submitted illicit exemptions would be banned from submitting medical exemptions in the future, and they would be reported to the California Medical Board for possible discipline.[47]

Reiss observed that the Medical Exemptions Bill was more politically palatable than the previous Nonmedical Exemptions Bill had been because "you don't have to finger the parents." Instead, the "baddie" was the "the doctors writing fake exemptions," and the Medical Exemptions Bill addressed "a few bad apples that are abusing the system."

However, Reiss argued that vaccine refusers were directly responsible for the new bill because it responded to their new behaviors after the implementation of the Nonmedical Exemptions Bill. In particular, Reiss reflected on vaccine opponents' backlash against Santa Barbara County's oversight of medical exemptions: "The people who are aggressively opposed to SB276 [the Medical Exemptions Bill] are exactly the people that worked really hard to undermine the other tools [e.g., oversight by county health officers]. So . . . if they don't want to have other ways of oversight, they're going to have this one."[48]

The vaccine-refusing parent activists who resisted the Medical Exemption Bill generally avoided discussing medical issues about vaccination but focused instead on ethical and political values. For example, they invoked rights to bodily autonomy and claimed that a right to refuse vaccines was tantamount to the right to receive an abortion.[49] Christina Hildebrand, founder of A Voice for Choice, claimed the mantle of "MLK [Martin Luther King] and the civil rights movement."[50] Another common trope was that the bill would place oversight of people's "health in bureaucrats' hands," which would constitute a "complete tyranny."[51] Along these lines, Barbara Loe Fisher, of the National Vaccine Information Center, claimed, "Forcing physicians to violate their professional judgment and their conscience is a form of state-sponsored tyranny that should not be part of public health law in any state."[52]

When critics of the Medical Exemptions Bill did invoke medical issues, they framed their objections in terms of parents' expertise about the medical needs of their children. For example, actress Jessica Biel posted the following on Instagram:

> I . . . support families having the right to make educated medical decisions for their children alongside their physicians. . . . That's why I spoke to legislators and argued against this bill. Not because I don't believe in vaccinations, but because I believe in giving doctors and the families they treat the ability to decide what's best for their patients.[53]

Others compared the Medical Exemptions Bill to the horrors of racist and extremist regimes. Some protesters carried signs saying "Welcome to Nazifornia."[54] Actor Rob Schneider tweeted about "the People's Republic of Chinafornia" and referred to Senator Pan as "Chairman Mao Jr." in a racist attack against the Asian American legislator.[55]

Pan had been the victim of racist attacks and death threats from the moment he started pushing for legislative reforms to California's immunization policies. He argued that his opponents' only methods are "deception, intimidation and bullying" because they "don't have science on their side." Pan described on-line attacks on other physicians who promoted vaccination, "not only in so-cial media—they'll go attack their reputations online too, right, give them bad reviews." For elected officials, it was even worse: "We regularly get death threats, which we have to report." Indeed, during Katie's interview, Pan was interrupted by a phone call from his wife on an unrelated matter. Reassuring Katie that he was fine to continue with the interview, he noted, "The anti-vaxxers also attack my family, so I have to keep an eye on that."[56]

Pan did not allow threats against him and his family to deter him. "It tells me that I'm doing the right thing. I've dedicated my life to the health of chil-dren, protecting children, so I'm not gonna be deterred by people who want to threaten my life and attack my family."[57] Referring back to his formative years as a physician, he asked the *San Jose Mercury News*, "How could I be deterred when I'd seen the danger of these diseases firsthand?"[58]

Only a few weeks after Katie's interview, Kenneth Austin Bennett, an anti-vaccine activist, physically assaulted Pan on the streets of Sacramento. Bennett "livestreamed it on Facebook—he was proud of what he did," Pan told the media.[59] Bennett later stated that "if Pan got what he deserved he would be hanged for treason for assaulting children, for misrepresenting the truth."[60] Bennett was not the only parent activist who embraced violence. In the days before Governor Newsom signed the Medical Exemptions Bill, protesters blocked the entrance to the Capitol, harassed senators as they entered and left the building, and threw a menstrual cup full of blood onto senators. "It looked like it was thrown at me," Pan told *Sactown Magazine*, "because it was splattered around me."[61]

The Medical Exemptions Bill had mainstream critics, too. Governor Newsom and members of the Medical Board of California questioned whether the state had the authority to regulate medical exemptions, and Newsom stated that he was reluctant to have the state interfere with the relationship between doctors and patients.[62] Even more significant was that Senator Ben Allen, who had sponsored the Nonmedical Exemptions Bill, refused to support the Medical Exemptions Bill and abstained from voting on it, in light of his earlier promise to protect medical exemptions.

Pan and Newsom eventually reached a compromise that heightened the threshold for government oversight of medical exemptions, such that only schools with vaccination rates below 95%, or doctors who granted five or more exemptions per year, would be subject to review.[63] But just when Pan thought his bill was over the line, Newsom raised further demands, which had to be addressed in a separate bill. Senate Bill 714 grandfathered existing medical exemptions and removed a provision that made doctors liable to perjury charges for certifications of invalid medical exemptions.[64] The fraught Medical Exemptions Bill eventually passed, but political compromises had extracted several of its teeth.

A Turning Tide?

The bloody fights over the Medical Exemption Bill all took place in 2019, the same year that a novel coronavirus was identified in Wuhan, China. Further battles over routine childhood immunization were soon overshadowed by the disruptions of the COVID-19 pandemic and by battles over pandemic control measures. Before we turn our attention to this COVID-19 coda, we first reflect on how significantly the 2015 Nonmedical Exemptions Bill and the 2019 Medical Exemptions Bill changed the social meaning of vaccine refusal in California and throughout the the U.S.

In 1998, Andrew Wakefield's now-retracted *Lancet* paper attempted to link vaccination to autism. It received a sympathetic, if critical, reception. Early 2000s celebrity couple Jenny McCarthy and Jim Carrey were regular features on television programs, where they spouted falsehoods about vaccines and about the health of McCarthy's child. The couple were often criticized, but they continued to be welcomed into mainstream media spaces. Even if they were thought to be misguided, they were generally considered to be well-meaning. There was a sense that scientists and public health officials had a responsibility to try to reach and reassure people such as McCarthy and to show them that vaccines were safe and effective. Few people were protesting McCarthy's appearances on television shows, and no one was staging demonstrations at her public talks. She remained a member of "polite society."

Contrast McCarthy's experience with the treatment of Rob Schneider and Jessica Biel after they attacked the Nonmedical Exemptions Bill and Medical Exemptions Bill. Schneider lost a lucrative sponsorship deal, and Biel was roundly rejected and de-platformed. No major television personality invited these actors on air to talk about vaccines. The mainstream community's patience for vaccine refusal seemed to have been exhausted.

Similar things happened in families and friendship groups across the country and the world. Parents who vaccinated their children may have tolerated their anti-vaccine friends or relatives in the early 2000s, but they were less willing to do so by the late 2010s.[65] Even the libertarian tech overlords who control our internet spaces—such as Google, Facebook, Twitter, and Instagram—appeared to be losing patience with vaccine refusers, or at least they started to align their platforms' policies with reduced tolerance among the general public. They took steps to limit anti-vaccine speech and anti-vaccine community organizing.[66]

During the period between California's Nonmedical Exemption Bill and its Medical Exemption Bill, other political communities around the world implemented new vaccine mandates, expanded their existing mandates, or made existing mandates more stringent.[67] In 2016, Australia's Federal Government withdrew NMEs to vaccine mandates associated with family assistance payments and child care subsidies. (This was called "No Jab, No Pay.") Later, several Australian states required vaccination for enrollment in child care and early education. They also chose to withdraw or not offer NMEs.[68] (This was called "No Jab, No Play.") These policy changes drew on community sentiment in much the same way that California's policy changes did, although the media played a much stronger role in Australian efforts to diminish and stigmatize vaccine refusal.[69] When Italy introduced preschool mandates for an expanded suite of vaccines in 2017, the government was also acting in response to recent disease outbreaks and parent advocacy. France implemented a more expansive and stricter mandatory vaccination scheme in 2018.[70] In 2019, Germany introduced preschool and child care vaccine mandates for measles.[71] Germany's immunization policy developments mirrored California's: Its new vaccine mandate replaced a clinician counselling requirement that it had introduced only in 2015.[72] In the United States, Maine, New York, and Washington passed legislation in 2019 to eliminate NMEs for school-entry vaccine mandates in the wake of measles outbreaks.[73] In 2021, Connecticut did likewise.

But in California, vaccine refusers continued to find new ways to avoid vaccinating their children. Some children were able to remain in school because SB714, the companion bill to the 2019 Medical Exemptions Bill, "grandfathered" their medical exemptions. Other families turned to homeschooling.[74] Also, despite the state's new oversight of egregious cases, physicians retain the right to consider family history in granting medical exemptions. Still other parents were able to have their children diagnosed with learning disabilities, which then entitled them to remain in school under the Individuals with Disabilities Education Act.[75]

The activism of California's vaccine refusers likely influenced efforts to protect NMEs in other states. For instance, Maine's NME elimination bill passed by only one vote in the face of a mobilized opposition. The defenders of Maine's NMEs collected enough signatures to force the state to hold a referendum on overturning the law. The referendum failed, and Maine's law eliminating NMEs was upheld.[76]

COVID-19 Politics

Many of the parents and organizations that mobilized to resist changes to California's childhood vaccine mandates also objected to COVID-19 pandemic control measures and COVID-19 vaccines.[77] Freedom Angels, a group that protested the Medical Exemptions Bill, organized "Operation Gridlock" at California's Capitol to protest COVID-19 lockdowns and restrictions.[78] In June 2020, protesters harassed health officials at their homes and denounced mask mandates and COVID-19 vaccines, even though COVID-19 vaccines were not yet available.[79] On January 30, 2021—an early date in the public rollout of new COVID-19 vaccines—anti-vaccine protestors swarmed the entrance to the Dodger Stadium mass vaccination clinic and shut it down.[80] The participants in the Dodger Stadium protest included many of the same people—and featured much of the same rhetoric—from the fights over the Clinician Counselling Bill, the Nonmedical Exemptions Bill, and the Medical Exemptions Bill. These groups continued to mobilize against COVID-19 vaccine mandates.

Familiar faces led California's efforts to promote COVID-19 vaccination and to overcome refusal of the new COVID-19 vaccines. For example, in February 2021, Senator Pan introduced Senate Bill 742 to protect vaccination centers from intimidation and physical obstruction by anti-vaccination activists. This bill, signed by Governor Newsom in October 2021, aimed to prevent disruptions such as the one at Dodger Stadium.[81] In August 2021, a Public Health Order from the California Department of Public Health required school staff to show proof of full vaccination against COVID-19 or submit to weekly COVID-19 testing, building on similar policies imposed on other government employees and health care workers.[82] In October 2021, Governor Newsom announced that middle school and high school students would need to be vaccinated against COVID-19 once vaccines for their age groups received full U.S. Food and Drug Administration approval.[83] In January 2022, Pan introduced Senate Bill 871 (SB871), which would expand COVID-19 vaccine mandates for all children in child care and in public or private schools. He also introduced Senate Bill 866, with fellow Democratic senator Scott Weiner, to grant children ages 12 years or older the legal right to consent to vaccination, even without parental permission. The legislature fast-tracked both bills, but there was substantial opposition, both from vaccine refusers and from the Capitol Resource Institute, a civil society organization "working to preserve and advance a culture of traditional family values," which meant that it opposed sex education, critical race theory, and, apparently, vaccine mandates.[84]

In April 2022, Pan decided to abandon SB871, stating that COVID-19 vaccination rates for children remained "insufficient" and that "the state needed to focus its efforts on increasing access to COVID vaccinations for children through physicians and other health providers who care for children and on education efforts to give families accurate information about the COVID vaccine."[85] Four months later, Weiner abandoned SB866 when he could not secure sufficient votes to proceed; Weiner blamed "months of harassment and misinformation—including death threats against me and teen advocates—by a small but highly vocal and organized minority of anti-vaxxers."[86]

Conflicts over California's COVID-19 mitigation measures accelerated and transformed political fights about immunization policy in ways we are only now beginning to understand. The attachment-parenting moms who were once the public face of California's vaccine resistors have, in the COVID-19 era, been joined by gun-toting White-supremacist men in camouflage militia uniforms. It is beyond the scope of this book to try to trace this recent evolution of California's—and America's—public health governance battles. But there are nonetheless clear connections between the conflicts over governing routine childhood immunization in the late 2010s and the COVID-related struggles that began in 2020.

6

Drawing the Wrong Lessons from the History of Mandates

Introduction

The California case demonstrates the centrality of nonmedical exemptions (NMEs) to contemporary debates about vaccine mandates. The "mandates & exemptions" regime that U.S. state governments adopted during the 1960s and 1970s nudged parents toward vaccination, but—in almost all cases—it allowed committed refusers to send their undervaccinated children to school. Contemporary efforts to eliminate NMEs therefore aim to transform America's vaccine mandates into coercive policies, with substantial ethical and political consequences.

It is common for today's advocates of eliminating NMEs—or of otherwise tightening vaccine mandates—to invoke much earlier instances of vaccine mandates to support their cause. Some point out that General George Washington compelled the Revolutionary Army to be variolated against smallpox during the Valley Forge encampment of 1777–1778, or that England mandated the smallpox vaccine within a few decades of its development in the 19th century.[1] Others observe that the U.S. Supreme Court has long upheld vaccine mandates and has determined that the Constitution does not require states to offer NMEs.[2] According to such claims, the elimination of NMEs—and the increased coercion of school vaccine mandates—restores well-established and legal forms of public health governance. It follows that California's Nonmedical Exemption Bill and other efforts to eliminate NMEs are *not* watershed moments for public health. They merely reinstate an earlier—and a *better*—kind of immunization policy.

We disagree. The history of coercive vaccination does not lend support to contemporary efforts to eliminate NMEs because the political, legal, social, and ethical contexts are so different and also because history illustrates that long-lasting and effective coercive immunization policies can generate immense backlash. In the U.S. context, the earliest coercive vaccine mandates were usually enforced only during major outbreaks and against targeted "captive" populations. In contrast, today's school-based vaccine mandates are sustained government efforts to ensure the vaccination of all children. Also, governments enforced previous eras' coercive vaccine mandates on populations whose ethical and political sensibilities were very different from those today. In particular, recent decades have witnessed the institutionalization of many limits on public health powers, the development of conscientious objector provisions, and new kinds of ethical ideals for medicine and public health. Accordingly, even when today's more coercive vaccine mandates superficially resemble those from the distant past, their legal, political, and ethical implications are vastly different. Finally, the historical record provides little reason to be complacent about the stability or long-term effectiveness of large-scale coercive immunization programs. In particular, England's experience with coercive vaccine mandates in the 19th century illustrates that increasing the coerciveness of immunization governance can undermine public health by generating powerful forms of social and political backlash.

The 19th-Century State

Early vaccine mandates in the United States do not represent a "pure" form to which contemporary efforts to eliminate NMEs aim to return. Instead, the early history of America's experience with coercive immunization governance consisted of isolated episodic emergency extensions of nascent state power, rather than ongoing exercises of comprehensive governance.

We can illustrate this difference by attending to the development of modern state capacities in Western societies from the end of the 18th century to the beginning of the 20th century. During this time, states increased their abilities to improve people's lives using the natural and social sciences; new techniques for investigation, information collection, and distribution; and the power of nationalism and patriotism. Governments pursued ambitious public projects and recruited new bureaucratic armies of public workers to promote the welfare of individuals and communities. In particular, a core ideal of the Enlightenment period was to use new scientific and medical knowledge to promote the efficiency of economic production and the well-being of the public, often by expanding the state's capacity to promote health.[3] French scholars often led the way in the period prior to the Revolution—for example, much of Diderot's *Encyclopédie* project focused on health issues—but Jeremy Bentham and other English reformers took up the lead by the early 19th century.[4]

Bentham and his followers, known as the Philosophical Radicals, developed the "theoretical underpinning for British social and health policy" and helped "create the modern public health movement" in many other countries, including the United States.[5] Bentham intended his philosophical work to help justify and inform his practical reforms, many of which aimed at public health goals or goals that were public health adjacent. The public projects pursued by Bentham and his followers (including the great sanitarian, Edwin Chadwick) included supplying clean water, constructing sewers, instituting public sanitation inspectors, funding education, expanding police services, and investing in new kinds of prisons and workhouses. These efforts required substantial increases in the size of the state and of its power over people.

States have always been powerful, but the modern state that began to emerge in the mid-19th century had a new kind of power. It could observe, regulate, and intervene in many more parts of people's lives. It could push into people's homes and into what may have previously been private decisions. Premodern governments were hardly respectful of individual liberty or peasants' privacy rights: On the contrary, royal prerogatives in medieval England and France often saw leadership as an opportunity to pillage from the population. But premodern states usually lacked the capacity to intervene systematically in the practices of social life. They had relatively little funding and few direct employees. Modern states have many more resources at their disposal. As historian Richard Pipes put the point, the French King Louis XIV (d. 1715) was an absolute monarch who was *allowed to do* whatever he wanted to his people, but Russian Tsar Nicolas II (d. 1918) was a totalitarian ruler who *could do* whatever he wanted to his people.[6] And although many modern governments used their new powers to help their populations, even beneficent uses of state power manifested as newfound intrusions into parts of people's lives that the state had not previously governed. Vaccination and vaccine mandates would take their place among such interventions.

America's Public Health Governance: From the Sanitarians to the Progressive Era

As we discussed in Chapter 2, the U.S. Constitution reserves most police powers to individual states and their municipalities. Accordingly, most U.S. vaccine mandates have operated at the state or local levels. Boston was the first U.S. city to mandate smallpox vaccination for children (in 1827), and Massachusetts was the first state to introduce statewide mandate statutes (in 1855).[7] By the 1880s, city health departments were supervising or conducting mandatory in-school smallpox vaccination across Massachusetts.[8] In addition to vaccine mandates, states and major cities, including New York City, passed laws to promote, supervise, and subsidize smallpox vaccination, especially for poor people and racial minorities.[9] Importantly, the most coercive public health laws—including vaccine mandates—were usually created during emergencies and were rarely enforced after outbreaks passed.[10]

Through the middle of the 19th century, public health was thought to be mostly an urban problem, which is to say largely unimportant in an American society that was predominantly rural.[11] But the latter half of the 19th century saw the rise of the sanitary movement both within government and among elites. Its aim was to regulate food and water supplies and to improve health in schools and in workplaces.[12] In the United States, the public health portfolios of the late-19th century's sanitarian period (clean water, sewers, and food) expanded during the Progressive Era (roughly the 1890s to the 1920s). New movements aimed to promote the well-being of immigrants, factory workers, children, and other disadvantaged members of society, in the context of urbanization, industrialization, and mass immigration.[13] Novel public and private associations sprang up to undertake this work.

Of particular interest to Progressive Era reformers was the health and well-being of children. The 1910s and 1920s saw an expansion of federal and state legislation to protect children from early deaths, the ravages of factory work, and abuses in underfunded orphanages.[14] Many reforms targeted schools, including the creation of school health inspectors, student health exams, hearing and vision tests, the promotion of school nursing, and the introduction of community and school social workers.[15] The workers behind these public programs were often employees of new federal, state, and local departments for public health, sanitation, and education.[16] They were educated members of a new middle class, and they had faith in government's ability to improve the well-being of disadvantaged persons through effective institutions and investments in infrastructure. Duffy observes that "professionalization and efficiency were the key methods by which the Progressive Movement . . . hoped to create a brave new society."[17]

Vaccine mandates were one part of the Progressive Era's broader public health mission.[18] Objection to vaccine mandates was often a potent symbol of a broader resistance to the expansion of state powers during the Progressive Era.[19] Anti-vaccination societies putatively focused on vaccine mandates as an organizing principle, but their members usually objected to other government policies, too. Social workers, visiting nurses, and professionally trained public schoolteachers were pushing themselves into family spaces. Scientific progress multiplied the state's tools to make people's lives better, but Colgrove notes that "these advances also produced an anti-modernist backlash against the paternalistic and potentially coercive uses to which scientific medicine might be put."[20] The public protested against medical inspections of children in schools, and against the forced removal of asymptomatic tuberculosis-infected children from their families.[21]

In the early 20th century, leading progressives—such as Jane Addams, Louis Brandeis, and John Dewey—often prioritized social well-being over individual rights.[22] This ethos animated the new bureaucratic armies of public health workers. But their progressive programs conflicted directly with a powerful American tradition of freedom from government intrusion into private life and civil society. As Michael Willrich puts the point,

> Many antivaccinationists had close intellectual and personal ties to a largely forgotten American tradition and subculture of libertarian radicalism. That tradition took on a feverish new life as industrial capitalism, progressive reform, and the professionalization of knowledge fostered the rise of a distinctly modern interventionist state during the Progressive Era. The same men and women who joined antivaccination leagues tended to throw themselves into other maligned causes of their era, including anti-imperialism, women's rights, antivivisection, vegetarianism, . . . and opposition to state eugenics.[23]

These early 20th-century struggles reveal a deeper cultural cleavage between advocates of progressive state policies and defenders of the Jeffersonian conviction that "government is best which governs least."[24]

Bringing the Empire Home

America's early 20th-century debates about the priority of liberty in its domestic politics stood in tension with its imperial ambitions. America developed its most powerful public health capacities during its violent and oppressive efforts to control disease in the territories it added to its empire in the late 19th century: Cuba, the Philippines, Puerto Rico, and the Panama Canal Zone.[25] American occupiers showed little concern for the liberty of the people whose countries they occupied. Instead, they relied on force to promote public health, both to bolster economic activity in their colonies and also to fulfill what Kipling called the White Man's Burden: "Take up the White Man's Burden / The savage wars of peace— / Full the mouth of Famine / And bid the sickness cease."[26]

America's public health institutions brought home their often-successful experiences of mandatory vaccination and other disease control mechanisms and imposed them on America's own marginalized ("captive") populations. Late-19th-century and early 20th-century coercive immunization programs focused on poor and oppressed populations as the primary vectors of disease and conceived of vaccine mandates as a way to govern the health of people who could not govern themselves (similar to the "beneficent paternalism" that supposedly justified colonialism). These attitudes prevailed among other colonial powers, too. A *Lancet* article from 1894 bemoaned that poor itinerant workers in England were impeding progress in public health because they were "parasite[s] upon the charity and good nature of the community" but were also "vehicle[s] for the spread of other parasites"; the author complained that "no compulsory steps have been taken to curtail seriously the vagrant's movements or to promote his elementary cleanliness."[27]

Paternalistic ideas likewise informed U.S. public health, where the relevant "vagrants" were immigrants or itinerant Black workers—that is, formerly enslaved persons or their descendants. The focus on marginalized people as disease vectors is illustrated by the list of names by which smallpox was known in the early 1900s, including "Cuban itch," "Porto Rico scratch," "Manila scab," "Filipino itch," "Mexican bump," and "Italian itch."[28] These names denote the various peoples of America's empire and the immigrants languishing in urban slums.

The cultivation of U.S. public health capacities grew in the context of Jim Crow restrictions on Black Americans in the American South. For example, Mississippi experienced outbreaks of yellow fever in 1878 and 1897, leading to quarantines enforced by armed guards at town borders, the destruction of railroad lines, the lynching of Black persons believed to be evading quarantine, and the burning of makeshift hospitals. This fiercely coercive response was the combined creation of both government activity and the paranoia of White mobs.[29] Oppressive public health measures were imposed disproportionately and sometimes exclusively on Black populations to supposedly protect White people and to ensure that Black laborers would remain economically productive.[30]

The experiences of public health officials in the Jim Crow regimes of the American South informed similarly coercive interventions in the immigrant slums of America's northern cities. In the early 20th century, many immigrants were not vaccinated against smallpox, and there were frequent outbreaks in overcrowded and under-resourced immigrant neighborhoods. In response, the health departments of major U.S. cities, including New York, Boston, and Chicago, became increasingly militarized, and conducted police-accompanied inspections and raids of tenements, often resulting in forcible vaccination.[31] Early 20th-century urban vaccine mandates were enforced on the poorest and most vulnerable members of society by what Michael Willrich has called "paramilitary vaccination squads."[32] Mandatory vaccination of these populations had an epidemiological justification: They were more likely to carry and spread disease. The Southern work camps of previously enslaved Black Americans and the urban slums of new immigrants were crowded, unsanitary, and offered little access to resources for healthy living.[33]

Jacobson and Public Health Police Powers

Jacobson v. Massachusetts (1905) crystallized early 20th-century American legal views about the relationship between public health police powers and individual rights.[34] Supporters of vaccine mandates frequently invoke this case, but the dark legacy of *Jacobson* complicates its usefulness in contemporary immunization policy.[35] After outlining the case, we consider its relationship to another 1905 case (*Lochner v. New York*) that prevented state government from protecting workers' health by regulating labor contracts. We then trace the legacy of *Jacobson* to the Bioethics Revolution of the 1970s and 1980s via the history of eugenics and Nazi medical atrocities.

Jacobson concerned the constitutionality of a Massachusetts law that allowed municipalities to mandate smallpox vaccines. Refusers such as Reverend Henning Jacobson could be fined $5 unless they qualified for a medical exemption. Jacobson believed that he was especially vulnerable to vaccine complications, although he did not submit evidence to claim a medical exemption (which was available under the Massachusetts law). He instead objected that the law unjustly deprived him of his liberty. Writing for a 7–2 majority in *Jacobson*, Justice Harlan rejected Jacobson's claim that members of society retain a natural liberty from external constraint of their bodies. Instead, Justice Harlan argued that modern political life involves a

> social compact that the whole people covenants with each citizen, and each citizen with the whole people, that all shall be governed by certain laws for the "common good," and that government is instituted "for the protection, safety, prosperity and happiness of the people, and not for the profit, honor or private interests of any one man."[36]

Here, we see the idea that liberty requires a "social conscience and a powerful interventionist state" in light of "the overwhelming social and economic forces of modern urban–industrial life."[37] This is a far cry from Jeffersonian liberalism and limited government. We applaud the Court's finding that the state may sometimes infringe on individual liberties to address social injustices, but we worry about the potentially unjust uses to which state power can be put in efforts to promote the "common good," as we discuss below.

With regard to vaccine mandates, *Jacobson* has been consistently upheld, and later decisions have expanded *Jacobson* to cover school vaccine mandates (*Zucht v. King* [1922] and *Prince v. Massachusetts* [1944]).[38] Vaccine refusers have frequently attempted to limit *Jacobson*, *Zucht*, and *Prince*—by claiming that mandates should be permitted only during active epidemics or only for vaccines against high-risk diseases—but courts have maintained the sweeping powers those decisions granted.[39] A notable exception, and a telling sign that something big is shifting, has been the Court's recent reluctance to uphold restrictive public health laws during the COVID-19 pandemic.[40]

Two paths leading out from *Jacobson* are relevant to contemporary debates about vaccine mandates. First, there is a striking contrast between *Jacobson*'s toleration of state intrusion upon the bodies of *individual persons* for the sake of public health, and the Court's concurrent rejection of government interventions in *markets* for the same purpose. In the same 1905 term in which it decided *Jacobson*, the U.S. Supreme Court also decided *Lochner v. New York*, which overturned a New York law that aimed to protect the well-being of workers by limiting the number of hours they could labor each week.[41] Justice Peckham's majority opinion found that such laws unjustly violate the liberty of individual workers and owners to contract for the exchange of labor for wages. *Lochner* therefore upheld a near absolute liberty of contract in economic matters, even though (as Justice Harlan noted in his *Lochner* dissent) economic regulations can be part of public health policy and, therefore, *Lochner* appears to be in tension with *Jacobson*.

The combination of these two holdings—the permissibility of coercive public health laws for individuals, but the illegitimacy of corporate regulation, even in the name of public health—informed the development of public health policy in the early 20th-century United States. Even though the New Deal of the 1930s erased much of the anti-regulatory spirit of *Lochner*, a broad reluctance to interfere with economics for the sake of health continued, both in the U.S. legal system and among medical elites. For example, the American Medical Association (AMA) frequently resisted public health measures—and was usually successful in doing so—on the grounds that "socialized medicine" unjustly interfered with physicians' and patients' liberty to contract for services. In contrast, the AMA and other elite medical institutions often supported public health laws that restricted individual liberty—a discussion we return to in Chapter 7.

Legal commentators have sometimes remarked that we appear to be entering a new "*Lochner* era."[42] Conservative justices seem increasingly willing to strike down laws that regulate industry.[43] And the current conservative majority on the Supreme Court appears especially skeptical of executive agencies' powers to create and enforce health and welfare regulations.[44] So, at the same time that many states have been making their immunization mandates more coercive, America's courts may be preventing governments from regulating industry to protect people, or from passing laws to promote health. The early 20th-century marriage of *Jacobson* and *Lochner* seems to be finding new life.

A second trajectory from *Jacobson* is the role it played in authorizing a broad set of early 20th-century public health measures, many of which enacted oppressive violence on society's most vulnerable members. Legal, social, and ethical responses to these abuses—abuses that were validated by *Jacobson*—generated new protections for individual rights. Among these protections were conscientious objector rights, including rights to NMEs to vaccine mandates. The legacy of *Jacobson* makes it difficult to re-create the coercive vaccination policies that *Jacobson* affirmed.

Of particular importance is the direct role *Jacobson* played in authorizing the eugenics movement of the early 20th century, which aimed to promote public health and broader social vitality by improving the "genetic quality" of the population.[45] Notable among state eugenics policies were coercive sterilization measures that attempted to reduce reproduction among "unfit" persons, including the poor, people with mental illness or cognitive disability, people of color, and immigrants. The constitutionality of coercive sterilization was challenged in *Buck v. Bell* (1927), which focused on whether a state law that permitted the sterilization of intellectually disabled persons violated their 14th Amendment due process rights.[46] By an 8–1 majority, the court affirmed the constitutionality of such laws. The majority opinion was written by Justice Oliver Wendell Holmes, Jr., generally a defender of civil liberties and constitutional democracy against the overreaches of state power.[47] However, Holmes endorsed a large exception to individual liberty rights when it came to public health. In *Buck v. Bell*, Holmes argued,

> It is better for all the world, if instead of waiting to execute degenerate offspring for crime, or to let them starve for their imbecility, society can prevent those who are manifestly unfit from continuing their kind. The *principle that sustains compulsory vaccination* is broad enough to cover cutting the Fallopian tubes. . . .Three generations of imbeciles are enough.[48]

Holmes drew a direct line from *Jacobson* to *Buck*. Even though *Jacobson* had affirmed the constitutionality of only a small financial penalty for vaccine refusal, the principle it established was that individual rights to bodily autonomy can be overpowered in the name of public health goals. From an all-things-considered perspective, the trivial fine *Jacobson* validated may seem wildly disproportionate compared with coercive sterilization. But, as Jamal Greene argues in *How Rights Went Wrong*, the Supreme Court often addresses rights in all-or-nothing frameworks, rather than by applying context-specific proportionality tests.[49] Accordingly, *Jacobson*'s affirmation of the priority of public health over individual bodily autonomy provided a sufficient precedent for upholding the constitutionality of coercive sterilization.

America's sterilization programs and its broader eugenics efforts diminished after World War II.[50] The postwar decline of the eugenics movement was no coincidence. It had been a great embarrassment to U.S. public health leaders that the Nazi regime modeled many of their eugenics policies on the American experience.[51]

Jacobson provided legal legitimacy to both mandatory vaccination and forced sterilization. But efforts to grapple with the legacy of *Jacobson*—including horrific Nazi medical abuses—have transformed domestic and international thinking about the rights of patients and research subjects. This new way of thinking about the ethical, social, and legal dimensions of medicine and medical research has been called the Bioethics Revolution. We explore its significance for immunization policy in the following section. For now, we note that NMEs and other medical liberty rights are a fundamental part of *Jacobson*'s legacy.[52] Even though *Jacobson* has not been overturned, it has undermined itself by bequeathing us a world that is much less tolerant of coercive medical interventions.

Postwar Bioethics

Reflection on Nazi abuses transformed the way that ethicists and policymakers thought about the rights of individuals, including as patients, both in the United States and in many other countries. The 1947 Nuremburg Code and the 1964 Declaration of Helsinki stated that patients and research subjects have the right to informed consent. However, physicians and public health officials in the United States initially resisted attempts to incorporate these guidelines in domestic law and in standards for physicians and researchers. U.S. doctors declared that the Nuremberg Code was "a good code for barbarians but an unnecessary code for ordinary physicians."[53] By their own estimation, U.S. physicians were incapable of the abuses Nazi caused, and protection of informed consent was therefore unnecessary. Unfortunately, U.S. physicians wildly overestimated their capacity or propensity to protect their patients and research subjects, and they would not support institutional protections for the rights of patients and research subjects until the 1970s and 1980s.[54]

The AMA's founding Code of Medical Ethics (1847) did not address the ethics of research on human subjects, since organized biomedical research studies on human subjects did not become widespread until the late 19th century.[55] The AMA's code instead emphasized beneficence backed by authority, which permitted physicians to withhold information or use deception for therapeutic benefits. This continued in the 1902 revision, which resisted endorsing patient consent rights to avoid "burden[ing] patients and their doctors" and "interfer[ing] with the vital progress of medical knowledge."[56] Physicians also attempted to avoid ethically charged questions by conducting medical research on institutionalized children and other vulnerable populations (prisoners, the poor, immigrants, and Black persons).[57]

An especially horrific example of postwar American complacency about medical abuses was the U.S. Public Health Service's syphilis experiment in Tuskegee, Alabama, which it named the "Tuskegee Study of Untreated Syphilis in the Negro Male." This 40-year study (1932–1972), funded by the U.S. Public Health Service and Center for Disease Control (CDC), aimed to observe the progression of untreated syphilis. Research participants were Black men with and without syphilis. Participants were deceived about the nature of the study, and those with syphilis were not offered effective treatments even after antibiotics became widely available. Regional hospitals conspired to prevent participants from receiving treatment, and there was clear evidence that the study design and implementation were racist.[58] For example, researchers believed their subjects possessed out-of-control sexual appetites, such that they would willingly have sex with infected women, and it was hence inevitable that all would eventually acquire syphilis.[59]

The Tuskegee Study and its methods were not secret. The study generated dozens of publications, and federal bureaucrats renewed its funding multiple times. In the 1960s, people inside the U.S. Public Health Service occasionally raised ethical objections, but high-ranking officials in the CDC maintained their support until media exposure—and subsequent political pressure—made it unfeasible. The study ended in 1972 only because Peter Buxton, an employee of the U.S. Public Health Service, blew the whistle to the press. Even then, it is likely that the Tuskegee Study became national news only because the civil rights movement had focused national attention on abuses experienced by Black Americans.

The Tuskegee Study was not an isolated incident. It was common to use members of vulnerable populations as medical research subjects. Such subjects were easier to manipulate or deceive into participating because of their social deprivation and relative powerlessness, and powerful social groups showed them little concern. Indeed, throughout the 1950s and 1960s, the U.S. Army directly or indirectly engaged in nonconsensual research on a wide variety of populations, from hospital patients to prison populations and people in crowded airports.[60] Furthermore, some polio vaccine trials were conducted on institutionalized children in New York and on adults in a prison in New Jersey.[61] Jonas Salk, the heroic inventor of the inactivated polio vaccine, also experimented on institutionalized children, although this was not widely reported at the time.[62]

In response to Tuskegee and other abuses, the United States finally codified protections for participants in medical research. It later extended similar protections for patients receiving medical care. In both cases, individuals were now invested with a right to make their own decisions. New institutional protections to regulate research and medical care were implemented by the 1970s and early 1980s, during the same period that the vaccine mandates & exemptions regime was established in U.S. states.[63]

One of the core values that emerged in the 1970s and 1980s as part of the Bioethics Revolution was *informed consent*.[64] This is the idea that patients have the right to decide among potential options for their treatment based on their voluntary, capacitated, and informed choices. Patients who have the capacity to express preferences, understand their condition, and appreciate the outcomes of possible interventions should make their own medical decisions.[65] The prominence of this idea is illustrated by the institutionalization of ethics protections for patients in U.S. hospitals, as well as by the ethics education that medical students and physicians receive.[66]

Advocates of creating and tightening vaccine mandates—for example, by eliminating NMEs—frequently argue that coercive public health policies are consistent with free choice about medical interventions or that restrictions on individual liberty are otherwise justified.[67] Furthermore, pediatric ethicists frequently observe that *childhood* vaccine mandates do not implicate informed consent because children are the patients, not parents, and because young children cannot provide informed consent.[68] (We elaborate on this argument in Chapter 8.) However, it matters very little if philosophers, physicians, and policymakers can convince *themselves* that vaccine mandates do not violate informed consent—or that such violations are justified. What matters is whether members of *the public* agree. At least a substantial minority appears to think that "informed consent" names an expansive right to be left alone when it comes to vaccination decisions for themselves and for their children. It will not solve problems for public health policy to tell such people that they misunderstand what "informed consent" means.

We are perhaps only now starting to consider the threat that the Bioethics Revolution—or at least its popular understanding—poses for public health.[69] George Annas notes that "almost 100 years after *Jacobson*, both medicine and constitutional law are radically different. We now take constitutional rights much more seriously, including the right of a competent adult to refuse any medical treatment, even life-saving treatment."[70] There is at least the appearance of a mismatch between the ethics of medical treatment, which emphasizes patient autonomy and informed consent, and public health ethics, which tolerates coercion to protect the public.[71] It is not clear how to respond when these two frameworks come into conflict, as they seem to do in the case of vaccination. As Bayer and Fairchild put the point, "Compulsion and, indeed, coercion—so anathema to this tradition of bioethics—are central to public health."[72] It is even less clear how we ought to navigate this tension when a nontrivial component of the population embraces expansive conceptions of health care liberty.

Historian Mark Largent suggests that the values of clinical bioethics and public health ethics should be balanced against each other: "In considering vaccine policies, we must recognize that individual American's [sic] rights to privacy, voluntary consent, and personal autonomy need to be similarly weighed against broad public health concerns."[73] However, Largent's solution is feasible in the United States only if Americans are actually committed to *both* sets of principles and are willing to find compromises between them. But there is reason to think that most people "often think of health largely as an individual matter rather than a societal responsibility"[74] and that this is especially true when it comes to their children's health. We return to this theme in Chapter 9.

It is striking that the Bioethics Revolution of the 1970s and 1980s—a period marked by increased commitments to individual rights in medical decision-making, and skepticism about the use of coercion in public health—overlapped substantially with the time period (1960s to 1970s) in which U.S. states implemented modern vaccine mandates. Why would U.S. states implement more coercive immunization measures around the same time that people were demanding more medical liberty? This supposed tension evaporates when we reflect on the details of the mandates & exemptions regime (as we did in Chapter 2): These state-based school vaccine mandates came with NMEs *built in*. Efforts in the 1960s and 1970s to create new vaccine mandates were therefore consistent with a commitment to free choice in health care decision-making. In contrast, contemporary efforts to eliminate NMEs defy recent trends toward greater clinical authority for patients and parents.

So far, this chapter has illustrated a long and indirect journey by which the state's development and use of public health powers generated limits on those same powers: While *Jacobson* affirmed early vaccine mandates, it also supported horrific abuses, and responses to those abuses—including the Bioethics Revolution—have undermined the potential for coercive vaccine mandates today. Accordingly, the NMEs that are the focus of today's reformers are not temporary aberrations, and we should not think of NME elimination efforts as the reestablishment of pure, authentic immunization policies. Instead, we have inherited NMEs as the consequence of previous generations' public health policy abuses.

We now discuss a more direct and powerful way in which coercive immunization policies can undermine themselves, by drawing on England's experience with coercive smallpox vaccine mandates in the 19th century.[75]

The Rise and Fall of Mandatory Vaccination in England

America's earliest vaccine mandates in the 19th and early 20th centuries did not directly generate backlash that was sufficiently powerful to establish widespread NMEs or to eliminate mandates. This is likely because coercive vaccination in America was a local or state matter and because authorities often created and enforced mandates only during outbreaks. In contrast, England's earliest vaccine mandates were long-lasting national policies that generated broad-based resistance.[76] Among the consequences of that resistance was the introduction of NMEs and, soon thereafter, a national commitment to voluntarism in vaccination (which holds to this day). Accordingly, we look to England's 19th-century smallpox vaccine mandates as a case study about how increasing the coercive-ness of broad-based vaccine mandates can, via a backfire effect, generate stark limits on immunization policy.

The English government's formal promotion of vaccines began with the Vaccination Act of 1840, which funded the vaccination of the poor. It also outlawed an earlier practice of cultivating immunity, variolation, on the grounds that vaccination was safer.[77] This law kept vaccination voluntary, but it marked the entry of government into the regulation of immunization. In 1853, the government updated the Vaccination Act to require all children to be vaccinated, introduce state record keeping, and impose a one-pound fine for noncompliance.[78] Taken together, the 1840 and 1853 laws governed a previously ungoverned behavior for *everyone*, seeking to directly increase immunization rates and to instill a new social norm. However, these mid-century English laws disproportionately impacted the poor. A one-pound fee would be more than a fortnight's wages for a laborer, but it would be trivial for an aristocrat.[79]

In 1867, the English government required that children be vaccinated within 7 days of birth and implemented additional inspections to ensure compliance. It also claimed vaccination as a government affair, imposing a 1-month prison term if anyone other than government-authorized agents delivered vaccines.[80] So, two kinds of freedom were under attack: the freedom to refuse vaccines and the freedom to provide vaccines outside of government supervision. In 1871, the state appointed a set of new vaccination officers, empowering them to compel noncompliers to appear in court.[81] Finally, in 1873, vaccination was rendered "compulsory" by changing the penalty for noncompliance from a one-off fine to a schedule of escalating fines. Whereas refusers could previously avoid vaccination (and further penalty) if they had one pound to spare, now the most recalcitrant refusers faced bankruptcy. This made England's post-1873 vaccine mandate much more coercive.

Even though the 1873 update harnessed immense state power to ensure vaccination, it did not initiate a new era of high immunization rates. Instead, it generated a popular backlash that ultimately led to the abolition of vaccine mandates. Anti-vaccine groups mounted massive public protests, often resulting in violence. They circulated pamphlets that painted sympathetic pictures of the vaccine-refusing parents who lost their homes and other possessions. Anti-vaccine candidates ran for Parliament and often won, reaching a high point of 100 members (out of 666) in 1906.[82]

Mass resistance to compulsory vaccination was a political problem for England's emergent administrative state. Parliament created a commission to study this resistance and to propose solutions. In 1896, the Royal Commission on Vaccination's report expressed eminently *modern* ideas about the purpose and limits of state public health powers:

The penalty [a fine, with increased fines for non-payment or non-vaccination] was not designed to punish a parent who may be considered misguided in his views and unwise in his action, but to secure the vaccination of people. If a law less severe, or administered with less stringency, would better secure this end, that seems to us conclusive in its favor . . . [I]t would conduce to increase vaccination if a scheme could be devised which would preclude the attempt (so often a vain one) to compel those who are honestly opposed to the practice to submit their children to vaccination, and, at the same time, leave the law to operate, as at present, to prevent children remaining unvaccinated owing to the neglect or indifference of the parent.[83]

There are some instructive points here for thinking about vaccine mandates in contemporary contexts:

1. The goal of a vaccine mandate is to bring about a population-level increase in vaccination rates.
2. It is *not* the goal of a mandate to punish parents who refuse vaccines (or to punish their children).
3. If there are ways to promote population-level health without trying to compel committed refusers to vaccinate, then those should be attempted first.
4. Vaccine mandates are primarily about overcoming parental indifference or neglect, and not about overcoming committed vaccine refusal.

This final point is especially important. Regardless of the actual intentions of the legislators who created and tightened England's vaccine mandates between 1853 and 1873, the 1896 Commission decided that the mandate's purpose was to nudge indifferent or otherwise preoccupied parents toward vaccination, rather than to compel committed vaccine refusers to accept vaccination for their children.

In 1898, a new Vaccination Act implemented the recommendations of the Royal Commission.[84] It created conscientious objector rights (NMEs), although it made them difficult to obtain. An applicant had to convince two different magistrates, and there were many delays built into the system's administration.[85] In 1907, Parliament made it easier to receive these NMEs, and in 1946 it repealed vaccine mandates altogether, establishing a principle of vaccination voluntarism that continues in England, and throughout the United Kingdom, to the present day.[86]

In 2004, the British Medical Association concluded that vaccine mandates remained inappropriate,[87] affirming a 2003 Scottish Executive Report that claimed vaccine mandates run counter to the "core principle that vaccines should be administered on a voluntary basis."[88] During the COVID-19 pandemic, when many other nations made new COVID-19 vaccines mandatory for health workforces, the British Medical Association supported delaying this measure.[89] In 2022, the UK government scuttled its potential mandates, arguing that the burdens associated with coercive measures outweighed the benefits. By this time, vaccines offered limited protection against dominant strains of COVID-19, and many people had become resigned to "living with COVID."[90] It appears, then, that England's commitment to vaccine voluntarism—and its rejection of coercive vaccine mandates—remains stable, even in the face of new contagions and pandemic conditions.

There were unfortunate short-term costs associated with England's abolition of its vaccine mandates. Lower immunization rates in the following decades led to preventable death and disease for many children.[91] Daniel Salmon and colleagues suggest that "compulsory laws would have remained effective" had they stayed in place.[92] However, this observation misses a broader point about policymaking: The government had to respond to the growth of political resistance to vaccine mandates, especially as English society democratized in the late 19th and early 20th centuries. From a merely epidemiological point of view, it was a mistake for England to introduce conscientious objector rights and later abolish vaccine mandates. But, from a political point of view, the mixed approach recommended by the Royal Commission identified a reasonable pathway for promoting public health in a democratic society.

England's experience with coercive immunization in the 19th century illustrates the self-limiting potential of coercive state power in democratic societies. Even as England was developing and mobilizing its capacities to push vaccination on its people, the state was concurrently, although unintentionally, generating conditions to limit those capacities.

Conclusion

This chapter has focused on social, legal, and ethical differences between the contexts of previous generations' coercive vaccine mandates and those of their contemporary analogs. Even while America has a history of coercive vaccination—affirmed in *Jacobson* and other Supreme Court decisions—that history has a terrible legacy. Responses to that legacy, including the creation and institutionalization of NMEs, have undermined the potential for coercive vaccination today. Relatedly, England's 19th- and 20th-century national smallpox vaccine mandates—which were more punitive and long-lasting—created an overwhelming backlash and led to the creation of NMEs, the repeal of mandates, and a national commitment to voluntarism in vaccination that has lasted more than 70 years.[93]

It is notable that England was eventually able to generate high immunization rates without relying on mandates. This is because it used state power to fund vaccines, deliver them to populations, and persuade individuals of their importance. Contemporary analysis suggests that frequent encounters with England's ubiquitous public health machine—the National Health Service—played a significant role in normalizing and routinizing vaccination.[94] In contrast, America did not develop a national health care system or provide public funding for health insurance or robust public health services. In the absence of these mechanisms for promoting voluntary vaccination, American states had to turn to vaccine mandates to promote the new vaccines of the 1960s and 1970s, as we explored in Chapter 2. Chapter 7 explores America's vaccine mandates as perverse consequences of its grossly underfunded health care systems. It focuses also on the power of physicians and physician organizations to both undermine public health and to advocate for coercive health measures.

7

Powerful Doctors and Underfunded Public Health

Introduction

In Chapter 6, we argued that legacies of coercive public health measures make it difficult to impose coercive vaccination today. In particular, the Bioethics Revolution empowered patients and research subjects, and it set new limits on physicians and medical researchers. However, even as law and medicine now protect greater autonomy for patients and research subjects, the immense trust that people place in their doctors remains a source of substantial professional power for physicians.[1] Furthermore, physicians possess significant professional autonomy to advocate for the health of their patients and communities, which makes them among the most forceful spokespeople on health-related issues.[2] Many American physicians are in private practice and are therefore their own bosses, but even employed physicians usually have a large domain of discretion, both for the practice of their profession and for their participation in community outreach and political advocacy.

Physicians can be especially politically powerful when they work collectively to lobby governments on health-related laws. Organizations such as the American Medical Association (AMA) and the American Academy of Pediatrics (AAP) may be among the most effective independent voices for health-related political advocacy. Since the mid-2010s, the AMA, AAP, and other physician groups (including the American Academy of Family Physicians and the American College of Physicians) have been among the loudest voices calling for more coercive immunization policies.[3] They have been joined by high-profile physicians, such as Paul Offit, and physician–politicians, such as Richard Pan.[4] Hence, in the fight for California's Nonmedical Exemption Bill, we should view the central roles played by Pan, the California chapters of the AAP and AMA, and the Health Officers Association of California as part of broader efforts by physicians and their organizations to shape America's public health system and its immunization policies. Indeed, in the aftermath of the Nonmedical Exemptions Bill, the AMA and the AAP called for state chapters across the country to replicate California's policy changes in their own communities.[5]

Physicians have also sometimes used their power to scuttle America's public health institutions, including those that could have increased access to vaccines and promoted vaccine acceptance. In this work, physicians have often been joined by other powerful institutions, including health insurance companies. We argue that physicians' recent strong support for vaccine mandates should be understood as part of a historical pattern in which physicians agitate for coercive public health measures but simultaneously work to prevent funding for public health services that could facilitate more voluntary means for achieving community health. Importantly, although both the AMA and AAP have recently lobbied for more coercive immunization policies, the AMA is a much older and more powerful institution, and its role in resisting the development of public health capacities has been more pronounced. Accordingly, we focus more on the AMA's history of resisting meaningful investments in public health.

Professionalism and Power

We trust physicians because we believe they are experts about medicine and because we believe they will advocate for what is best for us.[6] A primary reason why physicians can advocate so effectively for their patients is because their work is largely free from outside interference. They are *professionals*, whose "occupations are characterised by a high degree of autonomy or self-regulation."[7] For example, the UK's 1858 Medical Act established the General Council of Medical Education and Registration, which "limit[ed] efforts by government or private corporations to control their work."[8] In the United States, each state generally defers to independent medical boards (staffed by physicians) to set licensing requirements and to impose discipline on their members.[9] These medical boards effectively exercise monopoly power on the practice of medicine within their jurisdictions.

The drive toward professionalization in medicine had an ethical motivation: to set high standards for patient care and to protect medicine from undue political interference or market incentives. There were also clear economic motivations. In the 19th century, physicians who had attended medical school and completed residencies faced stiff competition in the health care marketplace. Unqualified practitioners, who did not constrain their diagnoses and prognoses to the scientific evidence, often competed with physicians for patients. The professionalization of medicine aimed to delegitimize these practitioners, both to promote better patient care and to protect and grow physicians' market share.[10] Such market considerations played a powerful role in motivating physicians' efforts to remove other practitioners (e.g., midwives) from mainstream health care delivery.

California's history is instructive. The mid-19th-century explosion of the state's population—including its physicians—corresponded with the period of intense professionalization of medicine. When Americans flocked to California's gold mines in the 1850s, physicians followed.[11] In 1856, they founded the California Medical Association (CMA), which is a constituent organization of the AMA. The CMA's founding purpose was to protect "patients and the profession" from "the challenges of rampant quackery, epidemics of contagious disease," but also to "establish standards for the profession."[12] State and national medical organizations have consistently shaped the practice of clinical medicine, and they have also led political efforts to promote community health in ways that advanced the interests of their members.

Physician Professional Power Versus Public Health Provision

The recent engagement of physician organizations in vaccine mandate advocacy continues a long history of physician organizations fighting for the integrity of medicine and for the health of their communities. We should also understand these efforts in light of a similarly robust tradition of physician organizations fighting against meaningful public investments in health infrastructure. While physician organizations have often demanded that the state use its power to impose health on individuals, they have frequently opposed efforts by the government to build institutions that would promote healthy communities.

In *Pox: An American History*, Michael Willrich argues that the underfunding of health institutions is a central theme in the history of American immunization policy.[13] In particular, a primary motivator for vaccine mandates in the 19th and 20th centuries was that individuals were often unwilling or unable to purchase vaccines. At the same time, states and local communities usually lacked the funding and infrastructure to provide free vaccines, to educate and persuade the population, or to cultivate public trust in the medical system. When America's municipalities could not build trust and social solidarity around vaccination—because they lacked institutional means to provide robust stable public health services—they sometimes fell back on state police power to coerce people to get vaccinated.

Physicians and their professional organizations deserve substantial blame for America's long-standing failures to create and fund effective public health institutions. While states create and enforce public policies, physicians and their organizations have often played prominent roles in medical and health policymaking. We are fortunate that many physicians and some of their professional societies have *recently* advocated for expanded public health investments and for universal health insurance coverage.[14] This sudden and dramatic shift—among physicians, and among Americans, more generally—is one reason to hope for the future of American immunization policy. But broad-based physician support for such policies is very recent, though it has some historical antecedent.

In the earliest periods of U.S. public health, elite physicians and their professional societies had expressed strong support for government investments in public health capacities.[15] An 1883 article in *Journal of the American Medical Association (JAMA)* reported that "public health ever goes hand in hand with true liberty, and is the companion of orderly habits and pure morals."[16] This confident statement obscured a divide among physicians. Wealthy physicians with degrees from leading institutions, and often European training or experience, embraced government's role in health care, seeing an alliance with public health as an opportunity to increase the prestige of medicine. Ordinary medical providers were often less well trained and were more vulnerable to market forces; they saw the expansion of public spending on health care as destructive to their livelihoods.[17] Why would ordinary people pay physicians for medical care if it were available for free from the government?

The cosmopolitan elites controlled the AMA until the end of World War I, after which ordinary physicians led a successful revolt. After more run-of-the-mill physicians took control of the AMA and other medical associations, these groups began to fight against government investments in health. They argued that patients would be better served by private physicians or would be more likely to comply with medical advice that they had paid for.[18] But physicians' organizations were not shy about the fact that their resistance to public provision of health resources was also motivated by economic interests. By the 1920s and 1930s, the conflict between individual physicians—through their professional societies—and public health officials was an explicit and central part of the emerging American health care system. Physicians saw vaccination as part of preventive care, a service within the realm of private practice physicians, whereas public health officials saw vaccination as a population-level intervention that government should provide directly. This conflict focused on how vaccines should be *financed* (patients paying their doctors or citizens receiving free vaccines from the government), *who* should provide vaccines (private practice physicians or public health nurses), *where* vaccines should be offered (in private clinics or in community health centers), and how they should be *advertised* (only through private communication with one's physician or through public-facing messaging).[19]

When public health officials tried to set up free clinics in poorer neighborhoods, local physicians often attacked those efforts as intrusions on their territory and livelihoods.[20] A 1920 article in *New Orleans Medical and Surgical Journal* illustrates the frustrations some physicians felt about investments in government health capacities: "The idea of government wet nursing is socialistic rot of the most dangerous type, is destructive of the very fundamentals of liberty, and in my opinion has no place in a free country," and it also makes the physician "little more than [a] stool pigeon, a clerk for the health boards."[21] In *The Sanitarians: A History of American Public Health*, John Duffy writes that it was "not at all uncommon" for physicians of that time to view public health programs as threats to their social status, to their economic well-being, and to the American "way of life."[22] Much of the struggle was about power and money, but some of it was cultural. Community physicians did not like the flashy advertising campaigns and emotional persuasion attempts of public health officials, since these methods resembled the techniques used by medical quacks, while mainstream medicine was at pains to adopt a more detached professional manner.[23]

The development of pediatrics as its own medical specialty was, among other things, an effort by physicians to reassert control over early childhood preventive health measures.[24] Progressive Era reformers advocated robust wrap-around services to promote children's health—often anchored in schools, community centers, or public health clinics—but physicians usually wanted children's health to remain under their control. Physicians won the battle. By the middle of the 20th century, public health institutions had largely retreated from providing vaccines, and immunization was once again delivered mostly in the offices of private practice physicians, especially pediatricians. Physicians had succeeded in preventing, restricting, or eliminating public funding for vaccination, and they had created other barriers to the government providing vaccines to the American people.[25]

Leadership of the AMA, including Morris Fishbein (longtime editor of *JAMA*), was fiercely opposed to "socialized medicine" and saw public vaccination campaigns as part of broader efforts by the state to take over health care.[26] The AMA was often at odds with the American Public Health Association (APHA). The latter wanted to establish national health insurance and to better fund public health institutions, both of which would promote vaccination. Two dramatic battles between these foes—both won by the AMA, a much larger and wealthier organization—were fought over federal government efforts to create national health insurance, which would cover vaccinations. The AMA opposed efforts by President Franklin Delano Roosevelt (1939) and then President Harry Truman (1949) to create federal health insurance systems, efforts that were supported by the APHA.[27] After his defeat in 1949, Truman decided to focus only on health insurance for older Americans since the battle over federal funding for childhood health, including vaccines, was clearly lost. It took until 1965 for Medicare to be passed, and the AMA forcefully fought against that program, too.[28]

Physicians' professional organizations consistently rejected institutional reforms or investments that could promote vaccine acceptance and uptake.

Instead, they responded to low vaccination rates by blaming their patients for being too cheap to pay for vaccines.[29] Physicians pushed back when state and federal governments attempted to use tax dollars to purchase vaccines or to empower public health institutions to persuade people to vaccinate.[30] This resistance continued through the fights about national health insurance in the 1940s and past the battles over Medicare in the 1950s and 1960s. For example, in the 1950s, the AMA resisted federal involvement in distribution of scarce polio vaccine on the grounds that physicians should provide it only to paying customers.[31]

The AMA's destructive resistance to investment in public health infrastructure continued into recent decades. In the early 1990s, the AMA successfully opposed the Clinton administration's plans to create national health insurance.[32] In 2009 and 2010, the AMA chose not to attack the Obama administration's Patient Protection and Affordable Care Act (Obamacare) only because this law bolstered *private* health insurance. The AMA fought vigorously to ensure that Obamacare did not include a "public option" (i.e., a publicly run health insurance plan), and it succeeded. To this day, the AMA continues to reject single-payer health insurance on the grounds that it may decrease physician income and power.[33] Of course, the AMA's rejection of single-payer health insurance is also often accompanied by common refrains about protecting "patient choice."

A notable exception to the narrative we have been developing was the creation of the Vaccines for Children (VFC) program through the Omnibus Budget Reconciliation Act of 1993.[34] This law authorizes the Centers for Disease Control and Prevention (CDC) to purchase vaccines approved by the Advisory Committee on Immunization Practices for distribution to children whose families may be otherwise unable to afford them. Eligible children include those who are Medicaid-eligible, uninsured, underinsured, American Indian, or Alaska Native.[35] The AMA and AAP support the VFC program.[36] The program also appears to be popular among individual pediatricians, with a poll showing that 86% of responding pediatricians participate.[37] However, the VFC program does not challenge the role of pediatricians in administering vaccines because it provides free vaccines to both to private practice physicians and public health clinics. Furthermore, the VFC program allows physicians to bill for the office visit and the administration of the VFC vaccine,[38] although Obamacare requires most private health insurance plans to cover childhood immunizations at no cost (i.e., no co-pays) to families.[39]

Physicians and Elite Anti-Democratic Power

In *State of Immunity*, James Colgrove highlights the central role of elite power in America's history of vaccine mandates. In the early 20th century—the same time that physicians were jealously guarding and consolidating their independence and power vis-à-vis government and competitor providers—the AMA argued for medical leaders to have broad powers over the public, and it ridiculed those who resisted. In particular, physicians and their organizations demanded that smallpox vaccine mandates be considered a matter of expert judgment, and therefore not put to popular votes, just as we would not ask everyday citizens for their opinions about "the feasibility of producing transparent lead, or steel-hard aluminum, or synthetic proteins."[40] The political engagement of physicians therefore expressed a popular aphorism about conservatism: "There must be in-groups whom the law protects but does not bind, alongside out-groups whom the law binds but does not protect."[41] Throughout much of the 20th century, physicians insisted that the government not interfere with or disrupt their private economic relationships with patients, but at the same time they demanded that the government marginalize or criminalize alterative practitioners (e.g., chiropractors and naturopaths). Physicians reserved the right to govern themselves, but they wanted the government to restrict both the medical marketplace and the medical choices of everyday citizens (e.g., via vaccine mandates).

Physicians' political power in the 20th century rose in tandem with their cultural power. In *Anti/Vax: Reframing the Vaccination Controversy*, Bernice Hausman discusses the emergence of *biological citizenship*, according to which individuals have an ethical duty to care for their personal health. Recent manifestations of biological citizenship include individual obligations to be an active participant in decision-making about diet, exercise, and personal medical choices. Biological citizenship places responsibility on individuals for their health, and it constructs that self as an agent for decisions about diet, tattoos, cosmetic surgery, pharmaceutical consumption, and genetic tests.[42]

Biological citizenship was encouraged during the 20th century by cultural messaging that framed physicians as beneficent prophets of well-being. A combination of "significant medical breakthroughs, coupled with technological refinements and the expansion of mass media" elevated physicians and medical scientists above other professions.[43] Throughout the 20th century, medical researchers were portrayed as "enlightened protectors of humankind" and "scientific nobility."[44] Popular culture often valorized physicians and medical researchers as heroes and potential philosopher kings; many movies, radio shows, comic books, and television programs celebrated their accomplishments.[45] In the context of their ascendant cultural clout and their emerging political power, physicians played a primary role in creating and enforcing biological citizenship's new idea of the self.

The cultural power of physicians and the individualistic ethos of biological citizenship contributed also to *medicalization*, which is the idea that ever larger domains of human life contain medical problems that should be solved by physicians. A high point of medicalization was the expansive conception of mental illness embraced by mainstream psychiatry in the 1960s and early 1970s. At that time, many nonconforming behaviors were constructed as psychiatric disorders, and psychiatrists diagnosed everything from feminist commitments to civil rights activism as pathologies that psychiatry should treat and correct.[46] Given the social and political upshot of a diagnosis of psychiatric illness— including the possibility of coercive hospitalization—the expansive nature of postwar psychiatric diagnosis significantly expanded the power of physicians. Expressing this worry, Irving Zola wrote in 1972 that the institution of medicine

> is becoming the new repository of truth, the place where absolute and often final judgments are made by supposedly morally neutral and objective experts. And those judgments are made, not in the name of virtue or legitimacy, but in the name of health.[47]

Scholars of vaccine hesitancy have often focused on the specter of medicalization. For example, Mark Largent and Elena Conis have argued that disagreements about whether vaccines are safe and effective are often *proxy debates* for more fundamental arguments about who should have the power to shape society and which kinds of well-being we ought to care about.[48] Hausman situates medicalization within "a pervasive and ongoing theme of modern discontent—concerns about specialized expertise, loss of individuality and control, and increased surveillance and managerial regulation that characterize medicine in late modernity."[49]

Challenges with Physicians Promoting Public Health

Recent changes in medical education and in the political commitments of physicians provide reasons to hope that the medical establishment will advocate for greater investments in public health. Medical education now focuses substantial attention on the social determinants of health, including how the underfunding of today's health institutions prevents patients from receiving the care they need. Medical students also explore ethical issues raised by the power that physicians exercise in clinic and society.[50] Furthermore, physicians increasingly support public financing of health care, and young physicians are increasingly left-leaning in their voting and political contributions.[51] These recent reorientations may give us reason to believe that physicians will soon be promoting—or at least not undermining—efforts to build up America's public health infrastructure. Such investments will be necessary for cultivating community protection through the voluntary vaccination of free citizens.

But a word of caution is necessary. Physicians can be effective in their public advocacy and their political lobbying only if they continue to be highly trusted. This trust derives from the perception that physicians are motivated by their commitment to patient and population health, rather than to partisan political goals. As we discuss in Chapter 9, political polarization is an acid for social trust, and physicians cannot escape from its corrosive effect simply by virtue of their dedication to patient or public health. If the public comes to perceive physicians as partisan actors in fights about public health, then physicians may squander much of the trust they currently possess. Of particular concern is the dramatic recent decline in the trust that Republicans place in public health institutions. In 2009, there was no difference between Democrats and Republicans in their (equally high) degree of trust in organizations such as the CDC, the National Institutes of Health, the U.S. Food and Drug Administration, and state health departments.[52] However, by 2021, Republicans were much less likely to trust these institutions, even as they continued to trust their personal physicians.[53]

A further worry is about actual or perceived conflicts of interest for high-profile physician advocates for vaccines. In an environment of suspicion and declining trust, it is essential that the motives of physicians who advocate for vaccines on television and other media be beyond suspicion. There is a long history among vaccine refusers of arguing that physicians advocate for vaccines to drum up business for themselves or have other conflicts of interest.[54] Along these lines, the pediatrician ethicist, Douglas Diekema, has argued that two of America's most prominent public advocates for vaccines and vaccine mandates may appear to have financial conflicts of interest and are therefore not ideal spokespeople:[55]

> The two pediatricians most recognized as pro-vaccination messengers work in the field of vaccine development, and at least one has made significant amounts of money from a vaccine patent. Both frequently speak at the request of the American Academy of Pediatrics, a professional organization with the explicit goal of improving the health of children. The issue is not whether these spokespeople possess integrity, speak honestly, or make factually accurate proclamations. Since there are literally hundreds of other pediatricians who could speak passionately on behalf of vaccinating children who have no actual or perceived conflicts of interest, it is curious that the primary spokespeople on behalf of vaccination are those whose involvement will raise questions about trustworthiness.[56]

We do not know who Diekema is referring to, but his description applies well to Paul Offit[57] and Peter Hotez.[58] There is no reason to think that Drs. Offit and Hotez have been anything other than honest and ethical. It seems likely that neither one has an *actual* conflict of interest. (Offit sold his interest in RotaTeq years ago, and Hotez claims to receive no financial benefits from his role in developing vaccines.)[59] But given the central importance of maximizing public trust about vaccines, it is not enough for spokespeople *to be* ethical and honest. The public must also *believe* that they are. A core thesis of this book is that good intentions by experts are not sufficient to solve the political problems associated with public health policies. We have pushed that point in the context of debates about vaccine mandates, but it applies equally to questions about which physicians should be public spokespeople for vaccines.

Punishing Versus Providing

Other liberal democratic countries have succeeded in governing vaccination through voluntary means.[60] But those other countries also provide robust, stable, and high-quality public health services to their communities.[61] It costs money to build and run institutions to deliver health care and to promote vaccination. Accordingly, communities that make such investments treat immunization policy primarily as a matter of *public expenditure*. In contrast, communities that govern vaccination through mandates make immunization policy more a matter of *private penalties*.[62] Vaccine mandates are often inexpensive for governments to implement, but they can be very costly for the people they govern.

We wish that America's health institutions had developed along the former model of immunization governance. It would have been better if America had committed to delivering high-quality, free-at-point-of-service health care to everyone in society. But America's health institutions developed in a different direction, one that shifts costs onto vulnerable individuals when they need health care.

For an illustration of how large this background fact looms in immunization policy, consider a December 27, 2021, tweet from the CDC that encouraged people to vaccinate: "Hospital stays can be expensive, but COVID-19 vaccines are free. Help protect yourself from being hospitalized with #COVID19 by getting vaccinated."[63] The CDC makes explicit what is usually only implicit in American health policy: If you get sick, you're on your own to pay for it. Even before the COVID-19 pandemic, one in six Americans had a medical debt in collections.[64] Given that a hospital stay from COVID-19 costs between $38,000 and $73,000,[65] even a 5% or 10% co-insurance payment could be a major hardship for many American families, making this privatized economic hardship of COVID-19 a central part of America's experience of the pandemic.[66] Nearly one-fourth of patients hospitalized for COVID-19 report that their illness exhausted their savings.[67]

The California politicians, activists, and civil society actors we examine in this book were working in a broken health care system. They wanted to build a better, fairer, healthier California, but America's history of underfunded public health institutions left them with few tools to use, other than mandates that shifted costs and risks onto individuals and their children. The Clinician Counselling Bill, the Nonmedical Exemptions Bill, and the Medical Exemptions Bill made substantial progress within a deeply flawed system. We admire the work of Senator Pan and his allies, but our admiration cannot distract us from the failures of America's institutions.

Previous chapters devoted substantial attention to the political conflict that California's Nonmedical Exemptions Bill generated, but it is also worth recalling that the law imposed few financial costs on the state. By contrast, the abolition of nonmedical exemptions created immense social and economic costs for vaccine refusers. It is a sign of the injustice of American society—and the fundamental brokenness of its institutions—that California's only real short-term option for increasing vaccination rates was to threaten families with excluding their children from child care and education.

8

The Ethics and Public Acceptability of Mandates

Introduction

Debates about immunization policy—including debates about the elimination of nonmedical exemptions (NMEs)—are often highly *moralized*. They invoke ethical ideas about the scope of government power, the interests of children, and the responsibilities of parents. This chapter reconstructs aspects of these policy debates in terms of ethics arguments. We begin by examining three arguments that critics often raise against coercive vaccine mandates. Then we turn to four arguments that proponents offer in defense of such policies. Finally, we express skepticism that today's coercive vaccine mandates could lead to disease *eradication* or *elimination* (or even to stable disease *control*). And in Chapter 9, we consider what the (at best) modest benefits of NME elimination mean for efforts to justify these policy interventions.

In discussing ethical arguments for and against mandates, we consider also whether particular positions can be *endorsed* by the diverse communities of Americans who are governed by mandate policies. National—and, indeed, international—coordination is necessary to create and maintain community protection. Therefore, arguments about vaccine mandates need to be acceptable to people who have diverse conceptions of ethics and justice. Otherwise, mandates cannot be stable forms of immunization governance. We further develop worries about the stability and legitimacy of coercive vaccine mandates in Chapter 9.

Arguments Against Coercive Mandates

Critics offer many arguments against coercive vaccine mandates, and we illustrated some of them in our discussion of California's Clinician Counselling Bill, Nonmedical Exemption Bill, and Medical Exemptions Bill. Frequently, those arguments focus on false claims about high risks associated with vaccines and low risks of vaccine-preventable diseases. For example, Brandy Vaughan's "Learn the Risk" campaign—and the existence of the Vaccine Injury Awareness League—illustrates that worries about vaccine safety motivate many people's commitment to maintaining NMEs. We set these objections aside because they are based on false empirical claims and because responding to them requires education and persuasion rather than moral argument. Instead, we focus on arguments against coercive vaccine mandates that invoke moral or political values, including parents' rights, informed consent, and children's rights to care and education.

Parental Freedom

The parents who protested California's Nonmedical Exemptions Bill—including those who offered comments during the meeting of the Senate Health Committee—often objected that the bill would restrict parental freedom.[1] (This "freedom framing" was also ubiquitous during earlier fights over the Clinician Counselling Bill.) Recall actor Jenna Elfman's claim that "parents should vaccinate their children as much as they wish to." And consider the names of relevant vaccine resistant organizations: Our Kids Our Choice, Voice for Choice, and Freedom Angels.

Parents who bemoaned restrictions of their freedoms were correct: Dr. Richard Pan's legislative reforms, and the Nonmedical Exemptions Bill in particular, restricted their ability to make decisions for their children. If a parent agreed to have their child vaccinated, rather than to have them removed from care or school, then that choice was not really their own. Under such systems of immunization governance, a parent may retain a *formal* liberty right to refuse vaccines, but families that lack the resources to educate or care for children at home will be substantively compelled to accept vaccination. By way of analogy, consider a thief who demands "your money or your life." They may leave you with a choice, but that does not mean that your decision to give them your money was made freely.[2]

The central question is not whether coercive mandates undermine parental freedom—they do—but whether parents have a right to make free vaccination decisions for their children, and when that right can be undermined to protect the health of children and the broader public.[3] Most people take for granted that parents should be free to make choices about their children's diet, schooling, social associations, leisure activities, and religious observance. But these parental rights are not absolute, and parents should not be allowed to undermine their children's basic interests or place them at significant risk of serious harms.[4]

Although parents are entitled to what Lynn Gillam has called a "zone of parental discretion," that zone is not unlimited.[5] Some ethicists have argued that the only rights parents have are to discharge their responsibilities to promote their children's interests.[6] However, others have argued that parents have further rights over children based in families' interests—for example, in having intimate relationships in the family—and also the interests of other members of the family.[7] The legislatures of many U.S. states have sometimes granted even more expansive authority to parents. For example, religious exemptions from medical neglect laws allow parents to choose forms of religious observance that may cause their children to die from medical neglect.[8]

We offer two brief conclusions about the relationship between parental rights and the justification of coercive childhood vaccine mandates. First, there is deep disagreement among ethicists and everyday Americans about which rights parents ought to possess over their children.[9] We will not try to resolve those debates here; even some pediatricians and pediatric ethicists disagree about whether refusing vaccines ought to fall within the scope of parental discretion.[10] Second, and more important, we believe that coercive vaccine mandates may be justified even if they violate parents' rights to choose for their children. This is because there is little reason to think that such rights are absolute, regardless of their content. Such rights are surely pro tanto, which is to say that they can be infringed upon in the face of more weighty moral considerations.[11] But this is to invite the question of whether there are more weighty considerations in favor of coercive vaccine mandates than there are against them. To address this question, we need to consider all of the ethically relevant impacts of coercive vaccine mandates, which we aim to do in this chapter and the next.

Informed Consent

Another prominent argument that critics use against coercive vaccine mandates is that they prevent informed consent.[12] Informed consent consists of a voluntary, informed, and capacitated choice.[13] It is widely acknowledged that adult patients should provide informed consent to medical interventions, and that they have the right to be free from nonconsensual medical treatment.[14] In the U.S. context, most public-facing messaging frames vaccination as a personal medical choice undertaken to promote the health of the vaccinated person. For example, the Vaccine Information Statements that clinicians must provide to patients prior to vaccination focus on benefits that accrue to vaccinated individuals, rather than on claims about the importance of community protection.[15]

Attorney and anti-vaccine activist Mary Holland focused on informed consent in her testimony against California's Nonmedical Exemption Bill at the April 28, 2015, meeting of the Senate Judiciary Committee:

> Parents will be vaccinating their children under duress, invalidating any notion of informed consent. In employment, lack of consent is forced labor. In military service, it's conscription. In contracts, it leads to invalid contracts. In intimate relations, it's called rape. And in medical treatment, it's battery.[16]

It was impolitic of Holland to list other contexts in which consent is important (including sexual activity), but she was correct that eliminating NMEs for daycare and school vaccine mandates could undermine informed consent to vaccination. Holland's mention of rape elicited a condemnatory response from Senator Hannah-Beth Jackson, and Dorit Reiss claimed that Holland lost her audience after drawing the rape analogy.[17] But regardless of whether NME elimination resembles rape, Holland's worries about informed consent are overstated and inapt.

First, informed consent is a more-or-less absolute requirement for personal medical choices, but vaccination is not merely a personal medical choice. It is also a core part of public health governance because it is a means by which people can prevent themselves from harming and being harmed by others.[18] As Margaret Battin and colleagues put the point, receiving a vaccine prevents you from becoming a *victim* of disease, but it also prevents you from being a *vector*.[19] We take for granted that the state may sometimes use coercion to prevent people from harming each other, as we discuss below. Accordingly, the absence of meaningful informed consent to vaccination may not be enough to make coercive mandates unjustified. We must also pay attention to the benefits those policies generate, including preventing harm to others and maintaining community protection.

Second, the value of informed consent is inapt in discussions about childhood vaccine mandates. Informed consent is something that capacitated *patients* provide. The relevant patients for childhood vaccine mandates are usually infants and very young children who are not capable of providing informed consent. Parents are usually decision-makers for pediatric vaccinations, but they are not the patients and therefore do not *consent*. Instead, parents provide what the American Academy of Pediatrics calls "parental permission."[20] Parental permission matters morally and pragmatically: Involving parents in decision-making can reveal facts and values relevant to their child's treatment, improve parents' adherence to treatment, nurture children's developing autonomy, and promote the interests of other family members.[21] But parental permission is not nearly as valuable as informed consent, and it may sometimes be bypassed to ensure that children receive necessary medical care. Accordingly, considerations about informed consent are not relevant to questions about coercive childhood vaccine mandates. What matters is whether parents should have the right to refuse vaccines, and that question is far from settled, as we noted above.

Children's Rights to Care and Education

Parents who objected to California's Nonmedical Exemption Bill often claimed that the bill would deprive their unvaccinated children of access to school or daycare, and that this violated their children's rights.[22] Recall the placards claiming "SB277 makes my child a truant." Advocates of eliminating NMEs may claim that they intend children to be vaccinated *and* to be in school or daycare. According to such a view, eliminating NMEs need not deprive children of anything, but can instead prompt their parents to help them be healthy and educated. However, unless the state is going to forcibly vaccinate children against their parents' wishes—something we certainly do not support—coercive vaccine mandates are going to leave the children of committed vaccine refusers both unvaccinated and excluded from school and daycare.

The important question is not *whether* coercive vaccine mandates infringe on children's rights to care and education—they do—but whether some children's rights may be infringed upon to promote public health goals. Some political communities do not allow vaccine mandates to prevent children from attending school. For example, the Italian constitution guarantees children's access to school, so Italy's vaccine mandates cannot exclude children merely because of their vaccination status.[23] America's Individuals with Disabilities Education Act guarantees children with disabilities a fundamental right to education, such that disabled children with an Individualized Education Program (IEP) must be exempted from vaccine mandates.[24] However, some other political communities allow children to be denied access to care and education as part of a broader effort to increase immunization coverage. Political leaders and policy scholars in such communities may be troubled by this unfortunate consequence but are willing to accept it.[25] Whether they are correct depends, again, on a deeper analysis of what is at stake.

Arguments for Coercive Mandates

We have so far focused on ethics arguments *against* coercive mandates advanced by scholars and parent activists. But *advocates* of coercive vaccine mandates also offer ethics arguments. These focus on the rights of children to receive vaccines, the importance of preventing disease transmission, the duty to make a fair contribution to herd immunity, and the overall social benefits associated with high rates of vaccination.

A Child's Right to Be Vaccinated

Vaccine advocates sometimes argue that children have an enforceable *right to receive* vaccines. It follows that eliminating NMEs would be justified if doing so were an effective means for ensuring that children got vaccinated. For example, philosopher Roland Pierik argues that the state ought to use its power to protect children's fundamental right to health, which includes vaccination. Just as the state does not allow parents to refuse life-sustaining blood transfusions for their children, Pierik thinks the state ought to ensure that children receive vaccines.[26]

We agree that vaccination is almost always in children's interests, and that it is generally morally wrong for parents to refuse it.[27] Interventions to prevent abuse or neglect are often justified, especially if those interventions are necessary to prevent death, disability, or substantial suffering.[28] To return to Pierik's example, parental refusal of emergency blood transfusions can impose a "significant risk of serious harm" on children, and it should not be permitted. In contrast, vaccine refusal usually does not impose a similar kind of risk on children.[29] Accordingly, it does not follow from the fact that children have a right to be vaccinated that the state may use coercion to promote their vaccination.

Perhaps U.S. courts and pediatric bioethicists are wrong about the threshold for intervening to protect children's interests. Perhaps government *should* act to guarantee children's vaccination. However, even if those suppositions were true, there is little reason to think many Americans would agree. The politics of such policies would be highly toxic in the contemporary United States. Consider that the U.S. is the only country not to ratify or accede to the United Nations Convention on the Rights of the Child, the core children's human rights document.[30] American opposition to children's human rights is motivated by a strong "parental rights" movement that defends parents' discretion to make decisions about children's education and health care and to impose physical discipline.[31]

Furthermore, U.S. law and culture protect and even valorize some parental choices that place children at substantial risks of harms. Most states do little to regulate firearms in the home, even in the context of an epidemic of children's deaths from gun violence.[32] Similarly, communities in the United States are typically lax about regulating youth tackle football, backyard trampolines, and all-terrain vehicles, which injure or kill hundreds of thousands of children each year.[33] In this context, it is unreasonable to expect Americans to recognize a supposed right of children to be vaccinated or to grant permission to the government to protect that right.[34]

Harm Prevention

Advocates often argue for vaccine mandates by invoking the risks that vaccine refusal imposes on others.[35] This harm prevention discourse powered the I ♥ Immunity coalition that we examined in Chapter 4, including its focus on vulnerable members of society. It is widely acknowledged that one purpose of government—perhaps its *core* purpose—is to prevent people from harming each other. The Vaccinate California parents were concerned that NMEs increased the odds that their young children, vulnerable family members, and other Californians would experience such harms. If vaccine refusal harms others, then government coercion in the form of vaccine mandates can be justified, as philosophers Jessica Flanigan and Jason Brennan have argued.[36]

We agree with Flanigan and Brennan—and with John Stuart Mill—that government should restrain individuals from harming others.[37] For example, governments may use force to prevent assaults and murders and thefts. Indeed, the prevention of these kinds of interpersonal harms is perhaps the strongest justification of the state's use of coercive measures against individuals. We also acknowledge that vaccine refusal can lead to death and disease for many people. But focusing on harms caused by individuals can offer only limited support for vaccine mandates.

Mill's harm principle justifies the restriction of an individual's liberty based on the harm that *that individual* would otherwise cause. Hence the justification of coercion to prevent murder, assault, theft, etc. Therefore, what is required for a Millian harm prevention defense of coercive vaccine mandates is that being unvaccinated risks "tangible and immediate harm to others"[38] or that coercive mandates be "essential to prevent a serious concrete harm."[39] However, one unvaccinated person does very little to increase other people's risks of infections above the baseline. Instead, only a collection of unvaccinated people can impose significant risks of serious infection-related harms on others. To use a term from moral philosophy, a single unvaccinated person is almost always *causally impotent* with respect to the risks of infection that other people experience. This means that a Millian harm prevention argument does not back up government use of coercion to compel individuals to vaccinate.[40]

Finally, even if a Millian harm principle could provide support for vaccine mandates, this would be unlikely to be persuasive in the contemporary United States. Americans tolerate substantial third-party risk impositions. Their country allows people to carry loaded firearms in public, even though doing so substantially increases risks of injuries and death for third parties. It permits oversized trucks and sport utility vehicles on the road even though they are much more likely to kill or maim pedestrians or people in other vehicles. Some American states have recently *increased* speed limits, with the predictable outcome of additional deaths and injuries. The magnitude and risk of these kinds of harms far exceed those associated with an individual person's decision to refuse vaccines. If Americans will not curtail the liberty of gun owners or drivers, then it seems unlikely that they will be willing to restrict vaccination choices for smaller amounts of harm reduction.

Fairness

Advocates sometimes argue for mandates to ensure that people contribute to community protection.[41] Sometimes these "fair share" arguments also assert that vaccine refusers are "free riders" on herd immunity, since they benefit from community protection without contributing to it.[42] At this point, we need to "uncouple" our perspectives: Katie is persuaded by fairness arguments for mandates. What follows is Mark's critique.

The mere fact that other people have created a good that I enjoy does not generate an enforceable obligation for me to contribute to that good.[43] To borrow an example from Robert Nozick, if someone in my neighborhood decides to mail a book to everyone on my block, this does not generate a duty on my part to help pay for the books.[44] The fact that I find the book to be valuable—perhaps it even changes my life—does not change the fact that I am morally free to compensate my neighbor or to pay them nothing at all. It would be highly inappropriate for my neighbor to use threats or force to try to get me to pay for the book. Of course, I owe something to my neighbors, perhaps including a duty of mutual aid, but such associative duties do not emerge from the fact that my neighbor provided a good from which I benefitted. This point can be extended to cover vaccination. The mere fact that you have benefitted from community protection does not generate an obligation to contribute to it.

Furthermore, even if *adults* have enforceable duties of fairness to contribute to projects from which they have benefitted, it is not as obvious that *children* have such obligations, or that they must discharge those duties while they are children. Children are consummate free riders on goods created by others—that is part of what it means to be a child—and that is not usually thought to be a moral failure on their part. Children do have *some* age-appropriate moral duties. For example, they should share their toys with siblings and demonstrate limited generosity toward members of their communities. But they do not have general duties to "pay their fair share" for goods they have enjoyed, including community protection.[45] Consider that adults should pay their taxes to support public goods, but it would be unreasonable to tax children's allowances just because children benefit from public roads and schools. Fairness-based arguments might be able to justify some kinds of vaccine mandates for adults,[46] but the value of fairness does not justify vaccine mandates for children.[47]

Overall Net Benefits

Returning to our "merged perspective," we note that advocates of coercive vaccine mandates often appeal to the "greater good" achieved by their policies.[48] They may acknowledge that coercive mandates have substantial costs, but they assume that the immense benefits such policies generate will outweigh their downsides. For example, we found that Australian policymakers often justified coercive immunization policies by claiming that the overall benefits of those policies outweighed the overall burdens they imposed.[49] Our fieldwork for this book elicited similar responses from some of the actors who helped eliminate NMEs in California.

This way of defending coercive vaccine mandates draws explicitly or implicitly on consequentialist moral theories, according to which the right acts are those that generate the most overall good consequences.[50] The most popular consequentialist moral theory is utilitarianism, according to which right acts are those that maximally promote human happiness or increase the net balance of pleasure over pain.[51]

In previous generations, it was common for ethicists to defend public health policies on the grounds that they promoted the overall good (utility).[52] For example, public health scholars have claimed that "the philosophical basis of modern public health is generally considered to be nineteenth century utilitarianism"[53] and that public health is "basically utilitarian in character."[54] When it comes to vaccine mandates, utilitarians claim that a world with higher immunization rates, but more coercion, contains more good (utility) than a world with less coercion but lower immunization rates. Even arguments that are not explicitly utilitarian often presume a utilitarian framework, focusing on the overall number of lives saved, diseases prevented, and quality-adjusted life years achieved.[55] Some have argued that the best accounts of COVID-19 pandemic ethics were utilitarian.[56]

We have two concerns with these arguments. First, it is important to avoid naïve utilitarian arguments that focus narrowly on the value of health but that ignore the many other values that health policies can implicate.[57] If coercive vaccine mandates achieve dramatic and long-lasting increases in immunization rates, that counts strongly in their favor. But an informed utilitarian analysis of immunization policy requires that due weight be given to all of the benefits that vaccination policies may create and all of the costs they may impose.[58] In particular, vaccine mandates can have negative consequences for children's education, the economic well-being of families, and the political fabric of the nation.[59] These considerations are not mere pragmatic side constraints but must be at the center of any utilitarian evaluation of immunization policy. We give greater attention to some of these issues in Chapter 9.

Second, utilitarianism can require trade-offs between the interests of individuals for the sake of a greater overall good. Because utilitarianism reduces all moral questions to questions about which acts best promote overall utility, this theory may therefore require unacceptable violations of individual rights and even serious harms to some individuals for the sake of a greater good.[60] This is because utilitarianism is indifferent to *distribution*. Whether the same amount of good (utility) is equally distributed or is hoarded by one person is equally acceptable from that theory's point of view.[61] Utilitarianism therefore permits or requires trade-offs that contemporary guidelines for public health policy, clinical medical encounters, and human subjects research explicitly prohibit (as we discussed in Chapter 6).[62]

The fact that vaccination benefits almost all children means that there do not need to be pernicious trade-offs in *voluntary* immunization policies. Ensuring that everyone gets vaccinated has expected benefits for every individual and is good for society as a whole. However, *coercive* vaccination policies may cause substantial harms to some people—for example, children who are denied access to care or education—so that society as a whole can enjoy higher immunization rates. Accordingly, we should be suspicious of utilitarian arguments for vaccine mandates that treat harms to some children as a means by which society will benefit. In a similar way, we should question utilitarian arguments that treat the imposition of harms on some social groups as a justified cost of creating substantial overall benefits. Poorer children—who are disproportionately children of color—are more likely to be deprived of adequate care or education if parents choose not to vaccinate them after NMEs are eliminated.[63] (Wealthier families are more likely to be able to hire private care providers, provide quality homeschooling experiences, or avoid enforcement of mandates altogether.) Accordingly, NME elimination aims to promote overall social welfare in ways that shift burdens disproportionately to members of society's most vulnerable groups. Many public health ethicists insist that it is unethical for public health policies to create or widen existing inequalities between social groups.[64] How exactly we should balance a commitment to equality with the desire to increase overall social welfare is a persistent problem for social philosophy, and not one we will attempt to resolve.[65] It is enough to note that eliminating NMEs may create or contribute to inequalities that many ethicists are eager to mitigate.

Elusive Long-Term Benefits: Eradication, Elimination, and Stable Control

So far, this chapter's discussion has focused primarily on the near-term benefits and costs of coercive vaccination. However, immunization policies often aim at longer term goals, including the eradication, elimination, or stable control of vaccine-preventable diseases.

A common master narrative of immunization history is that it aims at disease eradication. According to such a story, all our struggles—including today's political fights over coercive immunization policies—are justified because immunization will ultimately rid the earth of vaccine-preventable infections. At the very least, policymakers may be counting on local elimination of disease, or long-term disease control. The costs of these efforts may be high, but the benefits will be so substantial and long-lasting that it will be worth it.[66] Moreover, since the timeframe for a potential benefit is the indefinite future, cost–benefit analyses likely support doing whatever is necessary to achieve that perpetual benefit. (When you multiply large benefits over an indefinitely long period of time, they can offset very large costs today.) Even very coercive contemporary vaccine mandates may be justified if our communities will later be free of vaccine-preventable diseases and, therefore, also free of mandates.

Unfortunately, disease eradication—or even stable, long-lasting control of vaccine-preventable diseases—may not be a realistic goal for today's liberal democratic societies. Accordingly, today's vaccine mandates need to be justified by their short-term and often more modest benefits, and their potential harms need to be given greater comparative weight. Our task in this final section of the chapter is to generate skepticism about the eradication, elimination, or long-term control of vaccine-preventable diseases, so as to motivate a more critical examination of NME elimination policies in Chapter 9.

When Edward Jenner invented vaccination, he proclaimed "that the annihilation of the Small Pox, the most dreadful scourge of the human species, must be the final result."[67] On May 8, 1980, Jenner's hope was realized.[68] Humanity's efforts to eradicate smallpox were expensive and onerous, but the sacrifices rid the world of one of its greatest plagues. Advocates of vaccination, both in the academy and in public health, sometimes appeal to the history of smallpox eradication to justify immunization policies, including mandates. They hope that vaccine history will repeat itself, and they may even take for granted that vaccine history is somehow propelled toward the eradication of disease.

The idea that history has a goal or a destination is found in many philosophies and religions. For example, Christianity teaches that human history is ultimately oriented toward a Last Judgment, a time when Jesus Christ will return to send all of the living and the dead to either heaven or hell.[69] There are many secular ideas about the direction of history, too. For example, the political scientist, Francis Fukuyama, famously and erroneously argued that the world had reached the "end of history" after the collapse of Soviet Russia, and that humanity would thereafter enjoy stable free-market democratic societies. President Barack Obama and Reverend Martin Luther King, Jr. both claimed that "the arc of history bends towards justice," drawing on the work of 19th-century abolitionist Theodore Parker to express a faith that humanity was heading in the right direction. The public health analogue is that the arc of *vaccination history* bends toward *eradication*, and that arriving at this destination promises ongoing stability and prosperity.

Consider an often-reprinted figure of the "Natural History of an Immunization Program" by Robert Chen and colleagues (Figure 8.1).[70] This model provides vaccination history with a beginning, a middle, and an end. It begins with a new vaccination program, in the middle is a loss of confidence followed by renewed commitment, and the end consists of disease eradication and cessation of vaccination.

A chief virtue of this model is its attention to the middle parts: Chen and colleagues correctly note that vaccination history is not a straight unbroken line

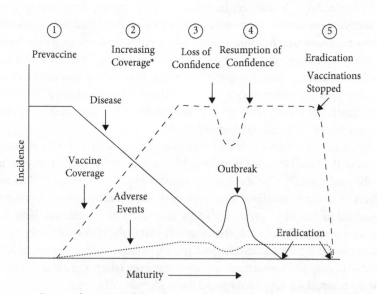

Figure 8.1 Potential stages in the evolution of an immunization program, showing the dynamics of the interaction between vaccine coverage, disease incidence, and incidence of vaccine adverse events.

Reproduced from Robert T. Chen et al., "The Vaccine Adverse Event Reporting System (VAERS)," *Vaccine* 12, no. 6 (1994): 542–50, https://doi.org/10.1016/0264-410x(94)90315-8.

of progress but, rather, a series of small steps forward combined with lurches backward.

Despite the virtues of the Chen model, we do not think it is appropriate to base our policies, programs, and individual behaviors on a speculative "natural history." More important, we should question whether vaccination history has an end.

According to the International Task Force for Disease Eradication, *eradication* is the "reduction of the worldwide incidence of a disease to zero as a result of deliberate efforts."[71] *Control* seeks to reduce the incidence of disease to a low level, but without ridding the world of it.[72] *Elimination* refers to the end of disease transmission in specific geographic areas. (So, for example, many countries have eliminated polio, but it has not been eradicated.) The modern idea of disease eradication originated at the Rockefeller Foundation during the interwar period, where it was invoked as the goal for its ambitious efforts to battle hookworm, yellow fever, and malaria. Later, The World Health Organization coordinated eradication campaigns against smallpox, malaria, and polio. The Bill and Melinda Gates Foundation has also recently focused on eradication of polio.

We worry that a commitment to eradication as a policy goal—*eradicationism*— may presume naïve ideas about human perfectibility, the ability of science to solve human problems, and political utopianism. Historian Nancy Leys Stepan suggests that for the prominent international epidemiologist, Fred Soper, eradication was "an all-or-nothing view of disease, a vision of the perfectibility of the human condition."[73] Eradicationism expresses hope for a new kind of human being: a persistently healthy organism. It also expresses a faith that scientific developments will help humanity escape many of the painful and burdensome parts of human life. The development of scientific technologies in the first half of the 20th century—new vaccines prominently among them—fueled widespread optimism that science would deliver a better future.[74] In addition to new vaccines, communities developed better epidemiological methods, surveillance technologies, understandings of disease, and a more robust state apparatus.[75] This contributed to the sense that medicine would soon be victorious in its millennia-long wars against infectious disease, and that it would extend that winning streak in new battles against chronic conditions such as cancer, heart disease, and stroke.[76]

Eradicationism expresses utopian ideas about the perfectibility of political institutions. As we noted in Chapter 2, America's health leaders in the 1960s embraced eradicationism because global victory over smallpox looked imminent, and because institutional investments provided hope for similar successes against other diseases.[77] The sustained push to increase measles vaccine uptake in the late 1960s underscored important legislative changes, including new and updated school entry mandates.[78]

Unfortunately, eradicationism faces substantial barriers. There is little reason to hope that any more than a handful of current vaccines *could* lead to disease eradication, and almost no reason to think any of them *will*. This is not because eradication is impossible from an immunological or epidemiological point of view, but because of stark limits on what human beings and our institutions can accomplish. Consider that 20 years ago philanthropists devoted immense sums of money to eradicating measles and polio, but those diseases persist. Polio remains endemic in Pakistan, Afghanistan, and Nigeria, with recent outbreaks in Malaysia. In September 2022, New York officials declared a State Disaster Emergency after an adult was paralyzed by polio and the polio virus was detected in 70 wastewater samples from around the New York City area.[79] Likewise, measles outbreaks are now common in many countries—including the United States—where they had previously been rare or nonexistent.[80]

Eradication requires that every person capable of spreading or incubating disease either voluntarily accept vaccination or be coerced into being vaccinated—and that is assuming that vaccination systems function well, with sufficient access and availability of vaccines. Achieving universal or near universal levels of vaccine uptake generally requires substantial and ongoing investments in public health institutions, in addition to political stability and high levels of public trust, or at least compliance.

Institutional insufficiencies and ideological commitments set limits on the amount of coercion that states can use for vaccine mandates when voluntary vaccination proves insufficient. As we discussed in Chapter 6, the United States learned to do public health through its experiences of empire building. But epidemiological victories in occupied territories did not translate well back home. As Fred Soper put the point, "If you have democracy, you cannot have eradication."[81] Soper was responding to reluctance in the United States to use enough DDT to drive yellow fever cases to zero in southern U.S. states, but his point generalizes. A society of people who think of themselves as free and equal will balk at many of the most effective forms of disease control, and eradication requires consistent and pervasive use of the most effective kinds of disease control. If people have democratic power, there is little reason to think that public health can overcome widespread resistance, and even less reason to think that it should. These issues echo in our contemporary experiences of the COVID-19 pandemic, and we return to them in Chapter 9.

Even if we could overcome the institutional and ideological forces that limit the effectiveness of vaccine mandates, human and political shortcomings will hamper eradication efforts. Consider the 21st-century setbacks to polio eradication caused by the Central Intelligence Agency (CIA). After the September 11, 2001 (9/11), attacks, the U.S. national security apparatus focused on locating Osama bin Laden.[82] They suspected he was hiding in Pakistan with extended family members. In an effort to locate bin Laden, the CIA infiltrated a regional hepatitis B vaccination drive, collected DNA from the people they vaccinated, and attempted to find matches to the bin Laden family. These efforts failed to locate the terrorist leader. But when word got out, communities across Pakistan and other countries resisted and sometimes attacked vaccination teams, setting back efforts to vaccinate against a long list of diseases, including polio.[83]

Finally, even the lone victory for eradicationism is less secure than it may appear to be. The last naturally occurring case of smallpox was in a Somali man in 1977.[84] But following 9/11, and after worries about post-Soviet labs and their unsecured smallpox vials, concerns about smallpox bioterrorism caused some countries to restart production of smallpox vaccine. Smallpox remains eradicated, but the United States is once again spending substantial sums to prepare for a world in which it is not.

The substantial long-term benefits of eradication could likely justify coercive immunization policies. But if eradication or elimination is not a reasonable policy goal, then we have a smaller set of benefits to invoke when we try to justify vaccine mandates. In particular, if vaccination history is more cyclical than linear—if it consists of patterns of higher and lower immunization rates, punctuated by different degrees of coercion in immunization policies—then we will have to try to balance the benefits of short- to medium-term higher immunization rates against what it costs to achieve those benefits. We take up that task in Chapter 9.

9

Policy Limitations and America's Institutions

Introduction

Chapter 8 surveyed ethics arguments about coercive vaccine mandates. But debates about California's Nonmedical Exemptions Bill—and whether other states should replicate it—cannot be resolved by moral philosophy and bioethics. Even if removing nonmedical exemptions (NMEs) could be justified in an abstract sense, America's efforts to eliminate NMEs are likely to produce underwhelming benefits and higher-than-expected costs, against the backdrop of emerging political polarization and declining social trust.

We advance this chapter's thesis by first arguing that efforts to eliminate NMEs in the United States are unlikely to result in substantially higher national immunization rates. In states whose legislatures can pass NME elimination laws, many of their local communities and schools will not fully enforce those laws. More important, only a handful of U.S. states are likely to eliminate NMEs, in light of new political polarization about immunization policy: Democratic state legislators favor NME elimination and Republicans oppose it. Ironically, the push to eliminate NMEs in states controlled by Democrats may indirectly contribute to efforts to make it easier to refuse vaccines in Republican-controlled states. Finally, we argue that neither the COVID-19 pandemic nor President Trump's response to it are responsible for partisan political fights about vaccine mandates for routine childhood vaccines. America's vaccine politics got worse during the pandemic; however, childhood immunization mandates were a source of partisan conflict through the 2010s, as illustrated by the fight over California's Clinician Counselling Bill, Nonmedical Exemptions Bill, and Medical Exemptions Bill.

Potentially Underwhelming Benefits

We have focused on a policy change that we assume many of our readers will find compelling. Our interviewees—people such as Hannah Henry and Leah Russin—believed that eliminating NMEs would improve their communities. Katie is perhaps more sympathetic with coercive public health measures than

Mark is, but we both identify with and admire the policymakers and activists who pushed for California's Nonmedical Exemptions Bill. Moreover, there is much to criticize about their *opponents*: The people who worked against the Nonmedical Exemptions Bill frequently expressed falsehoods about vaccines and made poor ethics arguments, as we discussed in Chapter 8. Furthermore, we are aghast at the sidewalk assault on Richard Pan and the blood-based vandalism of the California Senate chambers.

Coercive mandates can do substantial good. They aim to increase immunization coverage and reduce disease incidence, thereby protecting children, their families, and the community.[1] By this metric, California's elimination of NMEs was successful. It also promoted pro-vaccine social norms—and framed vaccine refusal as socially deviant—as we discussed in Chapter 5. More important, the percentage of fully vaccinated children enrolled in California's kindergarten classes increased substantially within 2 years of the policy being implemented, from 92.8% in 2015–2016 to 95.6% in 2016–2017.[2] California's Nonmedical Exemptions Law will likely pay dividends for children's health, their education and development, and their life prospects. It will make disease outbreaks less likely, which will benefit the state economy and its social and political stability.

Yet even as we recognize the success of California's Nonmedical Exemption Law, its benefits may be less substantial than its advocates may have hoped.[3] Eliminating NMEs increases vaccination rates only if mandates are properly enforced, if parents agree to vaccinate rather than to homeschool, and if other actors (e.g., doctors) agree not to undermine the intent of the law (e.g., by writing fraudulent medical exemptions).[4] Evidence from California invites some doubt about the success of these implementation efforts.[5]

California's schools are not universally excluding unvaccinated children. In the years after NMEs were eliminated, the number of children in the "overdue" category quadrupled.[6] These are students who were enrolled even though they had not completed required vaccines, yet they also did not have any official exemptions. Their presence is clear evidence of poor policy enforcement. New York state appears to be suffering from similar enforcement issues. It eliminated NMEs in 2019, but media reports during the 2022 polio emergency illustrate that some schools have vaccination coverage rates lower than 90%, while other schools have refused to submit any student vaccination data to state authorities.[7]

There are many explanations for the imperfect enforcement of vaccine mandates at schools. Some local school administrators may want to enroll unvaccinated students. As we have uncovered in other research, some school staff believe that it is inappropriate to exclude unvaccinated children, especially when communities are not experiencing outbreaks.[8] Such administrators may choose to prioritize a child's access to education over the marginal improvement to a school's immunization rates that would follow from excluding that child. Also, even school and daycare administrators who want to exclude unvaccinated children may face opposing pressure from parents and community members. New York state decided not to exclude unvaccinated Amish students from school after their parents sued, demonstrating that even governments can waver in the face of such pressures.[9] The COVID-19 pandemic has further illustrated that even small minority groups can prevent the enforcement of public health measures if they are sufficiently vocal and confrontational.[10] Furthermore, schools have financial incentives to keep students enrolled, even when administrators personally support excluding unvaccinated children and parents do not resist. In California, as in most U.S. states, a public school's funding is tied to enrollment, so schools have a strong incentive to enroll students. This incentive is even stronger for private schools, which often risk being shuttered by even moderate declines in tuition revenue.

Even when schools can enforce coercive mandates, doing so will produce higher immunization rates only when the threat of school exclusion is more salient than the burdens associated with vaccination. Parents who are only weakly committed to refusing vaccines, or who previously would have sought waivers out of convenience, will be pushed toward vaccination by the elimination of NMEs.[11] But parents who believe that vaccination is harmful or even deadly are highly unlikely to vaccinate their children, even if they also want their children to be in school.

In light of the imperfect enforcement of California's vaccine mandates after the elimination of NMEs, it is unclear how many parents in California decided to remove their children from care or school rather than have them vaccinated. Poor enforcement might have spared some parents from having to make that choice, but we suspect that at least some vaccine refusers elected to homeschool. Indeed, the percentage of California children being homeschooled increased substantially following the implementation of the Nonmedical Exemptions Law; however, homeschooling rates were rising before 2016, so much of the post-2016 increase could be part of a secular trend rather than a direct result of the legislation.[12] If refusers did choose to homeschool, this outcome would have increased the safety of schools, but it would have deprived unvaccinated children of education, without protecting the broader community from disease risks associated with these unvaccinated children.

Finally, even if schools are enforcing mandates, and even if parents are willing to vaccinate their children in the face of effective threats of exclusion, coercive vaccine mandate policies cannot work if other agents undermine them. In Chapter 5, we discussed how the number of *medical exemptions* went up dramatically in the aftermath of the Nonmedical Exemptions Bill. Parents demanded these exemptions, even when they did not qualify, and some physicians were willing to help. A small group of physicians even marketed their willingness to support fraudulent medical exemptions. California's subsequent Medical Exemptions Bill in 2019 cracked down on this abuse, but this project remains incomplete and state oversight of physicians is limited.

Moreover, it is unlikely that California will be able to enforce similar limits on other ways parents have found to keep unvaccinated children in school—for example, through diagnosis of a learning disability. As we outlined in Chapter 5, students with a learning disability and an Individualized Education Program (IEP) receive special protection under the Individuals with Disabilities Education Act (IDEA).[13] California's Nonmedical Exemption Bill proactively codified an exemption for such students, and the fact that IDEA compels such exemptions was confirmed in *Whitlow v. California* in 2016.[14] In light of strong incentives for schools and parents to keep children enrolled in school, it should be no surprise that California schools witnessed a dramatic increase in the number of children with IEPs after the Nonmedical Exemptions Law came into effect.[15]

The existence of many pathways for enrolling unvaccinated children demonstrates that the elimination of NMEs is not sufficient to prevent unvaccinated children from attending school, especially when parents possess enough social and economic capital. Privileged parents can pay doctors to provide fraudulent medical exemptions, or they can obtain a desired disability diagnosis in conjunction with school authorities. Thornton and Reich report an online comment from a mother about the role of social capital and privilege in such parents' flouting of the law: "All of my anti-vax California friends with kids have medical exemptions that they paid a hefty price for."[16] Families that can overpower, cajole, or gain the support of school officials can likewise keep their unvaccinated children enrolled.

Substantial, Unintended Costs of Eliminating Nonmedical Exemptions

As we noted above, it is not clear how many children have been denied care or education as a result of the Nonmedical Exemptions Law. However, even if only a very small number of children missed out, that would be a substantial loss. Daycare and school provide supervision, food, health interventions (e.g., vision and hearing tests), child abuse reporting, and many other important services.[17] Even missing one year of formal education can have a dramatic impact on a child's lifelong prospects for health, longevity, economic productivity, and other goods.[18] Also, there can be substantial spillover effects of denying a child access to care or school, with damaging consequences for families and across future generations. A child who is denied access to out-of-home care or formal education will need to be cared for by a family member, usually a mother. Hence, one likely outcome of effective efforts to eliminate NMEs will be a reduction in mothers' participation in the formal workforce, which is associated with worse outcomes for both mothers and their children.[19]

NME abolition may also generate much more serious downstream consequences. Efforts to make vaccine mandates more coercive may cause a backlash that will undermine public health. Indeed, other kinds of changes in immunization policy have sometimes had harmful unintended consequences.

Heidi Larson, the anthropologist founder of the Vaccine Confidence Project, cautions policymakers about unintended bad outcomes from their well-intentioned vaccine policies. In her book, *Stuck*, Larson describes a joint statement by the American Academy of Pediatrics (AAP) and the Centers for Disease Control and Prevention (CDC) from 1999, which called on vaccine manufacturers to remove thimerosal from vaccines, absent any evidence that this ingredient was harmful.[20] The statement aimed to improve perceptions about vaccine safety: Some parents worried that thimerosal was harmful, even though it was not. The announcement backfired because parents assumed that AAP and CDC would ask for thimerosal to be removed only if it were dangerous, which those organizations had long denied. Trying to cater to parents' worries ended up damaging public trust rather than increasing it. Furthermore, because thimerosal is a very effective preservative—and because there was not an alternative immediately available—its removal from vaccines decreased access to vaccines for people in poorer countries that lacked capacity for sustained cold storage along the supply chain.[21]

There were similarly bad unintended consequences associated with removing RotaShield rotavirus vaccine from circulation. This vaccine was very effective at preventing rotavirus infections, which disproportionately killed and otherwise harmed children in poor countries that lacked the ability to effectively treat the disease. A tiny number of vaccinated children developed intussusception— a dangerous kind of bowel blockage—and the vaccine was removed from the market in 1999, soon after that side effect was confirmed. However, no alternative rotavirus vaccine was immediately available, which led to additional and avoidable deaths in poorer countries, where the harms associated with a few cases of intussusception were far outweighed by the harms associated with rotavirus disease.[22]

We worry that efforts to eliminate NMEs will also have bad unintended consequences, particularly a politicized backlash against public health measures. The elimination of NMEs may generate political conflict about vaccine mandates that will undermine public support for immunization governance. This worry is especially pressing in light of Americans' current social and political divisions and the recent failures of public health governance during the COVID-19 pandemic.

The Perils of Politicizing Vaccines

We wish vaccination policy were boring, like policies for water and sewer systems usually are. We agree with Hannah Henry, who says,

> We don't go around saying . . . "I'm all for sewer systems! I think they're wonderful, and I have a position on sewer systems, and . . . everybody should continue to flush their toilets and celebrate their sewer system." Like, we just take it for granted. And that's what had been the case with vaccinations.[23]

Vaccine policy used to be less interesting, and we hope it can become boring again. We would love for governments to have largely technical debates about which vaccines to recommend and fund, while ordinary people voluntarily chose to vaccinate their children and themselves. But U.S. vaccination policy is not boring today. In the context of the COVID-19 pandemic, it is one of America's most contentious political issues. And we worry that efforts to eliminate NMEs heighten these conflicts in ways that are ultimately counterproductive for public health.

The COVID-19 pandemic nationalized conflicts about vaccine mandates. Political battles over lockdowns, school closures, and mask mandates raged across the country in 2020 and 2021, setting the stage for subsequent clashes over vaccine requirements. But we should not let national politics and elite messaging during the COVID-19 pandemic obscure the important role that state-level fights about vaccine mandates played over the past decade. America's most consequential vaccine policies are the school entry mandates that cue parents to vaccinate their children. Long after the COVID-19 pandemic is over—and when those political conflicts about the pandemic have decreased—America will still rely on school vaccine mandates to keep the country safe from infectious diseases. America needs these mandates to be uncontroversial.

Consider that it takes up to 95% of the childhood population to be immunized against measles to keep that disease out of our communities. This means that vaccination has to be not only a mainstream behavior but also near ubiquitous. And preserving that ubiquity requires extra care to avoid orienting even small groups of people toward refusal. Reaching 95% support or compliance is an enormous challenge. Fifteen percent of Americans believe "the government, media, and financial worlds in the U.S. are controlled by a group of Satan-worshipping pedophiles who run a global child sex trafficking operation."[24] Nearly one-third believe that our destinies are controlled by the stars,[25] while 26% believe that the sun revolves around the earth.[26] With significant percentages of Americans believing these absurd ideas, it is difficult to get 95% of Americans to believe vaccines are safe, effective, and necessary.

The only chance for vaccination to be a (nearly) unanimous behavior is for every significant social group to strongly support vaccination. We cannot hope to have a stable immunization social order if even small groups turn against vaccines. Instead, vaccination must be unremarkable, something that occurs in the background of our lives. When vaccination emerges to the foreground of our thinking, and when it becomes a topic of political conflict, it is no longer possible to adequately protect our communities.

School entry vaccine mandates have long been the backbone of the ubiquitous, boring, state-based project of getting Americans vaccinated. As we discussed in Chapter 2, one of the main reasons why states turned to school vaccine mandates was because of how easy, nonintrusive, and stable they were. NMEs were a central part of that boring, stable order in most states for many decades. Even in West Virginia, which never introduced NMEs, and in Mississippi, which got rid of its NMEs, poor and conservative voting populations remained vaccine-compliant for decades.[27] Furthermore, vaccine mandates remained politically anodyne even as they were modified over the final decades of the 20th century. States of both partisan stripes added new vaccines to their existing mandatory regimes—varicella, Hib, hepatitis B—with little or no controversy.[28]

Until the period of conflict that this book addresses, both major U.S. political parties supported vaccine mandates, even if they sometimes did so for different reasons. There is a long history of Democrats supporting more "big-state" public health interventions to protect communities from predictable harms.[29] Meanwhile, for Republicans, the economic benefits of vaccination are clear: The absence of infectious disease keeps businesses and schools open, workers employed, and health care services less dependent on state funds.

Emerging Polarization About Mandates and Vaccines

As we described in Chapter 3, state legislators have made many efforts to tinker with vaccine mandate exemption policies over the past 30 years, and especially in the past 15 years. Scholarship about this period provides little evidence that one political party was driving these efforts more than another. Indeed, as late as 2015, a bill to eliminate religious exemptions in North Carolina had Democratic and Republican co-sponsors.[30] One explicit political lens for examining state vaccine mandate polices is offered by Goldstein and colleagues, who focus on state legislation from 2011 to 2017, which covers the period of California's

Clinician Counselling Bill and Nonmedical Exemptions Bill.[31] They found that "anti-vaccination bills" (bills that made it easier to get exemptions) were more likely to be sponsored by Republican legislators. However, "pro-vaccination bills" (bills that made it more difficult to get exemptions) were equally likely to be introduced by Democrats or Republicans. The North Carolina bill is an example of the latter bipartisan tendency. Also, pro-vaccination bills were much more likely to get passed.

These findings align with other 2010s analyses of the relationship between political affiliation and vaccination. For example, in 2013, Dan Kahan found that attitudes toward vaccination were not subject to cultural cleavages. A person's voting habits could predict their views about environmental issues or gun control, but could not yet predict their opinions about the safety and efficacy of vaccines. Instead, Republicans were more supportive of exemptions to vaccine mandates than Democrats were, even while Democrats and Republicans were equally supportive of vaccination. Kahan cautioned policymakers that trying to further limit or eliminate exemptions could cause polarization about mandates to spill over into polarization about vaccines, if people came to view vaccination through the lens of political partisanship.[32]

Kahan's concerns were not heeded by the political activists and civil society organizations that promoted California's elimination of NMEs as a model for the country.[33] The American Medical Association, the American College of Physicians, the American Academy of Pediatrics, and the Safe Communities Coalition increasingly found themselves working with Democratic legislators to try to eliminate NMEs, while parent groups who wanted to protect NMEs found sympathetic allies among Republicans.[34]

Leah Russin from Vaccinate California explained that the activists who worked to pass the Nonmedical and Medical Exemptions Bills were aware of the political divide, and they tried to overcome it. "We worked hard to reach out to Republicans during these bills," she said, explaining that for the latter bill, "The Californian Medical Association even hired a Republican strategist."[35] Those efforts were in vain. California's Medical Exemptions Bill drew even fewer Republican votes than the Nonmedical Exemptions Bill had.[36]

Proof of further politicization can be found in the state-level policy changes following the measles outbreaks that shocked Americans in 2019.[37] All of the states that eliminated NMEs after that date—Washington, New York, Maine, and Connecticut—had Democratic Party trifecta governments.[38] That is, Democrats controlled the state's executive (the governor) and its legislature (often both houses of their legislature) in each of those states. As of the middle of 2022, only nine other states had Democratic trifecta governments, which limits the likely success of further bills.[39] There are 13 states in which neither party has a trifecta but instead have divided government. In the context of polarized conflict about vaccination policy, these states are unlikely to change their immunization laws. More troubling is that 23 states currently have Republican Party trifecta governments. Republican legislators in these states have introduced many bills to protect or expand NMEs, and even to eliminate school and daycare vaccine mandates.[40] It would be a public health disaster if these efforts were successful in weakening or removing existing mandates.

While all this state-level polarization was unfolding during the end of the 2010s, vaccination policy also featured in the national political discourse. During the 2016 Republican presidential primary, candidates Donald Trump, Ben Carson, and Rand Paul endorsed vaccine-critical views. Trump invoked an "epidemic" of autism supposedly caused by vaccinations. Carson and Paul, both physicians, expressed more moderate vaccine skepticism by signaling their support for slowed-down or alternative vaccine schedules.[41] Trump's victory in the primary and general elections planted vaccine skepticism deeply in the contemporary Republican Party, with consequences for that party's response to the COVID-19 pandemic.

The COVID-19 pandemic was accompanied by dramatic political polarization over lockdowns, masking, and school closures. In Chapter 5, we discussed the upshot of these developments for California. Nationally, President Trump and other Republicans led the rallying cry for "freedom" and for returning to normal life as soon as possible.[42] In contrast, Democrats generally supported restrictive pandemic measures.[43] The result was often stark differences in COVID-19 infection rates between states that were governed by Republicans and those governed by Democrats.[44] There were also heated and violent protests about state-level pandemic restrictions. For example, in Michigan, heavily armed right-wing terrorists occupied the state capitol building, and a small group attempted to kidnap the governor.[45]

As COVID-19 vaccines became available, President Trump's administration and other Republican leaders were often lackluster in promoting uptake.[46] President Trump sought credit for the speed of vaccine development, and did not speak *against* COVID-19 vaccines.[47] However, he received his vaccines in private and did little to promote vaccination.[48] Many Republican media voices (e.g., Tucker Carlson) actively opposed COVID-19 vaccination.[49] Following Trump's loss in the 2020 presidential election, many more Republican leaders and public figures spoke out against COVID-19 vaccines, including some who had previously made supportive statements.[50] Republican leaders also preemptively and consistently voiced their opposition to potential COVID-19 vaccine mandates.[51]

Republican leaders' resistance to COVID-19 vaccine mandates went well beyond public statements, and also included legislative and executive interventions.[52] We studied state-level efforts to create, modify, prevent, or eliminate vaccine mandates—whether for COVID-19 or for existing childhood vaccinations—from January 2020 to September 2021. Almost all efforts to loosen or prevent mandates were championed by Republicans, while almost all efforts to introduce or tighten mandates were led by Democrats.[53] While most of these policies focused on mandates, and most attention was paid to COVID-19 vaccinations, Republican efforts were sometimes obviously influenced by more general anti-vaccine sentiments. For example, Representative Ed Hill from the Montana State Legislature attempted (and failed) to have homeopathic remedies fulfill the requirements of his state's school entry vaccine mandates.[54] In 2021, Tennessee's Department of Health was briefly pressured by that state's legislature to stop promoting *any* vaccines for adolescents.[55] Florida's legislature also considered eliminating *all* school vaccine mandates,[56] before deciding to limit themselves to prohibiting COVID-19 vaccine mandates.[57]

Republican efforts to restrict state support for COVID-19 vaccination were wide ranging. They tried to prevent vaccination status from being used in driver's license renewal procedures, employment contexts, or school enrollment. Republicans proposed amending nondiscrimination laws to make "unvaccinated persons" a protected group, preventing state monies from being spent to promote COVID-19 vaccines, and outlawing any collection of information about people's vaccination status. Most alarmingly, some Republicans tried to make it illegal for public health authorities in their states to recommend COVID-19 vaccines.[58]

It is difficult to identify exactly how state-level political fights about COVID-19 vaccines and mandates relate to the views of ordinary Americans. Were Republican public figures leading their constituents toward vaccine skepticism, or were they following preexisting attitudes of the people they represented? Regardless, there is now a clear political polarization regarding COVID-19 vaccines and COVID-19 vaccine mandates.[59] Researchers at the Kaiser Family Foundation tracked partisan affiliation of unvaccinated Americans over time and asked them about their intentions to vaccinate. By October 2021, they found that Republicans were 3.5 times as likely to be unvaccinated compared to Democrats, reporting that "political partisanship is a stronger predictor of whether someone is vaccinated than demographic factors such as age, race, level of education, or insurance status."[60] Although the evidence from the historical record is incomplete, this appears to be the first time that a person's political partisanship has been the strongest predictor of their vaccination status.

Kahan's fears have come true, at least for COVID-19 vaccines. Americans are politicized not just about mandates but also about some vaccines.

Polarization about vaccine mandates continues apace. A February 2022 YouGov poll found that only 64% of Republicans and 83% of Democrats agreed with the statement, "Do you think parents should be required to have their children vaccinated against measles, mumps, and rubella [MMR]?"[61] The fact

that 64% of Republicans support MMR mandates is likely sufficient to maintain these mandates in most U.S. states. However, only 39% of Republicans and
77% of Democrats agreed with the statement, "Do you think parents should be
required to have their children vaccinated against infectious diseases?"[62] The
lower numbers here may reflect decreased support for mandating vaccines
other than MMR. (Notably, the survey questions do not ask about *exemptions* to
requirements, which lie at the heart of this book and which can greatly alter the
salience of a policy, or its likelihood of making parents vaccinate. But attitudes
toward mandates in general are nevertheless telling.) The YouGov poll also found
that only 24% of Republicans and 75% of Democrats agreed with the statement,
"Do you think parents should be required to have their children vaccinated
against COVID-19 if they are eligible for the vaccine?"[63] Clearly, polarization
about a COVID-19 vaccine mandate is most stark, even as polarization about
mandates for other vaccines also exists.

It is notable that among Democrats, levels of support for MMR vaccine
mandates, mandates in general, and COVID-19 vaccine mandates were similarly high: 83%, 77%, and 74%, respectively, for a range of only nine percentage
points. In contrast, Republican support for MMR vaccine mandates was much
higher (at 64%) than their support for mandates in general (39%) or for COVID-
19 mandates (24%), with a 40 percentage point difference between support for
MMR vaccine mandates and COVID-19 vaccine mandates. Whatever polarization commenced prior to the COVID-19 pandemic, there is little doubt that
experiences since 2020 have exacerbated it.

It is not clear how or whether we can undo political polarization about
COVID-19 vaccines. A more important question is whether we can prevent
views about routine childhood vaccines from becoming similarly polarized.
There is, so far, only limited and indirect evidence of this spillover effect—from
COVID-19 polarization to polarization about routine childhood vaccines.
Childhood immunization rates have recently declined in many communities,
but it is unclear how much of this decline is due to new politicization of vaccination behaviors, rather than to other pandemic-related barriers to immunization.[64] It will be vital to continue to research underimmunization and its causes
so that governments can respond. Government action to address access barriers
may prove costly and challenging for the moribund American health institutions
we have examined in this book; this is a *technical* and *financial* problem as well
as a political one. But the direct political threat of polarization remains pressing.

Political polarization surrounding acceptance of COVID-19 vaccines is very
recent—it appeared in force only in the second half of 2021—so we should not
be surprised to see little evidence of a spillover effect. However, some objections
to COVID-19 vaccines are also raised against routine childhood vaccines—for
example, that we should not trust the pharmaceutical companies that make

vaccines or the government agencies that approve them, or that vaccines commonly have dangerous side effects. Given the similarities between COVID-19
vaccine refusal and other kinds of vaccine refusal, it would be surprising if there
were no spillover effect at all. There has already been a substantial decrease in
the number of American children who are up-to-date with their vaccines, with
national rates among kindergartners falling by over 1% from 2020 to 2021.[65] The
overall vaccination rate declines may be steeper, as kindergartens experienced a
10% drop in enrollment during that period, and some of the children who did
not start kindergarten may not have received vaccines.[66] There have been similar declines around the world during this period,[67] so it is difficult to identify
how much of the American decline is connected to COVID-related spillover into
general vaccine rejection.

Even if there is not a substantial spillover, *vaccine mandates* are now thor-
oughly politically polarized and will likely continue to be for the foreseeable
future. The polarization of vaccine mandates—and of coercive public health
measures more generally—has eroded America's ability to implement effective policies to prevent and control outbreaks. There is little reason to think that
American life is going to become less politicized in the near future, or that the
two major political parties will become less polarized. And there is every reason
to believe that new outbreaks will be coming our way. Accordingly, we should
expect divisive politics to further undermine America's efforts to govern vaccine uptake. Chapter 10 explores the consequences of this dynamic. We suggest
strategies for government, private institutions, and individuals to adapt to the
likely decline of America's immunization governance.

10

Conclusion

Confronting Dystopia

Introduction

Yascha Mounk begins the concluding chapter of *The Great Experiment* by telling his readers that he faces a "Chapter Ten Problem."[1] When an author reaches the final chapter of their book—which is often Chapter 10, as in the case of Mounk's book and ours—they frequently transition from describing a problem to trying to solve it. This pivot can be perilous. Authors must navigate between proposing solutions that cannot work in our current world and advocating for feasible policy changes that will not make much of a difference.

We are not worried about the "Chapter Ten Problem" because we do not see any solutions to America's vaccine wars. The thesis of this book is that a recent history of *small tweaks* has not been able to reverse the rise of vaccine refusal, but also that *big policy changes* (e.g., eliminating nonmedical exemptions [NMEs]) face significant implementation challenges and may backfire, even if—like California's Nonmedical Exemptions Bill—they increase vaccine coverage rates.[2] Therefore, rather than attempt to solve the problems this book has discussed, we focus in this final chapter on the upshot of our thesis for the future of American public health. We first discuss why the future looks bleak and then we suggest how individuals and institutions can adapt to an America in which immunization rates decline and outbreaks of vaccine-preventable infections become more common.

Efforts to eliminate NMEs illustrate and contribute to institutional failure. Removing NMEs was the last change that policymakers could make to a half-century's worth of immunization policies. This step was more effective at increasing immunization rates than were the clever but more subtle tweaks that had preceded it. However, as we argued in Chapter 9, increasing the coercive capacities of mandates by removing NMEs is unlikely to be a successful policy model for the country as a whole. No more than a handful of additional states are likely to adopt such a policy, and it may not preserve community protection due to implementation failures and subversive efforts by vaccine-refusing parents. Our only hope for substantial improvements in immunization rates—or even for holding on to the rates we have now—relies on broad institutional changes in American society. And there is little reason to think those changes will happen any time soon.

The Decline of Public Health Citizenship?

Contemporary vaccine advocates often invoke solidarity by calling on people to care more about each other.[3] They talk about the inherently collective nature of building and maintaining community protection, and they sometimes speak of vaccination as an act of love. They appeal to many of the ethical ideas we addressed in Chapter 8. It is no coincidence that Hannah Henry's I ♥ Immunity campaign tried to make people feel good about contributing to vaccination's end goal. But coercive vaccination laws are unlikely to promote solidarity. Instead, our recent experience with COVID-19 pandemic governance illustrates that coercive public health measures can fray the fabric of society.

Vaccine advocates who invoke solidarity sometimes appeal to a "golden age" in American immunization history, a time when the people of this country were more committed to the public good. They point to the school-aged Polio Pioneers who participated in 1950s vaccine testing, and to Jonas Salk's supposedly altruistic refusal to patent the polio vaccine.[4] These myths are supposed to illustrate how immunization policy used to be driven by a greater sense of common purpose among Americans. They orient us toward a civic ideal that has deep roots in democratic theory, but which is all but absent in the history of America's democratic experiment.

Public health citizenship has been a component part of democratic theory from the time of the French and American Revolutions. For example, Thomas Jefferson and the Constituent Assembly of the French Revolution endorsed demanding duties of health citizenship as a matter of public obligation: Citizens of democracies had duties to be healthy for the benefit of all.[5] *The New York Times* applied these ideas to vaccination in 1900, arguing that it was a "public duty which every citizen owes to those with whom he comes in daily contact" and that this "common sense" idea "prevailed in all classes of the population," such that only "out-and-out savages" should resist vaccination or mandates.[6] Here we see the idea that it is a mark of modernity—and of civilization itself—that people should participate in public projects to promote the health of their communities.

Do the American people possess sufficiently communitarian commitments to support public health projects that restrict the liberty of dominant social groups? Have they ever? The history of American public health governance has been marked by periods of public resistance and political retreat, as public health leaders identified the limits of American compliance. A late-19th-century public health officer speculated about likely resistance to compulsory (i.e., forcible) vaccination of adults in the United States in light of his country's commitment to personal liberty:

> The people of this country are too thoroughly imbued with a sense of personal independence to submit patiently to personal compulsion. The attempt would excite hostility to vaccination that does not exist at present, and would hinder rather than promote the cause of vaccination.[7]

A health official in early 20th-century Louisiana provided a numerical estimate of public rejection of compulsory vaccination. Such a law

> would probably meet the passive resistance of one-third of our people, the violent opposition of another third, the unwilling compliance of most of the remaining third, and cheerful compliance by the small fraction comprising the intelligent and law-abiding class.[8]

We have little evidence to judge whether Americans are sufficiently committed to solidarity to accept compulsory (i.e., forcible) vaccination, because such measures have rarely been attempted. But when America *has* coerced vaccination, it has reinforced hierarchical and authoritarian social orders, rather than solidaristic commitments. As we discussed in Chapters 2 and 6, coercive vaccination has usually been imposed only on "captive" populations—for example, poor immigrants, itinerant workers, prisoners, and service members. And, even then, it often met with fierce resistance. America has never succeeded in systematically coercing the vaccination of comfortable White populations. It seems unlikely that contemporary White America is going to be *more* willing to accept this practice now than it has been in the more distant past.

We can cultivate further skepticism about a future of solidary-based American vaccine acceptance by critically reflecting on the mid-20th century's "golden age" of vaccination compliance. Consider again the example of the Polio Pioneers and Jonas Salk. Did parents line their children up at school for polio vaccine trials because they wanted them to benefit the community by participating in a medical experiment? Did they think of their children as instruments by which they could perform a civic obligation?[9] Some parents may have shared these sentiments, but we suspect that the primary motivation for most of the parents of the Polio Pioneers was the fear that their children would be infected with polio. It seems unlikely that a sense of community obligation led them to present their children for an experimental vaccine.[10] Moreover, Salk himself was not motivated by an altruistic civic commitment when he refused to patent the polio vaccine. His lawyers told him that the patent was unlikely to be upheld.[11] And the March of

Dimes Foundation had financed vaccine development, so he did not have any backers who needed to recoup their investments.

It is a self-congratulatory fiction to locate the public health compliance of earlier generations in their commitment to fairness and solidarity, or to suppose that America can resurrect a set of public virtues that it never possessed.

A truer story about earlier ages of American immunization governance would focus on the fact that previous generations of Americans often had more in common with each other than present generations of Americans do. The country was more White, more Christian, more racist, and more committed to imperialist foreign policies. Americans used to be more obedient to authority figures, and powerful social institutions enforced their compliance. This was especially true when super-majorities of Americans used to live in rural areas, whose cultures often tended toward greater conformity. In the period immediately after World War II, many people were bound together by a shared civil religion of American patriotism—including Cold War hatred of communism—and their private religious beliefs were more often connected to churches that occupied centrist positions in American political life.[12] Among White Americans, there used to be far greater economic equality, more optimism about improving standards of living, and greater trust in social institutions (including government, medicine, and science).[13] Racism and, more important, the institutions of White supremacy—in education, housing, and the workplace—shaped a shared experience for White Americans, even as they imposed common forms of oppression on non-White Americans.[14] That oppression also facilitated the vaccination of minority groups.

Today, many of the institutions that used to tie Americans to each other have eroded or disappeared.[15] In the case of White supremacy and racial segregation, that is for the best. But the erosion of its stabilizing institutions has also left America much less able to organize behavior through informal social means. This consequently leaves the country more reliant than ever on coercive state power to change recalcitrant behaviors. And that coercion has predictably bad consequences, in light of popular resistance and political polarization.

Medical Authority and Social Trust

In the early and mid-20th century, medical science generated substantial trust and good will for itself through its dramatic technological advances and its powerful propaganda machines. But this trust has eroded over subsequent decades, and it can no longer stabilize America's immunization social order. In particular, the period from the 1920s to the 1950s witnessed an explosion of scientific and medical breakthroughs, which led people to embrace novel interventions such as new vaccines.[16] But the speed of development for new life-changing medical technologies has slowed in recent decades. Most of our new drugs, devices, or interventions make at best marginal improvements on existing interventions. Even though some new and effective drugs are developed quickly—for example, COVID-19 vaccines—the golden age of medical advances has clearly passed.

Within this landscape, "anti-medicine" critiques point out problems with corruption, overtesting, and overtreatment in health care. They raise legitimate worries about the influence of pharmaceutical companies in creating "problems" for drugs to solve, convincing patients they have these problems, and getting physicians to prescribe their drugs.[17] In light of this, Bernice Hausman argues that broad-based trust in medicine is difficult to sustain and that a principle such as "stay away [from medicine] except for emergencies" is not unreasonable for ordinary people to adopt.[18] Hausman also argues that America's modern bureaucratic state performs vaccine approval and advocacy through mechanisms and institutions that are opaque and mysterious to most people, such that few people can directly assess whether the right questions are being asked and addressed. In the absence of broad deference towards or trust in the system, many people will have little reason to believe what they are told.[19] While large bureaucratic states and their institutions continue to be dominated by elite members of the managerial class, ordinary people are unlikely to see their own views and values well represented.[20]

The breakdown in trust is not confined to medical institutions, but extends broadly to America's social and political institutions. For example, the percentages of Americans who say they have "trust in Washington, DC" has fallen from over 70% during the 1960s to less than 40% in the 1990s and to just 17% in 2019.[21] People also increasingly distrust members of other political parties, and not just the opposing parties' politicians.[22] Americans are also less trustful of other people in general, with higher levels of distrust among younger people.[23] For example, more than 60% of Americans younger than age 65 years say that you cannot trust most people.

In *Trust in a Polarized Age*, Kevin Vallier argues that

> falling social trust can undermine democracy, economic growth, economic equality, the rule of law, the protection of minorities; it can foment tribalism and bigotry, weaken our capacity to form relations of love and friendship with others, and even negatively affect personal psychological well-being.[24]

Vallier's focus is on the relationship between decreased social trust, political polarization, and the functioning of liberal democratic societies. Immunization policy sits at the nexus between all of these phenomena. Decreased trust in elite medical opinions about vaccine safety and efficacy can lead to decreased immunization rates and to increased political polarization. Decreased trust erodes the potential for vaccination to be a consensus public project, and the decline of liberal democratic government undermines the legitimacy of vaccine mandates, even when low immunization rates force that policy choice upon communities, as the advocates of the Nonmedical Exemption Bill experienced.[25]

In Chapter 9, we discussed the emergence of political polarization around vaccine mandates. This political polarization will make it extremely difficult for America to create an immunization social order that can control vaccine-preventable infections.

Americans are more divided by political identity than ever before. That is not to say that the American past was free of conflict or strife.[26] Americans disagreed with each other about many things throughout the 20th century. We fought about communism, civil rights, women's rights, and school desegregation. But there were many more social institutions that held us together, and our *political parties* were generally centrist, such that our politics was not marked by polarized party conflict. As the political scientist Lilliana Mason puts the point in *Uncivil Agreement: How Politics Became Our Identity*, America's two major political parties used to have equal percentages of Blacks, Whites, Catholics, and union members.[27] There were many conservative Democrats and liberal Republicans. Given the central role of political parties in organizing American politics, there were often greater opportunities for agreement and compromise, even across substantial differences.

But our political parties have now become more distinct from each other, such that politicized conflict now draws on deeper cultural divides. As Mason observes,

> A single vote can now indicate a person's partisan preference as well as his or her religion, race, ethnicity, gender, neighborhood, and favorite grocery store. This is no longer a single social identity. Partisanship can now be thought of as a mega-identity, with all the psychological and behavioral magnifications that implies.[28]

Political partisanship motivates us to value in-group loyalty and to resist our "enemies," rather than to pursue our own interests and values. It is therefore a disaster for mega-identity polarized politics to infect immunization policy. After that happens, education, evidence, or moral appeals are unlikely to be persuasive.

The politicization of vaccination policy in the context of radical political polarization undermines vaccination efforts of all kinds, including the maintenance of community protection. This is because when we make decisions in which our identities are implicated, we rely on partisan affinities rather than on a balanced consideration of our interests. Partisans think of politics like team sports;[29] all they care about is winning.[30] When we view a problem through a political lens, we tend to think of people who disagree with us as our enemies, and our negative feelings about "the opposition" are often a much more powerful motivation than our own interests.[31] Given the power of partisan political thinking, it is especially troubling that our political identities are now contaminating how we think about immunization policy.

Adapting to Outbreaks

In light of decreased social trust and high levels of politicization and polarization in American life, we doubt that America's immunization social order will be able to control vaccine-preventable infections as effectively as it has in the past. Accordingly, we should consider how America can best adapt to lower immunization levels in the near future. A climate change analogy may be apt: For over 30 years, activists have called on governments to prevent or at least substantially mitigate climate change. Those efforts have largely failed. We are already living through destructive climate change, and its dangers are accelerating. Now we need to prepare for a world of substantially higher temperatures, greater frequency of extreme weather events, mass flooding, wildfires, and other disasters, even as we should continue to work to lower carbon dioxide emission levels. Our local and national communities are slowly adopting an adaptation orientation toward climate change in response to climate problems we already face.

In a similar way, individuals and communities may soon need to adopt an adaptation approach in response to vaccine refusal and lower immunization rates. If neither education, persuasion, nor coercion can preserve community protection across the country—if there is nothing left that could succeed, at least in the short term—then the best we can do is to prepare for a world in which outbreaks of previously well-controlled vaccine-preventable diseases are more prevalent. We have argued that immunization history is more like a cycle than a line of progress that ends with elimination and eradication of disease. If immunization history is cyclical, then how should we respond to the parts of cycle in which community protection falters?

Our local, state, and national governments are going to have to reorient themselves to new kinds of public health problems and solutions. COVID-19 has taught us some of these lessons. Mark witnessed the failure of America's institutions from the beginning of the pandemic. In contrast, Katie's state government implemented strict policies that protected West Australians for nearly 2 years, but now COVID-19 is endemic there, and Australia's federal and state governments have made most protective health measures voluntary. What can we learn from our governments' successes and failures during the COVID-19 pandemic?

First, we should continue to demand that government promote vaccination—through targeted education and persuasion efforts—even if such efforts are unlikely to keep immunization rates high enough to avoid outbreaks. Clever policy tweaks and social science interventions may no longer be able to sustain sufficiently high rates of immunization coverage, but we should still attempt to vaccinate everyone we can. Our communities will always have fence-sitters who can be oriented toward vaccination by the right kinds of education and outreach.[32] As social scientists uncover effective ways to promote vaccine acceptance, public health institutions should continue to put them into practice.

In the midst of diminishing community protection, public health should retreat strategically, so as to reduce harm as much as possible. In particular, communities should continue to "tinker" with existing NME processes to find ways to make NMEs more difficult to receive without eliminating them entirely. As we discussed in Chapter 3, these efforts can increase immunization rates without causing as much polarization and resistance as abolition does.[33] These small changes may not do *enough* to save community protection, but the good they can accomplish makes these efforts worth pursuing. One way to tinker with NME policies is to eliminate NMEs only for vaccines that protect against diseases for which there is a high risk of outbreaks. For example, physician researcher Douglas Opel and a team of co-authors advocated for abolishing NMEs only for the measles vaccine mandate—while preserving NMEs for other vaccine mandates—because measles is so contagious, the vaccine is so effective, and there has been a recent increase in outbreaks.[34] However, as we have argued elsewhere, disallowing exemptions only for specific vaccines can communicate to the public that some vaccines are more important than others, undermining acceptance of vaccines for which exemptions remain.[35]

Another lesson from the COVID-19 pandemic and more distant history is that public health institutions will need to develop new capacities. This may include greater investment in surveillance mechanisms, for example, by ramping up COVID-19 sewage sampling to cover other diseases, so that communities can predict and prepare for outbreaks.[36] We will also need to dedicate more time and resources to train nurses, physicians, and other medical professionals how to diagnose and treat previously well-controlled vaccine-preventable diseases. Governments should anticipate the resource needs associated with local and regional outbreaks. This may involve creating and funding mobile clinics and developing rapid response clinical and operations teams that can deploy quickly to outbreak areas.

Communities should also be planning for the impact of outbreaks on schools and other institutions. COVID-19 forced institutions to develop agile processes—for example, the capacity to move between in-person and online schooling. Future outbreaks of vaccine-preventable diseases may make the need for these kinds of institutional transitions more common. As educators and parents, we acknowledge the limitations of online learning and we deplore the lost opportunities and mental health consequences for children who have been locked down at home, away from their teachers and friends. But more of these measures may be necessary in the face of future outbreaks of measles, whooping cough, mumps, and other diseases. Furthermore, communities should be making contingency plans to provide daycare services for the children of essential workers, so the capacities of health care institutions can be protected during future outbreaks.

Finally, American state and national governments should communicate, transparently and on an ongoing basis, about their reorientation toward harm reduction in the context of diminished community protection against vaccine-preventable diseases. They will need to manage people's expectations about the likelihood of future outbreaks and about the impacts of those outbreaks on people's lives. We imagine that there will be substantial political pressure to avoid this kind of straight talk. But we should hope for leaders who will inform and inspire people to do the best they can in the difficult epidemiological and political conditions that appear to be coming our way.

Individuals and Private Institutions

Individuals and private institutions will also have to decide how to adapt to a world with more frequent outbreaks. If our political institutions and leaders are unable to govern vaccination, this will not *resolve* questions about immunization governance but will only *transfer* those questions to actors in private society.[37]

Private institutions, including businesses, cultural institutions, and civil society actors, will have to decide how to respond to the increase in outbreaks that will accompany lower immunization rates, especially in communities where governments cannot or will not implement mandates or other disease control measures.[38] In some cases, businesses may implement their own disease control measures or vaccine mandates to protect their economic interests: They will be able to keep the assembly lines going and the service counters staffed only if they can reduce the impact of disease on their workforce. Some companies may choose to implement such policies in response to the demands of their employees or customers, or because of the ideological commitments of corporate leadership. But other companies may adopt a lax attitude toward disease control measures because their employees, customers, or leadership are more tolerant of risks, less likely to believe facts about the dangers of disease or the safety and efficacy of vaccines, or because corporate leadership wants to take a stand in the public health culture wars. Also, private institutions in some communities will find their options restricted if their governments prohibit them from enforcing vaccine mandates and other disease control measures.

The COVID-19 pandemic has provided ample evidence of the different ways that private institutions can respond to workplace disease risks.[39] As government pandemic measures lapsed or were revoked, some private institutions continued to require masks or proof of vaccination. Others were quick to relax such measures after the state no longer required them. (Indeed, such businesses were unlikely to have been enforcing those measures in the first place.)

Individuals and families will also have to decide how to navigate the new immunization social order that emerges when community protection falters. For example, new parents will increasingly have to decide whether to let unvaccinated family and friends visit with their babies or otherwise be present in their lives. Already, some insist that visitors be vaccinated against seasonal influenza or have received a whooping cough booster.[40] Such practices seem likely to become more common as community protection diminishes.

More generally, we wonder what will happen to communities when some families adopt private kinds of immunization governance in their homes and associations. Will families want to be neighbors only with other vaccinated people? Will they want their children to go to school only with other vaccinated children? What about sports leagues, churches, and other associations? It is not clear whether these organizations will escape schisms along the battle lines that our new vaccine wars are creating.

These worries about vaccination-related segregation are not merely speculative. They respond to existing and accelerating tendencies toward ideological segregation in American life. More than at any other time in this country's history, Americans are living, working, and playing only with people who share their political and social values. Already, our ideological identities pick out whether we shop at Trader Joe's or eat at Cracker Barrel, and whether we go to church or listen to hip-hop.[41] Divergent vaccination decisions seem like new accelerants for these kinds of social divides.

Final Thoughts

We take some small comfort from the fact that people's beliefs and attitudes can change quickly and in unanticipated ways. Consider that the past 20 years have seen dramatic bipartisan shifts toward acceptance of gay marriage.[42] And the past 60 years have seen a slow but no less consequential transition toward America's embrace of interracial marriage.[43] So, perhaps Republicans will be able to reaffirm a commitment to immunization governance in a short period of time. This kind of hope hangs by a thin thread—the mere possibility of change—but we can see few other reasons to be optimistic.

America has many of the tools it needs to avoid the dystopia this chapter anticipates. It has created some of the best vaccines in the world, as illustrated by the country's role in developing both the Pfizer and Moderna COVID-19 vaccines. Notwithstanding the many deficiencies of its vaccination programs, America funds vaccines for all children in the country: directly through the Vaccines for Children program and indirectly through mandated no-cost insurance coverage. Also, while there are significant problems with America's health care system, skilled physicians, nurses, and other health professionals are available to promote and provide vaccinations. We possess many of the resources necessary to create and maintain nationwide community protection against vaccine-preventable diseases.

But resource scarcity did not cause the breakdown of America's immunization social order. New vaccines, more funding, and smarter clinicians will not end our polarizing fights about immunization policy. Until we can address this country's underlying social dysfunctions and political failures, we will not be able to make good use of our many immunization resources. Unfortunately, it appears that America's social and political conflicts are likely to intensify in the coming years. America's new vaccine wars are just getting started.

Notes

Preface

1. Mark C. Navin, *Values and Vaccine Refusal: Hard Questions in Ethics, Epistemology, and Healthcare* (New York: Routledge, 2016).
2. Katie Attwell and Melanie Freeman, "I Immunise: An Evaluation of a Values-Based Campaign to Change Attitudes and Beliefs," *Vaccine* 33, no. 46 (2015): 6235–40, https://doi.org/http://dx.doi.org/10.1016/j.vaccine.2015.09.092; Katie Attwell, "The Politics of Picking: Selective Vaccinators and Population-Level Policy," *SSM – Population Health* 7 (2019): 100342, https://doi.org/https://doi.org/10.1016/j.ssmph.2018.100342.
3. Jake Zuckerman, "Pandemic Brings Protests, and Guns, to Officials' Personal Homes," *Ohio Capital Journal*, January 27, 2021.
4. Stephen J. Flusberg, Teenie Matlock, and Paul H. Thibodeau, "War Metaphors in Public Discourse," *Metaphor and Symbol* 33, no. 1 (2018): 1–18, https://doi.org/10.1080/10926488.2018.1407992; Jing-Bao Nie et al., "Healing Without Waging War: Beyond Military Metaphors in Medicine and HIV Cure Research," *American Journal of Bioethics* 16, no. 10 (2016): 3–11; Daniel George, Erin Whitehouse, and Peter J. Whitehouse, "Asking More of Our Metaphors: Narrative Strategies to End the 'War on Alzheimer's' and Humanize Cognitive Aging," *American Journal of Bioethics* 16, no. 10 (2016): 22–24; Heidi Malm, "Military Metaphors and Their Contribution to the Problems of Overdiagnosis and Overtreatment in the 'War' Against Cancer," *American Journal of Bioethics* 16, no. 10 (2016): 19–21.

Chapter 1

1. William Sears, Martha Sears, and James Sears, *The Baby Sleep Book: The Complete Guide to a Good Night's Rest for the Whole Family* (New York: Little, Brown Spark, 2005); Martha Sears and William Sears, *The Attachment Parenting Book: A Commonsense Guide to Understanding and Nurturing Your Child* (New York: Little, Brown Spark, 2001).
2. Julia M. Brennan et al., "Trends in Personal Belief Exemption Rates Among Alternative Private Schools: Waldorf, Montessori, and Holistic Kindergartens In California, 2000–2014," *American Journal of Public Health* 107, no. 1 (2017): 108–112;

Kimiko de de Freytas-Tamura, "Bastion of Anti-Vaccine Fervor: Progressive Waldorf Schools," *The New York Times*, June 13, 2019.

3. "AMA Supports Tighter Limitations on Immunization Opt Outs," American Medical Association, last modified June 8, 2015, http://www.ama-assn.org/ama/pub/news/news/2015/2015-06-08-tighter-limitations-immunization-opt-outs.

4. "Senate Health Committee, Wednesday, April 8, 2015," California Senate, accessed December 31, 2021, https://www.senate.ca.gov/media/senate-health-committee-53/video.

5. Kat DeBurgh (Executive Director, Health Officers Association of California), interview with the author, June 13, 2019.

6. "AMA Supports Tighter Limitations on Immunization Opt Outs," American Medical Association, last modified June 8, 2015, http://www.ama-assn.org/ama/pub/news/news/2015/2015-06-08-tighter-limitations-immunization-opt-outs. "AMA Policy Advocates to Eliminate Non-Medical Vaccine Exemptions," American Medical Association, last modified June 13, 2019, https://www.ama-assn.org/press-center/press-releases/ama-policy-advocates-eliminate-non-medical-vaccine-exemptions; Alyson Sulaski Wyckoff, "Eliminate Nonmedical Immunization Exemptions for School Entry, Says AAP," AAP News, August 29, 2016, https://publications.aap.org/aapnews/news/8969?autologincheck=redirected?nfToken=00000000-0000-0000-0000-000000000000. American Academy Pediatrics, "Medical Versus Nonmedical Immunization Exemptions for Child Care and School Attendance," *Pediatrics* 138, no. 3 (2016): e20162145.

7. "COVID Data Tracker: COVID-19 Vaccinations in the United States," Centers for Disease Control and Prevention, accessed April 5, 2023, https://covid.cdc.gov/covid-data-tracker/#vaccinations_vacc-people-fully-percent-pop12. In Canada, 89.50% of the population older than age 5 years is fully vaccinated. "COVID-19 Vaccination in Canada—Vaccination Coverage," Government of Canada, accessed August 31, 2022, https://health-infobase.canada.ca/covid-19/vaccination-coverage. In Australia, more than 95% of the population older than age 16 years is fully vaccinated. "COVID-19 Vaccine Rollout Update," Government of Australia, accessed July 5, 2022, https://www.health.gov.au/sites/default/files/documents/2022/07/covid-19-vaccine-rollout-update-5-july-2022.pdf. In the United Kingdom, 87.2% of the population aged 12 years or older has received two doses, and 69.4% of the population aged 12 years or older has received a third/booster dose. "Coronavirus (COVID-19) in the UK—Vaccination in the United Kingdom," Government of United Kingdom, accessed July 5, 2022, https://coronavirus.data.gov.uk/details/vaccinations. In Singapore, 96% of the eligible population has received two doses, and 78% of the country's total population has received a third/booster dose. "COVID-19 Vaccination, Vaccination Statistics," Ministry of Health Singapore, accessed July 5, 2022, https://www.moh.gov.sg/covid-19/vaccination/statistics. In Germany, almost 76% of the total population has been fully vaccinated. "Vaccination Rate Against the Coronavirus (COVID-19) in Germany June 2022," Statistica, accessed July 5, 2022, https://de.statista.com/statistik/daten/studie/1196966/umfrage/impfquote-gegen-das-coronavirus-in-deutschland/#professional.

8. Rekha Lakshmanan and Jason Sabo, "Lessons from the Front Line: Advocating for Vaccines Policies at the Texas Capitol During Turbulent Times," *Journal of Applied Research on Children* 10, no. 2 (2019): Article 6.

9. Centers for Disease Control and Prevention, "National and State Vaccination Coverage Among Children Aged 19–35 Months—United States, 2010," *MMWR Morbidity and Mortality Weekly Report* 60, no. 34 (2011): 1157–63, retrieved September 23, 2022, from https://www.cdc.gov/mmwr/preview/mmwrhtml/mm603 4a2.htm.

10. Summer Sherburne Hawkins et al., "Associations Between Insurance-Related Affordable Care Act Policy Changes with HPV Vaccine Completion," *BMC Public Health* 21, no. 1 (2021): 304; Dina Fine Maron, "Improved Vaccination Rates Would Fall Victim to Senate Health Cuts," Scientific American, July 10, 2017.

11. Paul Offit, *Deadly Choices: How the Anti-Vaccine Movement Threatens Us All* (New York: Basic Books, 2011); Mark Largent, *Vaccine: The Debate in Modern America* (Baltimore, MD: Johns Hopkins University Press, 2012); Seth Mnookin, *The Panic Virus: The True Story Behind the Vaccine–Autism Controversy* (New York: Simon & Schuster, 2011); Melissa Leach and James Fairhead, *Vaccine Anxieties: Global Science, Child Health and Society* (London: Earthscan, 2007).

12. Andrew Wakefield et al., "Retracted: Ileal-Lymphoid-Nodular Hyperplasia, Non-Specific Colitis, and Pervasive Developmental Disorder in Children," *Lancet* 351, no. 9103 (1998): 637–41; Gregory A. Poland and Ray Spier, "Fear, Misinformation, and Innumerates: How the Wakefield Paper, the Press, and Advocacy Groups Damaged the Public Health," *Vaccine* 28, no. 12 (2010): 2361–62.

13. We discuss this scare in further detail in Chapter 10; Lillvis and colleagues explain it in detail (see note 14).

14. Denise Lillvis, Anna Kirkland, and Anna Frick, "Power and Persuasion in the Vaccine Debates: An Analysis of Political Efforts and Outcomes in the United States, 1998–2012," *Milbank Quarterly* 92, no. 3 (2014): 475–508; Douglas S. Diekema, "Personal Belief Exemptions from School Vaccination Requirements," *Annual Review of Public Health* 35, no. 1 (2014): 275–92; see also Lakshmanan and Sabo, "Lessons from the Front Line."

15. Saad B. Omer et al., "Vaccination Policies and Rates of Exemption from Immunization, 2005–2011," *New England Journal of Medicine* 367, no. 12 (2012): 1170–71; Nina Blank, Arthur Caplan, and Catherine Constable, "Exempting Schoolchildren from Immunizations: States with Few Barriers Had Highest Rates of Nonmedical Exemptions," *Health Affairs* 32, no. 7 (2013): 1282–90; W. David Bradford and Anne Mandich, "Some State Vaccination Laws Contribute to Greater Exemption Rates and Disease Outbreaks in the United States," *Health Affairs* 34, no. 8 (2015): 1383–90.

16. Robert A. Bednarczyk et al., "Current Landscape of Nonmedical Vaccination Exemptions in the United States: Impact of Policy Changes," *Expert Review of Vaccines* 18, no. 2 (2019): 175–90.

17. Paul Delamater, Timothy Leslie, and Tony Yang, "Changes in Medical Exemptions from Immunization in California After Elimination of Personal Belief Exemptions," *JAMA* 318, no. 9 (2017): 863–64.

18. Omer SB, Enger KS, Moulton LH, Halsey NA, Stokley S, Salmon DA, "Geographic Clustering of Nonmedical Exemptions to School Immunization Requirements and Associations With Geographic Clustering of Pertussis," *American Journal of Epidemiology* 168, no. 12 (2008): 1389–96. doi:10.1093/aje/kwn263

19. "Measles Cases and Outbreaks by Year (2010–2022)," Centers for Disease Control and Prevention, accessed August 19, 2022, https://www.cdc.gov/measles/cases-outbreaks.html; Mark Papania et al., "Elimination of Endemic Measles, Rubella, and Congenital Rubella Syndrome from the Western Hemisphere: The US Experience," *JAMA Pediatrics* 168, no. 2 (2013): 148–55; Christian Dimala et al., "Factors Associated with Measles Resurgence in the United States in the Post-Elimination Era," *Scientific Reports* 11, no. 1 (2021): 51.

20. Centers for Disease Control and Prevention, "Measles—United States, January 1–May 23, 2014," *MMWR Morbidity and Mortality Weekly Report* 63, no. 22 (2014): 496–99.

21. "Measles Cases and Outbreaks," Centers for Disease Control and Prevention, accessed December 21, 2021, https://www.cdc.gov/measles/cases-outbreaks.html.

22. Bernice Hausman, in *Anti/Vax*, presents evidence of a "decided increase in highly charged reporting on vaccination" (p. 37) in the years leading up to the Disneyland outbreak, according to which outbreaks of vaccine-preventable disease are no longer presented as merely regrettable social phenomena, or as evidence of systems failures, but as the result of blameworthy negligence by parents who are stupid and selfish, people who are worthy of mockery and scorn, and whose behaviors should be corrected by law. Bernice L. Hausman, *Anti/Vax: Reframing the Vaccination Controversy* (Ithaca, NY: Cornell University Press, 2019).

23. Centers for Disease Control and Prevention, "Measles Outbreak—California, December 2014–February 2015," *MMWR Morbidity and Mortality Weekly Report* 64, no. 6 (2015): 153–54, accessed December 31, 2021, https://www.cdc.gov/mmwr/prev iew/mmwrhtml/mm6406a5.htm.

24. Bryan D. Jones and Frank R. Baumgartner, *The Politics of Attention: How Government Prioritizes Problems* (Chicago: University of Chicago Press, 2005); Thomas Birkland, "Focusing Events, Mobilization, and Agenda Setting," *Journal of Public Policy* 18, no. 1 (1998): 53–74.

25. Owen Dyer, "Philippines Measles Outbreak Is Deadliest Yet as Vaccine Scepticism Spurs Disease Comeback," *BMJ* 364, no. 1739 (2019): 1739; Baffa Sule Ibrahim et al., "Burden of Measles in Nigeria: A Five-Year Review of Case-Based Surveillance Data, 2012–2016," *Pan African Medical Journal* 32, 1 (2019): 5; "Situation Report: UNICEF Somalia Monthly SitRep 9, September 2014," United Nations Office for the Coordination of Humanitarian Affairs, accessed December 17, 2021, https://relief web.int/report/somalia/unicef-somalia-monthly-sitrep-9-september-2014.

26. Gabrielle Canon, "'California Is America, Only Sooner': How the Progressive State Could Shape Biden's Policies," *The Guardian*, January 22, 2021, https://www.theg uardian.com/us-news/2021/jan/22/biden-administration-california-kamala-harris-gavin-newsom.

27. "California State Facts: The First Time the Golden State . . .," California.com, 2022, accessed October 26, 2022, https://www.california.com/california-state-facts-first-time-golden-state.

28. Rick Perlstein, *Nixonland: The Rise of a President and the Fracturing of America* (New York: Scribner, 2009); Rick Perlstein, *The Invisible Bridge: The Fall of Nixon and the Rise of Reagan* (New York: Simon & Schuster, 2015); Rick Perlstein, *Reaganland: America's Right Turn 1976–1980* (New York: Simon & Schuster, 2020).

29. John M. Sloop and Ken A. Ono, *Shifting Borders: Rhetoric, Immigration, and California's Proposition 187* (Philadelphia, PA: Temple University Press, 2002); Yueh-Ting Lee, Victor Ottati, and Imtiaz Hussain, "Attitudes Toward 'Illegal' Immigration into the United States: California Proposition 187," *Hispanic Journal of Behavioural Sciences* 23, no. 4 (2001): 430–43.

30. Philip Bump, "Nearly Half of Republicans Agree with 'Great Replacement Theory,'" *The Washington Post*, May 9, 2022, https://www.washingtonpost.com/politics/2022/05/09/nearly-half-republicans-agree-with-great-replacement-theory; Will Carless, "Month Before Buffalo Shooting, Poll Finds, 7 in 10 Republicans Believed in 'Great Replacement' Ideas," *USA Today*, May 5, 2022. https://www.usatoday.com/story/news/nation/2022/06/01/great-replacement-theory-poll-republicans-democrats/7461913001/?gnt-cfr=1; United States Studies Centre, "More Than a Quarter of Democrats Believe in Electoral Replacement Theory," May 18, 2022, Accessed July 5, 2022, https://www.ussc.edu.au/analysis/the-46th-more-than-a-quarter-of-democrats-believe-in-electoral-replacement-theory.

31. "Mary D. Nichols and State of California Receive Environmental Achievement Award—Special Award Presentation by Rep. Henry Waxman," Environmental Law Institute, 2014, accessed October 26, 2022, https://www.eli.org/mary-d-nichols-and-state-california-receive-environmental-achievement-award-special-award; Dian Kiser and T. Boschert, "Eliminating Smoking in Bars, Restaurants, and Gaming Clubs in California: Breath, the California Smoke-Free Bar Program," *Journal of Public Health Policy* 22, no. 1 (2001): 81–87, https://doi.org/10.2307/3343554; "San Francisco Bans Phthalates, Bisphenol A," Chemical and Engineering News, June 12, 2006, accessed October 26, 2022, https://cen.acs.org/articles/84/i24/San-Francisco-bans-phthalates-bisphenol.html.

32. Roni Caryn Rabin, "Eager to Limit Exemptions to Vaccination, States Face Staunch Resistance," *The New York Times*, June 14, 2019, https://www.nytimes.com/2019/06/14/health/vaccine-exemption-health.html; Bednarczyk et al., "Current Landscape of Nonmedical Vaccination Exemptions"; Neal D. Goldstein and Joanna S. Suder, "Towards Eliminating Nonmedical Vaccination Exemptions Among School-Age Children," *Delaware Journal of Public Health* 8, no. 1 (2022): 84–88.

33. Catherine Flores Martin (Executive Director of the California Immunization Coalition), interview with the author, June 13, 2019.

34. Northe Sanders, "Emerging Vaccine Legislation and its Impact on Access" (presentation, World Vaccine Congress, Washington D.C., April 3–6, 2023).

35. Katie Attwell and Adam Hannah, "Convergence on Coercion: Functional and Political Pressures as Drivers of Global Childhood Vaccine Mandates," *International*

Journal of Health Policy and Management 11, no. 11 (2022): 2660–71, https://doi.org/10.34172/ijhpm.2022.658.

36. Mark C. Navin et al., "COVID-19 Vaccine Hesitancy Among Healthcare Personnel Who Generally Accept Vaccines," *Journal of Community Health 47*, no. 3 (2022): 519–29, https://doi.org/10.1007/s10900-022-01080-w; Hilda Razzaghi et al., "COVID-19 Vaccination and Intent Among Healthcare Personnel, US," *American Journal of Preventive Medicine* 62, no. 5 (2022): 705–15; Feifan Chen, Yalin He, and Yuan Shi, "Parents' and Guardians' Willingness to Vaccinate Their Children Against COVID-19: A Systematic Review and Meta-Analysis," *Vaccines* 10, no. 2 (2022): 179.

37. Robert Stalnaker, "Common Ground," *Linguistics and Philosophy* 25, no. 5–6 (2022): 701–21.

38. Olúfẹ́mi O. Táíwò, *Elite Capture: How the Powerful Took over Identity Politics (and Everything Else)* (Chicago: Haymarket Books, 2022).

39. Katie Attwell and Mark C. Navin, "How Policymakers Employ Ethical Frames to Design and Introduce New Policies: The Case of Childhood Vaccine Mandates in Australia," *Policy & Politics* 50, no. 4 (2022), https://doi.org/10.1332/030557321X16476002878591; Mark C. Navin, Andrea T. Kozak, and Katie Attwell, "School Staff and Immunization Governance: Missed Opportunities for Public Health Promotion," *Vaccine* 40, no. 51 (2022): 7433–39, https://doi.org/https://doi.org/10.1016/j.vaccine.2021.07.061; Breanna Fernandes et al., "US State-Level Legal Interventions Related to COVID-19 Vaccine Mandates," *JAMA* 327, no. 2 (2021): 178–79; Mark C. Navin and Katie Attwell, "Vaccine Mandates, Value Pluralism, and Policy Diversity," *Bioethics* 33, no. 9 (2019): 1042–49; Katie Attwell and Mark C. Navin, "Childhood Vaccination Mandates: Scope, Sanctions, Severity, Selectivity, and Salience," *Milbank Quarterly* 97, no. 4 (2019): 978–1014; Katie Attwell et al., "Recent Vaccine Mandates in the United States, Europe and Australia: A Comparative Study," *Vaccine* 19, no. 36 (2018): 7377–84.

40. Tomas Rozbroj, Anthony Lyons, and Jayne Lucke, "Understanding How the Australian Vaccine-Refusal Movement Perceives Itself," *Health & Social Care in the Community* 30, no. 2 (2022): 695–705; Paul R. Ward et al., "Understanding the Perceived Logic of Care by Vaccine-Hesitant and Vaccine-Refusing Parents: A Qualitative Study in Australia," *PLoS One* 12, no. 10 (2017): e0185955; Julie Leask et al., "Communicating with Parents About Vaccination: A Framework for Health Professionals," *BMC Pediatrics* 12, no. 11 (2012): 154; Anat Amit Aharon et al., "Parents with High Levels of Communicative and Critical Health Literacy Are Less Likely to Vaccinate Their Children," *Patient Education and Counseling* 100, no. 4 (2016): 768–75.

41. See, e.g., Regina Rini, "Fake News and Partisan Epistemology," *Kennedy Institute of Ethics Journal* 27, no. 2 (2017): E-43–E-64.

42. Judith Petts and Simon Niemeyer, "Health Risk Communication and Amplification: Learning from the MMR Vaccination Controversy," *Health, Risk & Society* 6, no. 1 (2004): 7–23; Alina Salganicoff, Usha Ranjoi, and Roberta Wynn, "Women and Healthcare: A National Profile" (San Francisco: Kaiser Family Foundation, 2005); "Fact Sheet: General Facts on Women and Job Based Health," U.S. Department of Labor, n.d.; Naomi Smith and Tim Graham, "Mapping the

Anti-Vaccination Movement on Facebook," *Information, Communication & Society* 22, no. 9 (2019): 1310–27; Ashley Fetters and Gerrit De Vynck, "How Wellness Influencers Are Fueling the Anti-Vaccine Movement," *The Washington Post*, September 12, 2021; Stephanie Alice Baker and Michael James Walsh, "'A Mother's Intuition: It's Real and We Have to Believe in It': How the Maternal Is Used to Promote Vaccine Refusal on Instagram," *Information, Communication & Society* (online January 23, 2022).

43. Katie Attwell and David T. Smith, "Hearts, Minds, Nudges and Shoves: (How) Can We Mobilise Communities for Vaccination in a Marketised Society?" *Vaccine* 36, no. 44 (2018): 6506–08; Attwell and Navin, "Childhood Vaccination Mandates"; Sean T. O'Leary et al., "Policies Among US Pediatricians for Dismissing Patients for Delaying or Refusing Vaccination," *JAMA* 324, no. 11 (2020): 1105–07; Alexandra Sifferlin, "9 Ways Advertisers Think We Could Convince Parents to Vaccinate," *TIME*, February 6, 2015; Caitlin Jarrett et al., "Strategies for Addressing Vaccine Hesitancy—A Systematic Review," *Vaccine* 33, no. 34 (2015): 4180–90; Katie Attwell and Melanie Freeman, "I Immunise: An Evaluation of a Values-Based Campaign to Change Attitudes and Beliefs," *Vaccine* 33, no. 46 (2015): 6235–40; Mark C. Navin et al., "School Staff and Immunization Governance."

44. *Dobbs v. Jackson Women's Health Organization*, 597 U.S. __, 142 S.Ct. 2228 (2022).

45. Glenn C. Savage, "What Is Policy Assemblage?" *Territory, Politics, Governance* 8 (2019): 319–35; Sebastian Ureta, *Assembling Policy: Transantiago, Human Devices, and the Dream of a World Class Society* (Cambridge, MA: MIT Press, 2015).

46. Sidney Dekker, *Drift into Failure* (Burlington, VT: Ashgate, 2012).

47. See, e.g., Atul Gawande, *Checklist Manifesto* (London: Penguin, 2010).

48. John Rawls, *Theory of Justice*, rev ed. (Cambridge, MA: Harvard University Press, 1999), 4–10.

49. We could try to keep using a mechanistic model for immunization governance if we incorporated all of the background institutions of society into the model. But this idea of a "machine" would be so complicated as to undermine its usefulness for explaining and predicting social phenomena.

50. On punctuated equilibrium in policy change, see Frank Baumgartner and Bryan Jones, *Agendas & Instability in American Politics* (Chicago: University of Chicago Press, 1993); and Theda Skocpol and Paul Pierson, "Historical Institutionalism," in *Contemporary Political Science. Political Science: State of the Discipline*, eds. Ira Katznelson and Helen V. Milner (New York: Norton, 2002), 693–721.

51. See, e.g., Eric Schickler, *Disjointed Pluralism: Institutional Innovation and the Development of the U.S. Congress* (Princeton, NJ: Princeton University Press, 2001); Jacob S. Hacker, "Privatizing Risk Without Privatizing the Welfare State: The Hidden Politics of Social Policy Retrenchment in the United States," *American Political Science Review* 98, no. 2 (2004): 243–60; James Mahoney and Kathleen Thelen (eds.), *Explaining Institutional Change: Ambiguity, Agency, and Power* (Cambridge, MA: Cambridge University Press, 2010); Wolfgang Streeck and Kathleen Thelen (eds.), *Beyond Continuity: Institutional Change in Advanced Political Economies* (Oxford, UK: Oxford University Press, 2005)

52. See, e.g., Mahoney and Thelen, *Explaining Institutional Change*.

53. Robert M. Wolfe and Lisa K. Sharp, "Anti-Vaccinationists Past and Present," *BMJ* 325, no. 7361 (2002): 430–32.

54. Larry Pickering et al., "Protecting the Community Through Child Vaccination," *Clinical Infectious Diseases* 67, no. 3 (2018): 464–71.

55. Leah Russin (Advocate, Vaccinate California), interview with author, July 11, 2019; Dr. Richard Pan (California State Senator), interview with author, June 14, 2019; Dorit Reiss (law professor and parent activist), interview with author, June 15, 2019; Hannah Henry (Advocate, Vaccinate California), interview with author, June 12, 2019; Kris Calvin (CEO, AAP California), interview with author, August 1, 2019; Kat De Burgh (Executive Director, Health Officers Association of California), interview with author, June 13, 2019.

56. Eve Dube et al., "Vaccine Hesitancy: An Overview," *Human Vaccines and Immunotherapeutics* 9, no. 8 (2013): 1763–73; Noni E. MacDonald and SAGE Working Group on Vaccine Hesitancy, "Vaccine Hesitancy: Definition, Scope and Determinants," *Vaccine* 33, no. 34 (2015): 4161–64; Helen Bedford et al., "Vaccine Hesitancy, Refusal and Access Barriers: The Need for Clarity in Terminology," *Vaccine* 36, no. 44 (2018): 6556–58.

57. David E. McIntosh et al., "Vaccine Hesitancy and Refusal," *Journal of Pediatrics* 175 (2016): 248–49.e1; Eve Dubé, Maryline Vivion, and Noni E. MacDonald, "Vaccine Hesitancy, Vaccine Refusal and the Anti-Vaccine Movement: Influence, Impact and Implications," *Expert Review of Vaccines* 14, no. 1 (2015): 99–117; Bedford et al., "Vaccine Hesitancy, Refusal and Access Barriers."

58. Bedford et al., "Vaccine Hesitancy, Refusal and Access Barriers"; Katie Attwell, Adam Hannah, and Julie Leask, "COVID-19: Talk of 'Vaccine Hesitancy' Lets Governments off the Hook," *Nature* 602 (2022): 574–77.

59. Heidi J. Larson, "Negotiating Vaccine Acceptance in an Era of Reluctance," *Human Vaccines & Immunotherapeutics* 9, no. 8 (2013): 1779–81; Kristen A. Feemster, "Overview: Special Focus Vaccine Acceptance," *Human Vaccines & Immunotherapeutics* 9, no. 8 (2013): 1752–54; Eve Dubé, Dominique Gagnon, and Noni E. MacDonald, "Strategies Intended to Address Vaccine Hesitancy: Review of Published Reviews," *Vaccine* 33, no. 34 (2015): 4191–203.

60. Hausman, *Anti/Vax*.

61. Several scholarly articles tease out the distinction between access and acceptance barriers for vaccination. Angus Thomson, Karis Robinson, and Gaëlle Vallée-Tourangeau, "The 5As: A Practical Taxonomy for the Determinants of Vaccine Uptake," *Vaccine* 34, no. 8 (2016): 1018–24; Bedford et al., "Vaccine Hesitancy, Refusal and Access Barriers"; Attwell et al., "COVID-19"; Susan Thomas et al., "Structural and Social Inequities Contribute to Pockets of Low Childhood Immunisation in New South Wales, Australia," *Vaccine: X* 12 (2022): 100200.

62. Sachiko Ozawa et al., "Defining Hard-to-Reach Populations for Vaccination," *Vaccine* 37, no. 37 (2019): 5525–34; Alison F. Crawshaw et al., "Defining the Determinants of Vaccine Uptake and Undervaccination in Migrant Populations in Europe to Improve Routine and COVID-19 Vaccine Uptake: A Systematic Review," *Lancet Infectious*

Diseases 22, no. 9 (2022): E254–66; Amyn A. Malik et al., "Determinants of COVID-19 Vaccine Acceptance in the US," *EClinicalMedicine* 26 (2020): 100495; Ahmet Topuzoğlu et al., "The Barriers Against Childhood Immunizations: Qualitative Research Among Socio-economically Disadvantaged Mothers," *European Journal of Public Health* 17, no. 4 (2006): 348–52; Elisa J. Sobo, Diana Schow, and Stephanie McClure, "US Black and Latino Communities Often Have Low Vaccination Rates—But Blaming Vaccine Hesitancy Misses the Mark," The Conversation, July 7, 2021; Erin Schumaker, "Vaccination Rates Lag in Communities of Color, But It's Not Only Due to Hesitancy, Experts Say. Focusing on Hesitancy, Rather Than Access, Is Looking at the Problem Backward," ABC News, May 8, 2021; Thomas et al., "Structural and Social Inequities"; Samantha J. Carlson et al., "'Corona Is Coming': COVID-19 Vaccination Perspectives and Experiences Amongst Culturally and Linguistically Diverse West Australians," *Health Expectations* 25, no. 6 (2022): 3062–72.

63. Mandates also differ based on their selectivity: How does one get out of complying with the requirement—for example, exemptions—and on what grounds? Together, these features constitute the mandate's severity. In this book, we tell the story of policymakers and activists turning California's permissive mandate with a relatively easy opt-out into a restrictive mandate with far fewer possibilities for exemption. Navin and Attwell, "Vaccine Mandates"; Attwell and Navin, "Childhood Vaccination Mandates."

64. Mark C. Navin and Margie Danchin, "Vaccine Mandates in the US and Australia: Balancing Benefits and Burdens for Children and Physicians," *Vaccine* 38, no. 51 (2020): 8075–77; "Vaccine Recommendations and Guidelines of the ACIP," Centers for Disease Control and Prevention, accessed August 26, 2022, https://www.cdc.gov/vaccines/hcp/acip-recs/general-recs/contraindications.html.

65.. "State Immunization Policy Overview," National Conference of State Legislatures, accessed July 6, 2022, https://www.ncsl.org/research/health/immunizations-policy-issues-overview.aspx.

66. "States with Religious and Philosophical Exemptions from School Immunization Requirements," National Conference of State Legislatures, accessed July 6, 2022, https://www.ncsl.org/research/health/school-immunization-exemption-state-laws.aspx.

67. Anna Kirkland, *Vaccine Court: The Law and Politics of Injury* (New York: New York University Press, 2016).

68. Scott Anderson, "Coercion," in *Stanford Encyclopedia of Philosophy*, ed. E. N. Zalta (Stanford, CA: Stanford University, 2021).

69. See, e.g., Alan Wertheimer, *Coercion* (Princeton, NJ: Princeton University Press, 1987).

70. See e.g. Joel Feinberg, *Harm to Self* (New York: Oxford University Press, 1986).

71. Stephen Bell, Andrew Hindmoor, and Frank Mols, "Persuasion as Governance: A State-Centric Relational Perspective," *Public Administration* 88, no. 3 (2010): 851–70; Frank Mols et al., "Why a Nudge Is Not Enough: A Social Identity Critique of Governance by Stealth," *European Journal of Political Research* 54, no. 1 (2015): 81–98.

Chapter 2

1. Richard J. Altenbaugh, *Vaccination in America: Medical Science and Children's Welfare* (London: Palgrave Macmillan, 2018), 11; John Duffy, *The Sanitarians: A History of American Public Health* (Urbana: University of Illinois Press, 1990); Alexandra M. Levitt, Peter Drotman, and Stephen Ostroff, "Control of Infectious Diseases—A 20th Century Public Health Achievement," in *Silent Victories: The History and Practice of Public Health in Twentieth-Century America*, eds. J. Ward and C. Warren (Oxford, UK: Oxford University Press, 2006), 3–16.
2. Levitt et al., "Control of Infectious Diseases," 11.
3. Vincent J. Cirillo, "Two Faces of Death: Fatalities from Disease and Combat in America's Principal Wars, 1775 to Present." *Perspectives in Biology and Medicine* 51, no. 1 (2008): 121–33, https://doi.org/10.1353/pbm.2008.0005; Máire A. Connolly and David L. Heymann, "Deadly Comrades: War and Infectious Diseases," *Lancet* 360, no. 1 (2002): s23–s24, https://doi.org/10.1016/s0140-6736(02)11807-1.
4. James C. Riley, "Smallpox and American Indians Revisited," *Journal of the History of Medicine and Allied Sciences* 65, no. 4 (2010): 445–77; Jim Downs, "Reconstructing an Epidemic: Smallpox Among Former Slaves, 1862–1868," in *Sick from Freedom: African-American Illness and Suffering During the Civil War and Reconstruction* (Oxford, UK: Oxford University Press, 2012), 95–118.
5. Jennifer Calfas, "U.S. Covid-19 Deaths Top 800,000," *Wall Street Journal*, December 14, 2021; "U.S. Deaths Confirmed," Johns Hopkins University Coronavirus Resource Center, accessed August 26, 2021, https://coronavirus.jhu.edu.
6. Elena Conis, *Vaccine Nation* (Chicago: University of Chicago Press, 2015), 6.
7. James Colgrove, *State of Immunity: The Politics of Vaccination in Twentieth-Century America* (Berkeley: University of California Press, 2006), 174.
8. Colgrove, *State of Immunity*.
9. Colgrove, *State of Immunity*, 174.
10. Conis, *Vaccine Nation*, 7.
11. Colgrove, *State of Immunity*, 167.
12. Altenbaugh, *Vaccination in America*, 256–57 (citing Conis, *Vaccine Nation*).
13. Altenbaugh, *Vaccination in America*, 256–57; Conis, *Vaccine Nation*.
14. Altenbaugh, *Vaccination in America*, 259.
15. James G. Hodge and Lawrence O. Gostin, "School Vaccination Requirements: Historical, Social and Legal Perspectives," *Kentucky Law Journal* 90, no. 4 (2002): 831–90.
16. Colgrove, *State of Immunity*, 175.
17. Colgrove, *State of Immunity*, 175–76.
18. Colgrove, *State of Immunity*, 12.
19. Colgrove, *State of Immunity*, 11.
20. There is real merit to this approach. The current global obsession with vaccine hesitancy has distracted many governments from ensuring that their systems reach those with complex lives. The contemporary reframing of "poorly reached" helps refocus on the agent who should be doing that work: the state. For more on these arguments,

see Katie Attwell, Adam Hannah, and Julie Leask, "COVID-19: Talk of 'Vaccine Hesitancy' Lets Governments off the Hook," *Nature* 602 (2002): 574–77.

21. Shannon Pettypiece, "White House Warns of Covid Treatment, Vaccine Cuts Without Added Funding," NBC News, March 15, 2022, https://www.nbcnews.com/politics/white-house/white-house-warns-covid-treatment-vaccine-cuts-added-funding-rcna20097.

22. Jennifer Tolbert et al., "Implications of the Lapse in Federal COVID-19 Funding on Access to COVID-19 Testing, Treatment, and Vaccines," Kaiser Family Foundation, March 28, 2022, https://www.kff.org/coronavirus-covid-19/issue-brief/implications-of-the-lapse-in-federal-covid-19-funding-on-access-to-covid-19-testing-treatment-and-vaccines.

23. Conis, *Vaccine Nation*, 23.

24. Walter A. Orenstein and Alan R. Hinman, "The Immunization System in the United States—The Role of School Immunization Laws," *Vaccine* 17 (1999): 19–24.

25. Colgrove, *State of Immunity*, 72.

26. Altenbaugh, *Vaccination in America*, 2.

27. Alexis de Tocqueville, *Democracy in America: A New Translation by Arthur Goldhammer*, Ed. and Trans., Arthur. Goldhammer (New York: Library of America, 2004), 803.

28. "Historical Timeline of Public Education in the US," Race Forward: The Center for Racial Justice Innovation (2006), accessed July 19, 2022, https://www.raceforward.org/research/reports/historical-timeline-public-education-us.

29. Lawrence Cremin, *American Education: The National Experience, 1783–1876* (New York: HarperCollins, 1980).

30. Cody D. Ewert, *Making Schools American: Nationalism and the Origin of Modern Educational Politics* (Baltimore: Johns Hopkins University Press, 2022).

31. U.S. Constitution, Article 1, Section 8; Albert J. Rosenthal, "Conditional Federal Spending and the Constitution," *Stanford Law Review* 39, no. 5 (1987): 1103–64.

32. The landmark U.S. Supreme Court case of *Brown v. Board of Education* (1954) ruled race-based school segregation unconstitutional. Then the Civil Rights Act of 1964, the Education Amendments of 1972, and the Rehabilitation Act of 1973 clarified that the national Department of Education had a responsibility to prevent discrimination based on race, sex, or disability in schools throughout the country. Efforts to combat school-related poverty included the Elementary and Secondary Act of 1965, which directed federal financing to disadvantaged children.

33. Salmon et al., "Compulsory Vaccination"; James Colgrove, "Parents Were Fine with Sweeping School Vaccination Mandates Five Decades Ago—But Covid-19 May Be a Different Story," *The Conversation*, October 22, 2021.

34. Colgrove, *State of Immunity*, 177; Altenbaugh, *Vaccination in America*, 259–60; Kevin Malone and Alan Hinman, "Vaccination Mandates: The Public Health Imperative and Individual Rights," in *Law in Public Health Practice*, eds. Richard Goodman et al. (New York: Oxford University Press, 2007), 338–59.

35. Conis, *Vaccine Nation*, 100.

36. Colgrove, *State of Immunity*, 201; Los Angeles is a prominent example; see N. Anthony et al., "Immunization: Public Health Programming Through Law Enforcement," *American Journal of Public Health (1971)* 67, no. 8 (1977): 763–64.

37. Conis, *Vaccine Nation*, 100.

38. Conis, *Vaccine Nation*, 100.

39. Alan Hinman, "Position Paper," *Pediatric Research* 13, no. 5 (1979): 689–96.

40. Colgrove, *State of Immunity*, 177; Douglas S. Diekema, "Personal Belief Exemptions from School Vaccination Requirements," *Annual Review of Public Health* 35 (2014): 275–92; Robert M. Wolfe and Lisa K. Sharp, "Anti-Vaccinationists Past and Present," *BMJ* 325, no. 7361 (2002): 430–32.

41. Matthew D. Lassiter, "How White Americans' Refusal to Accept Busing Has Kept Schools Segregated," *The Washington Post*, April 21, 2021,https://www.washing tonpost.com/outlook/2021/04/20/how-white-americans-refusal-accept-busing-has-kept-schools-segregated; Sonya Ramsey, "The Troubled History of American Education After the Brown Decision," *The American Historian* (2017), accessed August 26, 2022, https://www.oah.org/tah/issues/2017/february/the-troubled-hist ory-of-american-education-after-the-brown-decision.

42. Richard Kluger, *Simple Justice: The History of Brown v. Board of Education and Black America's Struggle for Equality* (New York: Vintage, 1977); Samuel Wiggins, *Higher Education in the South* (Berkeley, CA: McCutchan, 1966); Thomas V. O'Brien, *The Politics of Race and Schooling: Public Education in Georgia, 1900–1961* (Lanham, MD: Lexington Books, 1999). Contemporary scholars locate the rise of the religious right in America in this issue; only later would the abortion issue become a "cover story" for the explicitly racist advocacy of segregated private schools that had initially mobilized evangelical Christians to enter U.S. politics in force; Randall Balmer, *Bad Faith: Race and the Rise of the Religious Right* (Grand Rapids, MI: Eerdmans, 2021).

43. "California's parent activists who worked to remove NMEs believed that their efforts would restore a 'pure' form of mandate policy". Leah Russin (Advocate, Vaccinate California), interview with the author, July 11, 2019.

44. James Colgrove and Abigail Lowin, "A Tale of Two States: Mississippi, West Virginia, and Exemptions to Compulsory School Vaccination Laws," *Health Affairs* 35, no. 2 (2016): 348–55; Elena Conis, "The History of the Personal Belief Exemption," *Pediatrics* 145, no. 4. (2020), https://doi.org/10.1542/peds.2019-2551.

45. In this way, the mandates of this period were similar to the "coercive voluntarism" that historian Christopher Capozzola identifies in the mobilization of America during World War I: Even as citizens remained formally free to refuse to comply, civic groups and local communities attempted to create cultures of obligation that made it more difficult, though not impossible, for individuals to choose to deviate from expected behaviors. Christopher Capozzola, *Uncle Sam Wants You: World War I and the Making of the Modern American Citizen* (Oxford, UK: Oxford University Press, 2008).

46. Ellis M. West, "The Right to Religion-Based Exemptions in Early America: The Case of Conscientious Objectors to Conscription," *Journal of Law and Religion* 10, no. 2 (1993): 367–401.

47. "Religious Exemptions Under the Free Exercise Clause: A Model of Competing Authorities," *Yale Law Journal* 90, no. 2 (1980): 350–76.

48. Douglas Laycock, "Regulatory Exemptions of Religious Behavior and the Original Understanding of the Establishment Clause," *Notre Dame Law Review* 81, no. 5 (2006): 1793–1842.

49. Some people have argued that determining eligibility for religious exemptions to vaccine mandates should be as simple as determining someone's denomination, from which they conclude that very few people are entitled to religious exemptions to vaccine mandates. See, e.g., Dorit Reiss, "Thou Shalt Not Take the Name of the Lord Thy God in Vain: Use and Abuse of Religious Exemptions from School Immunization Requirements," *Hastings Law Journal* 65, no. 6 (2014): 1551–1602; see also Hillel Levin, "Why Some Religious Accommodations for Mandatory Vaccinations Violate the Establishment Clause," *Hastings Law Journal* 68 (2016): 1193.

50. Courtney Miller, "Spiritual but Not Religious: Rethinking the Legal Definition of Religion," *Virginia Law Review* 102 (2016): 833.

51. Micah Schwartzman, "What If Religion Is Not Special?" *University of Chicago Law Review* 79, no. 4 (2012): 1351–1427.

52. In *Seeger*, the Court determined that a belief qualifies as sufficiently religious (for the purpose of a religious objection to military service) when it is "sincere and meaningful [and] occupies a place in the life of its possessor parallel to that filled by the orthodox belief in God." In *Welsh*, the Court further determined that a person's objection could be religious, from the point of view of exemptions laws, even if the person making the objection declares that their objection is secular. A "religious" objection may emerge from convictions that are part of a person's "ultimate concerns," even if the objector does not identify their ultimate concerns as religious.

53. Daniel A. Salmon and Andrew W. Siegel, "Religious and Philosophical Exemptions from Vaccination Requirements and Lessons Learned from Conscientious Objectors from Conscription," *Public Health Reports* 116, no. 4 (2001): 289–95.

54. Diekema, "Personal Belief Exemptions."

55. For a discussion of the pragmatic and legal limits to such inquiries, see Hillel Levin, "Private Schools' Role and Rights in Setting Vaccination Policy: A Constitutional and Statutory Puzzle," *William & Mary Law Review* 61 (2019): 1607.

56. This section draws on material from Mark C. Navin, "Prioritizing Religion in Vaccine Exemption Policies," in *Religious Exemptions*, eds. Kevin Vallier and Michael Weber (Oxford, UK: Oxford University Press, 2018), 184–202; and Mark C. Navin, "Privacy and Religious Exemptions," in *Core Concepts and Contemporary Issues in Privacy*, eds. Ann E. Cudd and Mark C. Navin (New York: Springer, 2018), 121–33.

57. Notably, three of the nine Supreme Court justices—Clarence Thomas, Neil Gorsuch, and Samuel Alito—did assert that the state was obligated to provide religious exemptions in these cases. So, it would require the difference of only two to make U.S. constitutional law much more friendly toward religious exemptions. See *Dr A, et al., v. Kathy Houchul, Governor of New York, et al.*, Supreme Court of the United States 595 U.S. (2021) 1, No. 21A145, https://www.supremecourt.gov/opinions/21pdf/21a145_gfbi.pdf. See also *John Does 1–3, et al. v. J. T. Mills, Governor of Maine, et al.*, Supreme Court of the United States 595 U.S. (2021) No. 21A90.

58. Joshua Kassner and David Lefkowitz, "Conscientious Objection," in *Encyclopedia of Applied Ethics* (2nd ed.), ed. Ruth Chadwick (San Diego, CA: Elsevier, 2012), 594–601.

59. Beth Bailey, *America's Army: Making the All-Volunteer Force*s (Cambridge, MA: Harvard University Press, 2009).

60. Jennifer S. Rota et al., "Processes for Obtaining Nonmedical Exemptions to State Immunization Laws," *American Journal of Public Health* 91, no. 4 (2001): 645–48; Saad B. Omer et al., "Legislative Challenges to School Immunization Mandates, 2009–2012," *JAMA* 311, no. 6 (2014): 620–21; Nina Blank, Arthur Caplan, and Catherine Constable, "Exempting Schoolchildren from Immunizations: States with Few Barriers Had Highest Rates of Nonmedical Exemptions." *Health Affairs* 32, no. 7 (2013): 1282–90; Neal D. Goldstein, Joanna S. Suder, and Jonathon Purtle, "Trends and Characteristics of Proposed and Enacted State Legislation on Childhood Vaccination Exemption, 2011–2017," *American Journal of Public Health* 109, no. 1 (2019): 102–107.

61. As examples, see Dorothy Bonn, "Texas Law Allows Conscientious Immunisation Exemptions," *Lancet Infectious Diseases* 3, no. 9 (2003): 525; Blank et al., "Exempting Schoolchildren from Immunizations"; Emily Thakar, "God Bless Texas: Should Texas Eliminate the Vaccine Exemption for Reasons of Conscience?" *Journal of Biosecurity, Biosafety and Biodefense Law*, 9, no 1 (2018).

62. Colgrove and Lowin, "A Tale of Two States," 350; Levin, "Why Some Religious Accommodations"; Allan J. Jacobs and Kavita Shah Arora, "When May Government Interfere with Religious Practices to Protect the Health and Safety of Children?" *Ethics, Medicine, and Public Health* 5, (2018): 86–93.

63. Colgrove and Lowin, "A Tale of Two States," 353.

64. Colgrove and Lowin, "A Tale of Two States," 353.

65. Interestingly, as we explain later, the medical exemption process in California would prove troublesome when the state removed NMEs, and this time West Virginia would become a reference point for reform advocates. Richard J. Pan and Dorit Rubinstein, "Vaccine Medical Exemptions Are a Delegated Public Health Authority," *Pediatrics* 142, no. 5 (2018): e20182009.

66. An important further question about the relationship between religious liberty and school vaccine mandates is whether religious schools have the right to be exempted from state laws that permit NMEs to vaccine mandates. That is, religious private schools that follow a faith tradition that is committed to personal and public health may thereby have a religious reason to reject state laws that tolerate unvaccinated children in schools. See Levin, "Why Some Religious Accommodations."

67. John Duffy, "School Vaccination: The Precursor to School Medical Inspection," *Journal of the History of Medicine and Allied Sciences* 33, no. 3 (1978): 344–55; Elena Conis and Jonathon Kuo, "Historical Origins of the Personal Belief Exemption to Vaccination Mandates: The View from California," *Journal of the History of Medicine and Allied Sciences* 76, no. 2 (2021): 167–90, 173.

68. Colgrove and Lowin, "A Tale of Two States."

69. Conis and Kuo, "Historical Origins."

70. Conis and Kuo, "Historical Origins," 171.
71. Conis and Kuo, "Historical Origins."
72. Conis and Kuo, "Historical Origins."
73. Conis and Kuo, "Historical Origins."
74. Conis and Kuo, "Historical Origins," 173.

Chapter 3

1. Saad B. Omer et al., "Legislative Challenges to School Immunization Mandates, 2009–2012," *JAMA* 311, no. 6 (2014): 620–21; Neal D. Goldstein, Joanna S. Suder, and Jonathan Purtle, "Trends and Characteristics of Proposed and Enacted State Legislation on Childhood Vaccination Exemption, 2011–2017," *American Journal of Public Health* 109, no. 1 (2019): 102.
2. Pawel Stefanoff et al., "Tracking Parental Attitudes on Vaccination Across European Countries: The Vaccine Safety, Attitudes, Training and Communication Project (VACSATC)," *Vaccine* 28, no. 35 (2010): 5731–37; Julie Leask, "Target the Fence-Sitters," Nature, 473, (2011): 443–5.
3. Jennifer Reich, "Neoliberal Mothering and Vaccine Refusal: Imagined Gated Communities and the Privilege of Choice," *Gender & Society* 28 (2014): 679–704.
4. Patrice M. Miller and Michael L. Commons, "The Benefits of Attachment Parenting for Infants and Children: A Behavioral Developmental View," *Behavioral Development Bulletin* 16, no. 1 (2010): 1–14, https://doi.org/10.1037/h0100514; Deborah L. Lee et al., "Associations Between Breastfeeding Practices and Young Children's Language and Motor Skill Development," *Pediatrics* 119, Supplement 1 (2007): S92–S98, https://doi.org/10.1542/peds.2006-2089; P. J. Quinn et al., "The Effect of Breastfeeding on Child Development at 5 Years: A Cohort Study," *Journal of Paediatrics and Child Health* 37, no. 5 (2001): 465–69; Brian G. Moss and William H. Yeaton, "Early Childhood Healthy and Obese Weight Status: Potentially Protective Benefits of Breastfeeding and Delaying Solid Foods," *Maternal and Child Health Journal* 18 (2014): 1224–32, https://doi.org/10.1007/s10995-013-1357-z; Cathal McCrory and Aisling Murray, "The Effect of Breastfeeding on Neuro-development in Infancy," *Maternal and Child Health Journal* 17 (2013): 1680–88.
5. Jennifer Reich, "Of Natural Bodies and Antibodies: Parents' Vaccine Refusal and the Dichotomies of Natural and Artificial," *Social Science & Medicine* 157 (2016): 103–10; Eve Dube et al., "Nature Does Things Well, Why Should We Interfere?" Vaccine Hesitancy Among Mothers," *Qualitative Health Research* 26, no. 3 (2016): 411–25.
6. Paul R. Ward et al., "Understanding the Perceived Logic of Care by Vaccine-Hesitant and Vaccine-Refusing Parents: A Qualitative Study in Australia," *PLoS One*, 12, no. 10 (2017): e0185955.
7. Katie Attwell et al., "'Do-It-Yourself': Vaccine Rejection and Complementary and Alternative Medicine (CAM)," *Social Science and Medicine* 196 (2018): 106–14.
8. Ward et al., "Understanding the Perceived Logic."

9. Parents of color who refuse vaccines may also pay a higher social and economic cost for their decisions than do parents who are White or otherwise privileged. Amelia M. Jamison, Sandra Crouse Quinn, and Vicki S. Freimuth, "'You Don't Trust a Government Vaccine': Narratives of Institutional Trust and Influenza Vaccination Among African American and White Adults," *Social Science & Medicine* 221 (2019): 87–94; Courtney Thornton and Jennifer Reich, "Black Mothers and Vaccine Refusal: Gendered Racism, Healthcare, and the State," *Gender & Society* 36, no. 4 (2022): 525–51.

10. Elisa J. Sobo, "Theorizing (Vaccine) Refusal: Through the Looking Glass," *Cultural Anthropology* 31, no. 3 (2016): 342–50.

11. Maya J. Goldenberg, *Vaccine Hesitancy: Public Trust, Expertise and the War on Science* (Pittsburgh, PA: University of Pittsburgh Press, 2021); Jennifer Reich, *Calling the Shots: Why Parents Reject Vaccines* (New York: New York University Press, 2016); Seth Mnookin, *The Panic Virus* (New York: Simon & Schuster, 2011); Eule Biss, *On Immunity: An Inoculation* (Minneapolis, MN: Greywolf, 2014); Mark A. Largent, *Vaccine: The Debate in Modern America* (Baltimore, MD: Johns Hopkins University Press, 2012); Bernice L. Haus, *Anti/Vax: Reframing the Vaccination Controversy* (Ithaca, NY: Cornell University Press, 2019); Heidi Larson, *Stuck: How Vaccine Rumors Start and Why They Don't Go Away* (Oxford, UK: Oxford University Press, 2020); Heidi Lawrence, *Vaccine Rhetorics* (Columbus, OH: Ohio State University Press, 2020); Paul A. Offit, *Deadly Choices: How the Anti-Vaccine Movement Threatens Us All* (New York: Basic Books, 2011); Jonathon Berman, *Anti-Vaxxers: How to Challenge a Misinformed Movement* (Cambridge, MA: MIT Press, 2020). Many journal articles also help explain parental vaccine refusal; some can be read without a paywall or can be found on the scholars' university research databases, and authors will often send you drafts if you email them. Katie Attwell and Melanie Freeman, "I Immunise: An Evaluation of a Values-Based Campaign to Change Attitudes and Beliefs," *Vaccine* 33, no. 46 (2015): 6235–40; Ward et al., "Understanding the Perceived Logic"; Katie Attwell, Samantha Meyer, and Paul Ward, "The Social Basis of Vaccine Questioning and Refusal: A Qualitative Study Employing Bourdieu's Concepts of 'Capitals' and 'Habitus,'" *International Journal of Environmental Research and Public Health* 15, no. 5 (2018): 1044; Katie Attwell, "The Politics of Picking: Selective Vaccinators and Population-Level Policy," *SSM – Population Health* 7 (2019): 100342; Sobo, "Theorizing (Vaccine) Refusal."

12. Nina R. Blank, Arthur L. Caplan, and Catherine Constable, "Exempting Schoolchildren from Immunizations: States with Few Barriers Had Highest Rates of Nonmedical Exemptions," *Health Affairs* 32, no. 7 (2013): 1282–90; Omer et al., "Legislative Challenges"; Goldstein et al., "Trends and Characteristics."

13. Rekha Lakshmanan and Jason Sabo, "Lessons from the Front Line: Advocating for Vaccines Policies at the Texas Capitol During Turbulent Times," *Journal of Applied Research on Children* 10, no. 2 (2019): Article 6, 8.

14. Dorothy Bonn, "Texas Law Allows Conscientious Immunisation Exemptions," *Lancet Infectious Diseases* 3, no. 9 (2003): 525, https://doi.org/10.1016/s1473-3099(03)00751-5.

15. Lakshmanan and Sabo, "Lessons from the Front Line," 8.

16. Denise F. Lillvis, Anna Kirkland, and Anna Frick, "Power and Persuasion in the Vaccine Debates: An Analysis of Political Efforts and Outcomes in the United States, 1998–2012," *Milbank Quarterly* 92, no. 3 (2014): 475–508.

17. G. Baird et al., "Measles Vaccination and Antibody Response in Autism Spectrum Disorders," *Archives of Disease in Childhood* 93, no. 10 (2008): 832–37, https://doi.org/10.1136/adc.2007.122937; Robert L. Davis, "Measles–Mumps–Rubella and Other Measles-Containing Vaccines Do Not Increase the Risk for Inflammatory Bowel Disease: A Case–Control Study from the Vaccine Safety Datalink Project," *JAMA* 285, no. 24 (2001): 3073; James A. Kaye, Maria del Mar Melero-Montes, and Hershel Jick, "Mumps, Measles, and Rubella Vaccine and the Incidence of Autism Recorded by General Practitioners: A Time Trend Analysis," *BMJ* 322, no. 7284 (2001): 460–63, https://doi.org/10.1136/bmj.322.7284.460; Heikki Peltola et al., "No Evidence for Measles, Mumps, and Rubella Vaccine-Associated Inflammatory Bowel Disease or Autism in a 14-Year Prospective Study," *Lancet* 351, no. 9112 (1998): 1327–28, https://doi.org/10.1016/s0140-6736(98)24018-9; Brent Taylor et al., "Autism and Measles, Mumps, and Rubella Vaccine: No Epidemiological Evidence for a Causal Association," *Lancet* 353, no. 9169 (1999): 2026–29, https://doi.org/10.1016/s0140-6736(99)01239-8.

18. Liam Smeeth et al., "MMR Vaccination and Pervasive Developmental Disorders: A Case–Control Study," *Lancet* 364, no. 9438 (2004): 963–69; Eric Fombonne et al., "Pervasive Developmental Disorders in Montreal, Quebec, Canada: Prevalence and Links with Immunizations," *Pediatrics* 118, no. 1 (2006): e139–e50; Annamari J. Makela, Pekka Nuorti, and Heikki Peltola, "Neurologic Disorders After Measles–Mumps–Rubella Vaccination," *Pediatrics* 110, no. 5 (2002): 957–63.

19. Jon Heron, Jean Golding, and the Alspac Study Team, "Thimerosal Exposure in Infants and Developmental Disorders: A Prospective Cohort Study in the United Kingdom Does Not Support a Causal Association," *Pediatrics* 114, no. 3 (2004): 577–83; Anders Hviid et al., "Association Between Thimerosal-Containing Vaccine and Autism," *JAMA* 290, no. 13 (2003): 1763–66; Cristofer S. Price et al., "Prenatal and Infant Exposure to Thimerosal from Vaccines and Immunoglobulins and Risk of Autism," *Pediatrics* 126, no. 4 (2010): 656–64; Robert Schechter and Judith Grether, "Continuing Increases in Autism Reported to California's Developmental Services System: Mercury in Retrograde," *Archives of General Psychiatry* 65, no. 1 (2008): 19–24.

20. Lillvis et al., "Power and Persuasion in the Vaccine Debates," 502.

21. See Mark Largent's personal example of this issue: "Annabelle's daycare notified me that she had not yet received a mandated vaccine, and I was unable to get an appointment with her pediatrician quickly enough to satisfy the daycare's requirements. So, I simply filled out an exemption form and wrote 'philosophically opposed to mandatory vaccinations' in the area labelled 'Reasons.' A couple months later . . . I was able to get her [Annabelle] into the pediatrician's office." Largent, *Vaccine*, 4.

22. Saad B. Omer et al., "Vaccination Policies and Rates of Exemption from Immunization, 2005–2011," *New England Journal of Medicine* 367, no. 12 (2012): 1170–71.

23. Mark C. Navin, Mark A. Largent, and Aaron M. McCright, "Efficient Burdens Decrease Nonmedical Exemption Rates: A Cross-County Comparison of Michigan's Vaccination Waiver Education Efforts," *Preventive Medicine Reports* 17 (2020): 101049, https://doi.org/10.1016/j.pmedr.2020.101049.

24. Mark C. Navin, Andrea T. Kozak, and Emily C. Clark, "The Evolution of Immunization Waiver Education in Michigan: A Qualitative Study of Vaccine Educators," *Vaccine* 36, no. 13 (2018): 1751–56.

25. Mark C. Navin, Jason Adam Wasserman, Miriam Ahmad, and Shane Bies, "Vaccine Education, Reasons for Refusal, and Vaccination Behavior," *American Journal of Preventive Medicine* 56, no. 3 (2019): 359–67, https://doi.org/10.1016/j.ame pre.2018.10.024.

26. Navin et al., "Efficient Burdens Decrease Nonmedical Exemption Rates."

27. Denise F. Lillvis, "Managing Dissonance and Dissent: Bureaucratic Professionalism and Political Risk in Policy Implementation," *Law & Policy* 41, no. 3 (2019): 310–35, https://doi.org/10.1111/lapo.12131; Denise F. Lillvis, Charley Willison, and Katia Noyes, "Normalizing Inconvenience to Promote Childhood Vaccination: A Qualitative Implementation Evaluation of a Novel Michigan Program," *BMC Health Services Research* 20, no. 1 (2020): 1–9.

28. Nina B. Masters et al, "Evaluating Michigan's Administrative Rule Change on Nonmedical Vaccine Exemptions," *Pediatrics* 148, no.2 (2021): e2021049942. A similar increase occurred in Oregon after a first-year dramatic decrease in nonmedical exemptions. See Natalie Pete, "Oregon Sees Increase of Nonmedical Vaccine Exemptions," *Statesman Journal*, June 7, 2018, https://www.statesmanjournal.com/story/news/education/2018/06/08/oregon-increase-nonmedical-vaccine-exempti ons/653460002.

29. Washington State Legislature. "2011 Senate Bill 5005: Certification of Exemption from Immunization," 2011, http://www.washingtonvotes.org/2011-SB-5005.

30. Saad B. Omer et al., "Exemptions from Mandatory Immunization After Legally Mandated Parental Counseling," *Pediatrics* 141, no. 1 (2018): e20172364.

31. Dan Ariely, *Predictably Irrational, Revised and Expanded Edition: The Hidden Forces That Shape Our Decisions*, rev. ed. (New York: Harper Perennial, 2010); Jonathon Haidt, *The Righteous Mind: Why Good People Are Divided by Politics and Religion* (New York: Penguin, 2012); Daniel Kahneman, *Thinking, Fast and Slow* (New York: Farrar, Straus & Giroux, 2011).

32. Ricard H. Thaler and Cass R. Sunstein, *Nudge: Improving Decisions About Health, Wealth, and Happiness* (New Haven, CT: Yale University Press, 2008).

33. Barbara A. Butrica and Nadia S. Karamcheva, "The Relationship Between Automatic Enrollment and DC Plan Contributions: Evidence from a National Survey of Older Workers," *Proceedings. Annual Conference on Taxation and Minutes of the Annual Meeting of the National Tax Association* 109 (2016): 1–33; James J. Choi et al., "Optimal Defaults," *American Economic Review* 93, no. 2 (2003): 180–85, https://doi.org/10.1257/000282803321947010.

34. Shlomo Cohen, "Nudging and Informed Consent," *American Journal of Bioethics* 13, no. 6 (2013): 3–11, https://doi.org/10.1080/15265161.2013.781704; Scott D. Halpern, Peter A. Ubel, and David A. Asch, "Harnessing the Power of Default Options to Improve Health Care," *New England Journal of Medicine* 357, no. 13 (2007): 1340–44, https://doi.org/10.1056/NEJMsb071595; Meng Li and Gretchen B. Chapman, "Nudge to Health: Harnessing Decision Research to Promote Health Behavior," *Social and Personality Psychology Compass* 7, no. 3 (2013): 187–98, https://doi.org/10.1111/spc3.12019; Jack Stevens, "Topical Review: Behavioral Economics as a Promising Framework for Promoting Treatment Adherence to Pediatric Regimens," *Journal of Pediatric Psychology* 39, no. 10 (2014): 1097–1103, https://doi.org/10.1093/jpepsy/jsu071. J. S. Swindell, Amy L. J. Halpern, and Scott D. McGuire, "Beneficent Persuasion: Techniques and Ethical Guidelines to Improve Patients' Decisions," *Annals of Family Medicine* 8, no. 3 (2010): 260–64, https://doi.org/10.1370/afm.1118; Katherine L. Milkman et al., "Using Implementation Intentions Prompts to Enhance Influenza Vaccination Rates," *Proceedings of the National Academy of Sciences of the United States of America* 108 no. 26 (2011): 10415–20. Noel T. Brewer et al, "Announcements Versus Conversations to Improve HPV Vaccination Coverage: A Randomized Trial," *Pediatrics* 139, no. 1 (2017): e20161764.

35. Douglas J. Opel and Saad B. Omer, "Measles, Mandates, and Making Vaccination the Default Option," *JAMA Pediatrics* 169, no. 4 (2015): 303–304; Douglas J. Opel, "The Architecture of Provider-Parent Vaccine Discussions at Health Supervision Visits," *Pediatrics* 132, no. 6 (2013): 1037–46, https://doi.org/10.1542/peds.2013-2037.

36. Another discrete clinical intervention that can increase vaccine acceptance is motivational interviewing. See, for example, Arnaud Gagneur et al., "A Postpartum Vaccination Promotion Intervention Using Motivational Interviewing Techniques Improves Short-Term Vaccine Coverage: PromoVac Study," *BMC Public Health* 18, no. 1 (2018): 811, https://doi.org/10.1186/s12889-018-5724-y; Arnaud Gagneur et al., "A Complementary Approach to the Vaccination Promotion Continuum: An Immunization-Specific Motivational-Interview Training for Nurses," *Vaccine* 37, no. 20 (2019): 2748–56, https://doi.org/10.1016/j.vaccine.2019.03.076.

37. Kathryn M. Edwards et al., "Countering Vaccine Hesitancy," *Pediatrics* 138, no. 3 (2016): e20162146.

38. Alison M. Buttenheim and David A. Asch, "Making Vaccine Refusal Less of a Free Ride," *Human Vaccines & Immunotherapeutics* 9, no. 12 (2013): 2674–75, http://www.landesbioscience.com/journals/vaccines/article/26676/; Navin et al., "The Evolution of Immunization Waiver Education in Michigan"; Opel and Omer, "Measles, Mandates, and Making Vaccination the Default Option."

39. Adam Oliver, "From Nudging to Budging: Using Behavioural Economics to Inform Public Sector Policy," *Journal of Social Policy* 42, no. 4 (2013): 685–700, https://doi.org/10.1017/s0047279413000299.

40. Hillel Y. Levin et al., "Stopping the Resurgence of Vaccine-Preventable Childhood Diseases: Policy, Politics, and Law," *University of Illinois Law Review* 2020, no. 1 (2020): 233–72.

41. For a review and discussion, see Kelly Levin et al., "Overcoming the Tragedy of Super Wicked Problems: Constraining Our Future Selves to Ameliorate Global Climate Change," *Policy Sciences* 45, no. 2 (2012): 123–52.

42. Alison Cutler, "Doctor Wrote Bogus COVID Vaccine Exemptions for Patients, Washington Officials Say," *The News Tribune*, December 30, 2021; Andrea Salcedo, "Retired Doctor's License Suspended After State Found She Mailed Fake Vaccine Exemption Forms: 'Let Freedom Ring!'" *The Washington Post*, September 29, 2021; Lauren Dunn and Linda Carroll, "Some Doctors Helping Anti-Vaccine Parents Get Medical Exemptions," NBC News, 2019, https://www.nbcnews.com/health/kids-hea lth/some-doctors-helping-anti-vaccine-parents-get-medical-exemptions-n963011.

43. For an informative discussion of the information deficit model and its defects, see Goldenberg, *Vaccine Hesitancy*.

44. Brendan Nyhan et al., "Effective Messages in Vaccine Promotion: A Randomized Trial," *Pediatrics* 133, no. 4 (2014): e835–e42.

45. Attwell and Freeman, "I Immunise"; Jennie Schoeppe et al., "The Immunity Community: A Community Engagement Strategy for Reducing Vaccine Hesitancy," *Health Promotion Practice* 18, no. 5 (2017): 654–61; Isabel Rossen, Mark J. Hurlstone, and Carmen Lawrence, "Going with the Grain of Cognition: Applying Insights from Psychology to Build Support for Childhood Vaccination," *Frontiers in Psychology* 7, no. (2016): 1483; Amanda Dempsey et al., "A Values-Tailored Web-Based Intervention for New Mothers to Increase Infant Vaccine Uptake: Development and Qualitative Study," *Journal of Medical Internet Research* 22, no. 3 (2020): e15800, https://doi.org/10.2196/15800.

46. Sean T. O'Leary et al., "Policies Among US Pediatricians for Dismissing Patients for Delaying or Refusing Vaccination," *JAMA* 324, no. 11 (2020): 1105–107.

47. Lillvis et al., "Power and Persuasion in the Vaccine Debates."

48. "1991: The Philly Measles Outbreak That Killed 9 Children," 6abc Action News—WPVI Philadelphia, 2015, https://6abc.com/1991-outbreak-faith-tabernacle-first-century-gospel-measles/504818.

49. Dr. Richard Pan (California State Senator), interview with the author, June 14, 2019.

50. Sara E. Oliver, Pedro Moro, and Amy E. Blain, "Haemophilus Influenzae Type B" in *Epidemiology and Prevention of Vaccine-Preventable Diseases The Pink Book: Course Textbook*, Centers for Disease Control and Prevention, 2021, accessed April 5, 2023, https://www.cdc.gov/vaccines/pubs/pinkbook/hib.html

51. Dr. Richard Pan (California State Senator), interview with the author, June 14, 2019.

52. "Dr. Richard Pan Pediatrician | California State Senator (D-Sacramento)," LinkedIn, 2022, accessed November 2, 2022, https://www.linkedin.com/in/drrichardpan.

53. Dr. Richard Pan (California State Senator), interview with the author, June 14, 2019.

54. Dr. Richard Pan (California State Senator), interview with the author, June 14, 2019.

55. State of California, "Bill Text—AB-2109 Communicable Disease: Immunization Exemption," California Legislative Information, 2012, https://leginfo.legislature. ca.gov/faces/billHistoryClient.xhtml?bill_id=201120120AB2109.

56. Dr. Richard Pan (California State Senator), interview with the author, June 14, 2019.

57. Kris Calvin (CEO, AAP California), interview with the author, August 1, 2019.

58. Kris Calvin (CEO, AAP California), interview with the author, August 1, 2019.

59. Naturopaths usually practice with either the Doctorate in Naturopathy (ND) or the Doctorate in Naturopathic Medicine (NMD). Neither is a medical degree. However, some physicians—who possess the MD or DO—have also trained as naturopaths.

60. Tim Jelleyman and Andrew Ure. "Attitudes to Immunisation: A Survey of Health Professionals in the Rotorua District," *New Zealand Medical Journal* 117, no. 1189 (2004): U769; Julie Leask et al., "Immunisation Attitudes, Knowledge and Practices of Health Professionals in Regional NSW," *Australian & New Zealand Journal of Public Health* 32, no. 3 (2008): 224–29, https://doi.org/10.1111/j.1753-6405.2008.00220.x.

61. Jon Wardle et al., "Complementary Medicine and Childhood Immunisation: A Critical Review," *Vaccine* 34, no. 38 (2016): 4484–500.

62. Governor Brown in fact had not entered seminary, but instead had lived at the Sacred Heart Novitiate, a training house for future members of the Society of Jesus (the Jesuits). See Robert Pack, *Jerry Brown, the Philosopher-Prince* (New York: Stein & Day, 1978).

63. Rene F. Najera, "An Adult Measles Patient Writes Home to Mom and Dad in the 1880s," *History of Vaccines*, October 5, 2022, accessed October 12, 2022, https://www.historyofvaccines.org/content/blog/california-immunization-exemption-legislation.

64. Dr. Richard Pan (California State Senator), interview with the author, June 14, 2019.

65. Dr. Richard Pan (California State Senator), interview with the author, June 14, 2019.

66. See Figure 1B in Omer et al., "Exemptions from Mandatory Immunization." Overall vaccination coverage rates increased after the physician counselling requirement was introduced, but they began to decline again after 2012. Also, the number of exemptions of any type (Washington state recorded only aggregate exemptions numbers) initially fell after the policy change, but it remained static in future years.

67. "Washington State School Immunization Slide Set, 2013–2014 School Year," Washington State Department of Health, updated April 2015, accessed July 28, 2022, https://doh.wa.gov/sites/default/files/legacy/Documents/Pubs//348-238-SY2013-14-ImmunizationGraphs.pdf.

68. Leah Russin (Advocate, Vaccinate California), interview with the author, July 11, 2019; Dr. Richard Pan (California State Senator), interview with the author, June 16, 2019; Catherine Flores Martin (Executive Director of the California Immunization Coalition), interview with the author, June 13, 2019.

69. "Washington State School Immunization Slide Set, 2013–2014 School Year."

70. Dr. Richard Pan (California State Senator), interview with the author, June 16, 2019.

71. Catherine Flores Martin (Executive Director, California Immunization Coalition), interview with the author, June 13, 2019.

72. Kris Calvin (CEO, American Academy of Pediatrics, California), interview with the author, August 1, 2019.

73. Catherine Flores Martin (Executive Director, California Immunization Coalition), interview with the author, June 13, 2019.

74. Alison Buttenheim et al., "Conditional Admission, Religious Exemption Type, and Nonmedical Vaccine Exemptions in California Before and After a State Policy

Change," *Vaccine* 36, no. 26 (2018): 3789–93; Paul L. Delamater et al., "Change in Medical Exemptions from Immunization in California After Elimination of Personal Belief Exemptions," *JAMA* 318, no. 91 (2017): 863–64.

75. See note 66 about Omer's Figure 1B.
76. Richard Pan, "Senate Bill 277 Introduced to End California's Vaccine Exemption Loophole," February 19, 2015, http://sd06.senate.ca.gov/news/2015-02-19-senate-bill-277-introduced-end-california%E2%80%99s-vaccine-exemption-loophole.
77. Malia Jones et al., "Mandatory Health Care Provider Counseling for Parents Led to a Decline in Vaccine Exemptions in California," *Health Affairs* 37, no. 9 (2018): 1494–1502.
78. Kris Calvin (CEO, American Academy of Pediatrics, California), interview with the author, August 1, 2019.
79. Jeremy White, "From Death Threats to Holocaust Warning, California Vaccine Bill an Extraordinary Fight," *The Sacramento Bee*, June 30, 2015, https://www.sacbee.com/news/politics-government/capitol-alert/article25909216.html.
80. This is another behavioral insights strategy. See Sabina Nuti et al., "Making Governance Work in the Health Care Sector: Evidence from a 'Natural Experiment' in Italy," *Health Economics, Policy and Law* 11, no. 1 (2016): 17–38, https://doi.org/10.1017/S1744133115000067.

Chapter 4

1. For example, there was no organized parent movement to push for California's earlier Clinician Counselling Bill or for similar bills during this period in other states.
2. Kerrie Wiley et al., "Parenting and the Vaccine Refusal Process: A New Explanation of the Relationship Between Lifestyle and Vaccination Trajectories," *Social Science & Medicine* 263, (2020): 113259; Jenifer Reich, *Calling the Shots: Why Parents Reject Vaccines* (New York: New York University Press, 2016); Katie Attwell et al., "Vaccine Rejecting Parents' Engagement with Expert Systems That Inform Vaccination Programs," *Journal of Bioethical Inquiry* 14, no. 1 (2017): 65–76; Courtney Gidengil et al., "Beliefs Around Childhood Vaccines in the United States: A Systematic Review," *Vaccine* 37, no. 45 (2019): 6793–6802.
3. Leah Russin (advocate, Vaccinate California), interview with the author, July 11, 2019.
4. Renée DiResta, "Renée DiResta," accessed December 31, 2021, http://www.reneediresta.com.
5. Renée DiResta, "How California's Terrible Vaccination Policy Puts Kids at Risk," 2014, accessed December 31, 2021, https://blog.noupsi.de/post/103050754027/vax viz; Renée DiResta, "Personal Exemptions From Reason," *Slate*, April 8, 2015; Renée DiResta, "A Win for Evidence-Based Policy, and California Kids," 2015, accessed December 31, 2021, https://blog.noupsi.de/post/123134500352/sb277.
6. DiResta, "How California's Terrible Vaccination Policy Puts Kids at Risk"; DiResta, "Personal Exemptions from Reason; DiResta, "A Win for Evidence-Based Policy."

7. New York State, "Title: Section 69-3.10—Religious Exemption from Immunization," in *New York Codes, Rules and Regulations: Volume A-1a (Title 10) SubChapter H—Maternal and Child Health Part 69—Family Health SubPart 69-3—Pregnant Women, Testing for Hepatitis B, Follow-Up Care Title: Section 69-3.10—Religious Exemption from Immunization,"* 1991; Aamer Imdad et al., "Religious Exemptions for Immunization and Risk of Pertussis in New York State, 2000–2011," *Pediatrics* 132, no. 1 (2013): 37–43, https://doi.org/10.1542/peds.2012-3449; Dorit Reiss, "Thou Shalt Not Take the Name of the Lord Thy God in Vain: Use and Abuse of Religious Exemptions from School Immunization Requirements," *Hastings Law Journal* 65, no. 6 (2014): 1551–602.

8. Arthur Allen, *Vaccine: The Controversial Story of Medicine's Greatest Lifesaver* (New York: Norton, 2008), 328. On the idea that disease contributes to spiritual growth, consider the following from Nietzsche's *Human, All Too Human*, Part 1, Section 224 (titled "Ennoblement Through Degeneration"): "The more sickly man, for example, will if he belongs to a warlike and restless race perhaps have more inducement to stay by himself and thereby acquire more repose and wisdom," Friedrich Nietzsche, *Human, All Too Human: A Book for Free Spirits*, translated by R. J. Hollingdale (Cambridge, UK: Cambridge University Press, 1996), 107.

9. Julia M. Brennan et al., "Trends in Personal Belief Exemption Rates Among Alternative Private Schools: Waldorf, Montessori, and Holistic Kindergartens in California, 2000–2014," *American Journal of Public Health* 107, no. 1 (2017): 108–12; Yun-Kuang Lai, Jessica Nadeau Louise-Anne McNutt, and Jana Shaw, "Variation in Exemptions to School Immunization Requirements Among New York State Private and Public Schools," *Vaccine* 32, no.52 (2014): 7070–76; D. Schmid et al., "An Ongoing Multi-State Outbreak of Measles Linked to Non-Immune Anthroposophic Communities in Austria, Germany, and Norway, March–April 2008," *Euro Surveillance* 13, no. 16 (2008); Elisa J. Sobo, "Social Cultivation of Vaccine Refusal and Delay Among Waldorf (Steiner) School Parents," *Medical Anthropology Quarterly* 29, no. 3 (2015): 381–399; Chris Cook, "Why Are Steiner Schools So Controversial," BBC Newsnight, August 4, 2014; Mathieu Foulkes, "How the Spiritual 'Waldorf' Movement Is Connected to German Vaccine Scepticism," *The Local*, November 23, 2021, https://www.thelocal.de/20211123/how-a-spiritual-movement-is-connected-to-german-vaccine-scepticism. Fran Rimrod, "Perth Alternative School Slammed for Refusing to Support to Vaccinate Students, August 9, 2017," WA Today, August 9, 2017, https://www.watoday.com.au/national/western-australia/perth-alternative-school-slammed-for-refusing-to-support-to-vaccinate-students-20170809-gxsv1h.html; Reich, *Calling the Shots*. For a discussion of the vaccination and exemption rates in Californian Waldorf Schools, see Sobo, "Social Cultivation of Vaccine Refusal and Delay."

10. Saad B. Omer et al., "Geographic Clustering of Nonmedical Exemptions to School Immunization Requirements and Associations with Geographic Clustering of Pertussis," *American Journal of Epidemiology* 168, no. 12 (2008): 1389–96; Ashley Gromis and Ka-Yuet Liu, "Spatial Clustering of Vaccine Exemptions on the Risk of a Measles Outbreak," *Pediatrics* 149, no. 1 (2022): e2021050971, https://doi.org/10.1542/peds.2021-050971.

11. Hannah Henry (advocate, Vaccinate California), interview with the author, June 12, 2019.

12. Hannah Henry (advocate, Vaccinate California), interview with the author, June 12, 2019.

13. Hannah Henry (advocate, Vaccinate California), interview with the author, June 12, 2019.

14. Hannah Henry (advocate, Vaccinate California), interview with the author, June 12, 2019.

15. Jennifer Zipprich et al., "Measles Outbreak—California, December 2014–February 2015," *MMWR Morbidity and Mortality Weekly Report* 64, no. 6 (2015): 153–54.

16. New York State, "Title: Section 69-3.10—Religious Exemption from Immunization"; Imdad et al., "Religious Exemptions for Immunization"; Reiss, "Thou Shalt Not Take the Name of the Lord Thy God in Vain."

17. Leah Russin (advocate, Vaccinate California), interview with the author, July 11, 2019.

18. Hannah Henry (advocate, Vaccinate California), interview with the author, June 12, 2019.

19. Leah Russin (advocate, Vaccinate California), interview with the author, July 11, 2019.

20. Leah Russin (advocate, Vaccinate California), interview with the author, July 11, 2019.

21. Ben Allen, "Senators Richard Pan and Ben Allen to Introduce Legislation to End California's Vaccine Exemption Loophole," 2015, accessed April 5, 2023, https://sd24. senate.ca.gov/news/press-release/senators-richard-pan-and-ben-allen-introduce- legislation-end-californias-vaccine.

22. Leah Russin (advocate, Vaccinate California), interview with the author, July 11, 2019.

23. Leah Russin (advocate, Vaccinate California), interview with the author, July 11, 2019.

24. Leah Russin (advocate, Vaccinate California), interview with the author, July 11, 2019.

25. Dr. Richard Pan (California State Senator), interview with the the author, June 14, 2019.

26. Dr. Richard Pan (California State Senator), interview with the author, June 14, 2019.

27. Vaccinate California, "People," 2022, accessed June 30, 2022, https://vaccinecalifor nia.org/about/people.

28. Leah Russin (advocate, Vaccinate California), interview with the author, July 11, 2019.

29. Hannah Henry (advocate, Vaccinate California), interview with the author, June 12, 2019.

30. Dorit Reiss (law professor and parent activist), interview with the author, June 15, 2019.

31. Leah Russin (advocate, Vaccinate California), interview with the author, July 11, 2019.

32. Hannah Henry (advocate, Vaccinate California), interview with the author, June 12, 2019.

33. Dorit Reiss (law professor and parent activist), interview with the author, June 15, 2019.

34. Dorit Reiss (law professor and parent activist), interview with the author, June 15, 2019.

35. Dorit Reiss (law professor and parent activist), interview with the author, June 15, 2019.

36. Jason Breslow and Chris Amico, "What Are the Vaccine Exemption Laws in Your State?" *Frontline*, March 24, 2015, accessed August 31, 2022, https://www.pbs.org/wgbh/frontline/article/what-are-the-vaccine-exemption-laws-in-your-state.

37. Andrew Gumbe, "US States Face Fierce Protests from Anti-Vaccine Activists," *The Guardian*, April 10, 2015.

38. "NC Lawmakers Behind Vaccine Bill Announce It's Dead," WFAE 90.7 Charlotte's NPR News Source, April 2, 2015.

39. Dorit Reiss, interview with the author, June 15, 2019.

40. Dorit Reiss, email to the author, February 2021.

41. Catherine Flores Martin (Executive Director of the California Immunization Coalition), interview with the author, June 13, 2019. In California, you have to register as a lobbyist if you spend more than 30% of your time on advocacy work with political actors. However, nonprofits such as the Immunization Coalition usually spent less than that percentage of time lobbying. Similarly, Kris Calvin from AAP California explained, "I don't do direct lobbying. I only do grass roots advocacy on behalf of the pediatricians who are my members." Instead, the organization made regular use of a registered lobbyist, and the California Medical Association also had a team of lobbyists at their disposal. Kris Calvin (CEO, AAP California) interview with the author, August 1, 2019.

42. Kat De Burgh (Executive Director, Health Officers Association of California), interview with the author, June 13, 2019.

43. Kat De Burgh (Executive Director, Health Officers Association of California), interview with the author, June 13, 2019.

44. Kat De Burgh (Executive Director, Health Officers Association of California), interview with the author, June 13, 2019.

45. Kris Calvin, "SB277 (Pan & Allen) Elimination of CA Personal Belief Exemption for School-Entry Vaccines," American Academy of Pediatrics–California, 2015, aap-ca.org/letter/sb-277-pan-allen-elimination-of-ca-personal-belief-exemption-for-school-entry-vaccines.

46. American Academy of Pediatrics Committee on Practice and Ambulatory Medicine, Committee on Infectious Diseases, Committee on State Government Affairs, Council on School Health, and Section on Administration and Practice Management, "Medical Versus Nonmedical Immunization Exemptions for Child Care and School Attendance," *Pediatrics* 138, no. 3 (2016): e20162145.

47. California Senate, "Senate Judiciary Committee, Tuesday, April 28, 2015," 2015, https://www.senate.ca.gov/media/senate-judiciary-committee-38/video (Holland's comments about coercion—and her analogy with rape—begin at 3:29:19). Also see Melody Gutierrez, "California's School Immunizations Bill Passes Another Committee," SFGate, April 28, 2015, https://www.sfgate.com/news/article/California-s-school-immunizations-bill-passes-6229962.php.

48. Dr. Richard Pan (California State Senator), interview with the author, June 14, 2019.

49. Kris Calvin (CEO, AAP California), interview with the author, August 1, 2019.

50. Hannah Henry (advocate, Vaccinate California), interview with the author, June 12, 2019.

51. "Understanding the Impact of Vaccines: A Conversation with the National Public Health Information Coalition," Milken Institute School of Public Health at The George Washington University, September 18, 2015, accessed August 5, 2020, https://cdn1.mha.gwu.edu/blog/niam-2015-NPHIC; Michelle Mello, David M. Studdert, and Wendy E. Parmet, "Shifting Vaccination Politics—The End of Personal-Belief Exemptions in California," *New England Journal of Medicine* 373, no. 9 (2015): 785–87, https://search.proquest.com/docview/1707845173?accountid=14681.

52. "Senate Bill 277 Clears Senate Judiciary Committee," California Medical Association, April 29, 2015, accessed August 7, 2020, https://www.cmadocs.org/newsroom/news/view/ArticleId/32506/Senate-Bill-277-clears-Senate-Judiciary-Committee.

53. Vanderslott, Samantha. "Exploring the Meaning of Pro-Vaccine Activism across Two Countries," *Social Science & Medicine* 222 (February 1, 2019): 59–66, at p. 62.

54. Mello et al., "Shifting Vaccination Politics."

55. Laurel Rosenhall, "Parents Lobby California Lawmakers from Both Sides of Vaccine Debate," *The Sacramento Bee*, February 25, 2015, https://www.sacbee.com/news/politics-government/capitol-alert/article11174378.htmlL; "SB277 (Pan & Allen) Elimination of CA Personal Belief Exemption for School-Entry Vaccines," American Academy of Pediatrics California, press release, March 30, 2015; Kat DeBurgh (Executive Director, Health Officers Association of California), interview with the author, June 2019.

56. Mansur Olson, *The Logic of Collective Action: Public Goods and the Theory of Groups* (Cambridge, MA: Harvard University Press, 1965).

57. See discussion of health social movements in Phil Brown et al., "The Health Politics of Asthma: Environmental Justice and Collective Illness Experience in the United States," *Social Science & Medicine* 57, no. 3 (2003): 453–64, https://doi.org/10.1016/s0277-9536(02)00375-1; see also Stuart Blume, "Anti-Vaccination Movements and Their Interpretations," *Social Science & Medicine* 62, no. 3 (2006): 628–42, http://dx.doi.org/10.1016/j.socscimed.2005.06.020.

58. Katie's own transition into becoming a vaccination activist, and eventually a vaccination policy expert in a tenured position at a university, was underpinned by similar fortunes in terms of time: paid parental leave, a partner whose earnings could help support the family, her own high-paid part-time work resulting from class and educational privilege, and extended family members helping to care for the couple's children. She also benefited from Australia's public health system providing free maternity and health care.

59. Email correspondence between Leah Russin (advocate, Vaccinate California) and the author, February 10, 2022.

60. Leah Russin (advocate, Vaccinate California), interview with the author, July 11, 2019.

61. Jennifer Reich, "Neoliberal Mothering and Vaccine Refusal: Imagined Gated Communities and the Privilege of Choice," *Gender & Society* 28, no. 5 (2014): 679–704.

62. Courtney Thornton and Jennifer A. Reich. "Black Mothers and Vaccine Refusal: Gendered Racism, Healthcare, and the State," *Gender & Society* 36, no. 4 (2022): 525–51, https://doi.org/10.1177/08912432221102150.

63. Email correspondence between Leah Russin (advocate, Vaccinate California) and the author, February 10, 2022.

64. Dorit Reiss (law professor and parent activist), interview with the author, June 15, 2019.

65. Hannah Henry (advocate, Vaccinate California), interview with the author, June 12, 2019.

Chapter 5

1. "Care of the Young Athlete Patient Education Handout—Football," American Academy of Pediatrics, 2010; American Academy of Pediatrics, "Tackling in Youth Football," *Pediatrics* 136, no. (2015): e1419–e30; "Care of the Young Athlete Patient Education Handout—Ice Hockey," American Academy of Pediatrics, 2011 ; American Academy of Pediatrics, "Incidence of Concussion in Youth Ice Hockey Players," *Pediatrics* 137, no. 2 (2016): e20151633.

2. Matthew Page, Kristin M. Lindahl, and Neena M. Malik, "The Role of Religion and Stress in Sexual Identity and Mental Health Among Lesbian, Gay, and Bisexual Youth," *Journal of Research on Adolescence* 23, no. 4 (2013): 665–77, https://doi.org/10.1111/jora.12025; Lauren J. Joseph and Stephen Cranney, "Self-Esteem Among Lesbian, Gay, Bisexual and Same-Sex-Attracted Mormons and Ex-Mormons," *Mental Health, Religion & Culture* 20, no. 10 (2017): 1028–41, https://doi.org/10.1080/13674676.2018.1435634; Ian Rivers et al., "LGBT People and Suicidality in Youth: A Qualitative Study of Perceptions of Risk and Protective Circumstances," *Social Science & Medicine* 212 (2018): 1–8, https://doi.org/10.1016/j.socscimed.2018.06.040.

3. Jean M. Twenge and W. Keith Campbell, "Associations Between Screen Time and Lower Psychological Well-Being Among Children and Adolescents: Evidence from a Population-Based Study," *Preventive Medicine Reports* 12 (2018): 271–83; Sheri Madigan et al., "Association Between Screen Time and Children's Performance on a Developmental Screening Test," *JAMA Pediatrics* 173, no. 3 (2019): 244–50, https://doi.org/10.1001/jamapediatrics.2018.5056; Yolanda Reid Chassiakos et al., "Children and Adolescents and Digital Media," *Pediatrics* 138, no. 5 (2016): e20162593, https://doi.org/10.1542/peds.2016-2593; Michelle O'Reilly et al., "Is Social Media Bad for Mental Health and Wellbeing? Exploring the Perspectives of Adolescents," *Clinical Child Psychology and Psychiatry* 23, no. 4 (2018): 601–13, https://doi.org/10.1177/1359104518775154; Lisa M. Cookingham and Ginny L. Ryan, "The Impact of Social Media on the Sexual and Social Wellness of Adolescents," *Journal of Pediatric & Adolescent Gynecology* 28, no. 1 (2015): 2–5, https://doi.org/10.1016/j.jpag.2014.03.001; Elia Abi-Jaoude, Karline Treurnicht Naylor, and Antonio Pignatiello, "Smartphones, Social Media Use and Youth Mental Health," *Canadian Medical Association Journal* 192, no. 6 (2020): E136–E41, https://doi.org/10.1503/cmaj.190434; Elena Bozzola et al., "The Use of Social Media in Children and Adolescents: Scoping Review on the Potential Risks," *International Journal*

of Environmental Research and Public Health 19, no. 16 (2022): 9960, https://www.mdpi.com/1660-4601/19/16/9960; Eric W. Owens, Richard J. Behun, Jill C. Manning, and Rory C. Reid, "The Impact of Internet Pornography on Adolescents: A Review of the Research," *Sexual Addiction & Compulsivity* 19, no. 1–2 (2012): 99–122, https://doi.org/10.1080/10720162.2012.660431; Megan Lim et al., "Young Australians' Use of Pornography and Associations with Sexual Risk Behaviours," *Australian and New Zealand Journal of Public Health* 41, no. 4 (2017): 438–43, https://doi.org/10.1111/1753-6405.12678.

4. Margaret M. C. Thomas, Jane Waldfogel, and Ovita F. Williams, "Inequities in Child Protective Services Contact Between Black and White Children," *Child Maltreatment* 28, no. 1 (2023): 42–54, https://doi.org/10.1177/10775595211070248; Alan J. Dettlaff et al., "Disentangling Substantiation: The Influence of Race, Income, and Risk on the Substantiation Decision in Child Welfare," *Children and Youth Services Review* 33, no. 9 (2011): 1630–37; Rachel Sanders, "The Color of Fat: Racializing Obesity, Recuperating Whiteness, and Reproducing Injustice," *Politics, Groups & Identities* 7, no. 2 (2019): 287–304.

5. The only reason for refusers to keep quiet would be because celebrity pediatrician Bob Sears warned them not to "share their fears with their neighbors, because if too many people avoid the MMR [measles, mumps, and rubella vaccine], we'll likely see the disease increase significantly." Although many vaccine refusers do not believe in herd immunity, Sears and some of his clients did. Blabbing too much about the dangers of vaccines might stop people vaccinating and ruin that precious resource of community protection generated by the vaccinating majority. Robert Sears, *The Vaccine Book: Making the Right Decision for Your Child* (New York: Little, Brown, 2007), 96–97.

6. Jonann Brady and Stephanie Dahle, "Celeb Couple to Lead 'Green Vaccine' Rally—Jenny McCarthy and Jim Carrey Talk About Autism March on 'GMA,'" ABC News, June 4, 2008, https://abcnews.go.com/GMA/OnCall/story?id=4987758; Mayim Bialik, *Beyond the Sling: A Real-Life Guide to Raising Confident, Loving Children the Attachment Parenting Way* (New York: Gallery Books, 2012); Alicia Silverstone, *The Kind Mama* (New York: Rodale, 2014); "Clueless: Celebrities Make Us Sick," *The Economist*, June 28, 2014, https://www.economist.com/united-states/2014/06/28/clueless.

7. Zoe Meleo-Erwin et al., "'To Each His Own': Discussions of Vaccine Decision-Making in Top Parenting Blogs," *Human Vaccines & Immunotherapeutics* 13, no. 8 (2017): 1895–901; Marina C. Jenkins and Megan A. Moreno, "Vaccination Discussion Among Parents on Social Media: A Content Analysis of Comments on Parenting Blogs," *Journal of Health Communication* 25, no. 3 (2020): 232–42; Anna Kata, "Anti-Vaccine Activists, Web 2.0, and the Postmodern Paradigm—An Overview of Tactics and Tropes Used Online by the Anti-Vaccination Movement," *Vaccine* 30 (2012): 3778–89.

8. Phillip J. Smith, Susan Y. Chu, and Lawrence E. Barker, "Children Who Have Received No Vaccines: Who Are They and Where Do They Live?" *Pediatrics* 114, no. 1 (2004): 187–95.

9. Jessica E. Atwell et al., "Nonmedical Vaccine Exemptions and Pertussis in California, 2010," *Pediatrics* 132, no. 4 (2013): 624–30.

10. Elisa J. Sobo, "Social Cultivation of Vaccine Refusal and Delay Among Waldorf (Steiner) School Parents," *Medical Anthropology Quarterly* 29, no. 3 (2015): 381–99.

11. Sobo, "Social Cultivation of Vaccine Refusal and Delay," 13.

12. Katie Attwell and David T. Smith, "Parenting as Politics: Social Identity Theory and Vaccine Hesitant Communities," *International Journal of Health Governance* 22, no. 3 (2017): 183–98; Katie Attwell, Samantha Meyer, and Paul Ward, "The Social Basis of Vaccine Questioning and Refusal: A Qualitative Study Employing Bourdieu's Concepts of 'Capitals' and 'Habitus,'" *International Journal of Environmental Research and Public Health* 15, no. 5 (2018): 1044. See also Mike Poltorak, Melissa Leach, James Fairhead, and Jackie Cassell, "'MMR Talk' and Vaccination Choice: An Ethnographic Study in Brighton," *Social Science and Medicine* 61, no. 3 (2005): 709–19; Jennifer Reich, "Neoliberal Mothering and Vaccine Refusal: Imagined Gated Communities and the Privilege of Choice," *Gender & Society* 28, no. 5 (2014): 679–704; Emily K. Brunson, "The Impact of Social Networks on Parents' Vaccination Decisions," *Pediatrics* 131, no. 5 (2013): e1397–404.

13. Alison M. Buttenheim, Malia Jones, and Yelena Baras, "Exposure and Vulnerability of California Kindergarteners to Intentionally Unvaccinated Children," *American Journal of Public Health* 102, 8, no. 1 (2012): 59–67.

14. Nationally, Lillvis et al. (2014) characterize the turn of the century as a "high point of vaccine criticism" following Andrew Wakefield's fraudulent study linking the MMR jab to autism, while Berezin and Eads' (2016) analysis of national media during this period shows an emphasis on vaccination's risks over its benefits, propagated by popular culture voices supplanting those of experts. Denise F. Lillvis, Anna Kirkland, and Anna Frick, "Power and Persuasion in the Vaccine Debates: An Analysis of Political Efforts and Outcomes in the United States, 1998–2012," *Milbank Quarterly* 92, no. 3 (2014): 475–508; Mabel Berezin and Alicia Eads, "Risk Is for the Rich? Childhood Vaccination Resistance and a Culture of Health," *Social Science & Medicine* 165 (2016): 233–45. See also Kata, "Anti-Vaccine Activists"; Paul J. Offit, "The Anti-Vaccination Epidemic," *The Wall Street Journal*, September 24, 2015.; Olga Khazan, "Wealthy L.A. Schools' Vaccination Rates Are as Low as South Sudan's," *The Atlantic*, September 14, 2014, https://www.theatlantic.com/health/archive/2014/09/wealthy-la-schools-vaccination-rates-are-as-low-as-south-sudans/380252.

15. Bob Sears, "California Bill AB2109 Threatens Vaccine Freedom of Choice," *Huffington Post*, March 22, 2012 (updated May 22, 2012), https://www.huffpost.com/entry/california-vaccination-bill_b_1355370R.

16. "CA: OPPOSE AB2109 Restricting Personal Belief Exemptions to Mandatory Vaccination," National Vaccination Information Centere, accessed August 30, 2022, http://nvicadvocacy.org/members/Resources/CAOPPOSEAB2109RestrictingVaccineExemptions.aspx. It is unclear why Fisher believed that parents would have to pay for visits to their pediatricians to discuss vaccines, as U.S. federal law requires all health insurance companies to provide a set of preventive care services—including immunization visits—without charging co-payment or co-insurance.

17. Sears, "California Bill AB2109."

18. David Gorski, "California Bill AB 2109: The Antivaccine Movement Attacks School Vaccine Mandates Again," Science-Based Medicine, March 26, 2012, https://scien cebasedmedicine.org/antivaccine-activists-attack-vaccine-mandates

19. "Petition: Concerned Americans for Parental Rights & Vaccine Exemptions Oppose CA AB2109," Change.org, 2012, accessed February 11, 2022, https://www.change. org/p/concerned-americans-for-parental-rights-vaccine-exemptions-oppose-ca-ab2109.

20. The Canary Party, a grassroots group suspicious of Western medicine, mobilized on their website: "Top 10 Reasons to Oppose California AB2109," The Canary Party, 2012, accessed December 28, 2021, https://canaryparty.org/commentary/top-10-reasons-to-oppose-california-ab2109; "Listen to the California Health Committee Hearings on AB2109," The Canary Party, 2012, accessed December 28, 2021, https:// canaryparty.org/commentary/listen-to-the-california-health-committee-hearings-on-ab2109. It also publicized Blumhardt's petition: "Shake, Rattle, and Roll Tuesday in Sacramento," The Canary Party, 2012, accessed December 28, 2021, https://cana ryparty.org/commentary/shake-rattle-and-roll-tuesday-in-sacramento. It then formed an alliance with the East Bay Tea Party (EBTP) to resist AB2109: "Tea Party Joins Canary Party in Opposing Vaccine Mandates in California," Age of Autism, 2012, accessed February 16, 2022, https://www.ageofautism.com/2012/06/tea-party-joins-canary-party-in-opposing-vaccine-mandates-in-california.html. The EBTP used Facebook to encourage AB2109 opponents to contact political actors and attend a rally opposing the bill: "Contact Political Representatives to Express Opposition to AB2109," The East Bay Tea Party, 2012, accessed February 16, 2022; "Details of Rally to Oppose AB2109," The East Bay Tea Party, 2012, accessed February 16, 2022.

21. "Rob Schneider and Tim Donnelly on Medical Freedom," September 28, 2012," Rob Schneider and Tim Donnelly, Vimeo, 2012, https://vimeo.com/60227438; "Actor Rob Schneider Joins in Protest Against Anti-Vaccination Bill," CBS Sacramento, September 5, 2012.

22. Ben Allen, "Senators Richard Pan and Ben Allen to Introduce Legislation to End California's Vaccine Exemption Loophole," February 4, 2015," news release, accessed April 5, 2023, https://sd24.senate.ca.gov/news/press-release/senators-richard-pan-and-ben-allen-introduce-legislation-end-californias-vaccine.

23. "SB-277 Public Health: Vaccinations (2015–2016), Bill History," California Legislative Information, 2015.

24. Anita Chabria, "California Mandatory Vaccination Bill Breezes Through Senate Committee," The Guardian, April 30, 2015.

25. Rebecca Plevin, "Discredited Vaccination Opponent Andrew Wakefield Crusades Against California SB 277," KPCC Southern California, May 1, 2015, accessed March 4, 2022, https://www.scpr.org/news/2015/05/01/51367/discredited-vaccination-opponent-andrew-wakefield

26. E. J. Dickson, "A Guide to 17 Anti-Vaccination Celebrities," Rolling Stone, 2019, https://www.rollingstone.com/culture/culture-features/celebrities-anti-vaxxers-jess ica-biel-847779.

27. Chabria, "California Mandatory Vaccination Bill."
28. Tracy Seipel and Jessica Calefati, "Controversial Mandatory Vaccine Bill Easily Clears California Assembly Committee," *The Mercury Times*, June 9, 2015, and *The Oakland Tribune*, June 10, 2015; "Lawmaker Against California Vaccine Bill SB277 Makes Comparison to Internment Camps," CBS News, June 9, 2015, https://sanfrancisco. cbslocal.com/2015/06/09/sb277-vaccine-bill-assemblyman-jim-patterson-internm ent-camps.
29. "SB-277 Public Health: Vaccinations, Bill History," California Legislative Information, https://leginfo.legislature.ca.gov/faces/billHistoryClient.xhtml?bill_id=2015201 60SB277.
30. Dorit Reiss, "California Court of Appeal Rejects Challenge to Vaccine Law," Bill of Health, July 30, 2018, accessed March 4, 2021, https://blog.petrieflom.law.harvard. edu/?s=California+Court+of+Appeal+Rejects+Challenge+to+Vaccine+Law&sub mit=Search.
31. Atwell et al., "Nonmedical Vaccine Exemptions and Pertussis"; Berezin and Eads, "Risk Is for the Rich?"; "Petition: Concerned Americans," Change.org; Lillvis et al., "Power and Persuasion"; "Rob Schneider and Tim Donnelly on Medical Freedom," Schneider and Donnelly; Sears, "California Bill AB2109."
32. "About Us," Circle of Mamas, 2020, accessed December 29, 2021, https://circleofma mas.com; "About Us," Learn the Risk, 2018, accessed December 29, 2021, https:// learntherisk.org/about-us.
33. "About A Voice for Choice," A Voice for Choice, 2020, accessed December 29, 2021, http://avoiceforchoice.org/about-avfc.
34. Paul L. Delamater, Timothy F. Leslie, and Y. Tony Yang, "Change in Medical Exemptions from Immunization in California After Elimination of Personal Belief Exemptions," *JAMA* 318, no. 9 (2017): 863–64.
35. Salini Mohanty et al., "Experiences with Medical Exemptions After a Change in Vaccine Exemption Policy in California," *Pediatrics* 142, no. 5 (2018): e20181051.
36. "Editorial: Anti-Vaxxers Have Found a Way Around California's Strict New Immunization Law. They Need to Be Stopped.," *Los Angeles Times*, November 8, 2017, http://www.latimes.com/opinion/editorials/la-ed-vaccine-exemption-crackd own-20171108-story.html; Karen Kaplan, "Here's What Happened After California Got Rid of Personal Belief Exemptions for Childhood Vaccines," *Los Angeles Times*, October 29, 2018.
37. Soumya Karlamangla, "Pushback Against Immunization Laws Leaves Some California Schools Vulnerable to Outbreaks," *Los Angeles Times*, July 13, 2018, https:// www.latimes.com/local/lanow/la-me-ln-sears-vaccines-fight-20180713-story.html; Rong-Gong Lin, Soumya Karlamangla, and Rosanna Xia, "Board Action Renews Vaccine Battle; Medical Panel's Bid to Pull the License of an O.C. Doctor Is Seen as Targeting Those Who Try to Skirt a New Law," *Los Angeles Times*, September 13, 2016.
38. Rong-Gong Lin, Soumya Karlamangla, and Rosanna Xia, "California Wants to Pull This Doctor's License. Here's How It's Sparked a New Battle over Child Vaccinations," *Los Angeles Times*, September 12, 2016, https://www.latimes.com/local/lanow/la-me-sears-vaccine-20160909-snap-story.html.

39. Soumya Karlamangla, "Why Hasn't California Cracked Down on Antivaccination Doctors?" *Los Angeles Times*, November 6, 2017.

40. Will Huntsberry, "One Doctor Is Responsible for a Third of All Medical Vaccine Exemptions in San Diego," *Voice of San Diego*, March 18, 2019; Will Huntsberry, "Medical Board Charges San Diego Doctor Who's Doled out Dozens of Vaccine Exemptions," *Voice of San Diego*, October 22, 2019; "San Diego Unified Vaccine Exemptions," *Voice of San Diego*, 2019, https://docs.google.com/spreadshe ets/d/e/2PACX-1vSVXxFMi1pgUGkhLzCzVXdKVZGBLoRN94I2gG3FN1- t18zZNUdWN8v0bdTt93_0criAClUQslXDgT78/pubhtml#.

41. Karlamangla, "Why Hasn't California Cracked Down on Antivaccination Doctors?"

42. Charity Dean, "Letter from Santa Barbara County Health Officer to County School and Childcare Administrators," 2016, https://www.avoiceforchoiceadvocacy.org/wp- content/uploads/2016/06/SBCPHD-MEPP-Letter.pdf.

43. "Staunch Pro-Vaccine Dr. Charity Dean Propels Herself up the CA Public Health Department Ranks to Become CA Vaccine Medical Exemptions Czar," Voice for Choice, 2019, https://avoiceforchoiceadvocacy.org/wp-content/uploads/2019/10/ AVFCA-Press-Release-Charity-Dean-self-propelled-Medical-Exemption-Czar.pdf.

44. Mohanty et al., "Experiences with Medical Exemptions."

45. Salini Mohanty et al., "California's Senate Bill 277: Local Health Jurisdictions' Experiences with the Elimination of Nonmedical Vaccine Exemptions," *American Journal of Public Health* 109, no. 1 (2019): 96–101.

46. Richard J. Pan and Dorit Reiss, "Vaccine Medical Exemptions Are a Delegated Public Health Authority," *Pediatrics* 142, no. 5 (2018): e2018200, 2.

47. "SB 276 Fact Sheet," Vaccinate California, 2019, accessed February 23, 2022, https:// vaccinatecalifornia.org/sb-276-fact-sheet.

48. Dorit Reiss (law professor and parent activist), interview with the author, June 15, 2019.

49. Lexington Howe, "Vaccine Bill SB 276 Is Not 'California for All,' Opposition Says," *The Village News*, September 17, 2019, https://www.villagenews.com/story/2019/09/ 12/news/vaccine-bill-sb-276-is-not-california-for-all-opposition-says/57425.html.

50. Mackenzie Mays, "Anti-Vaccine Protesters Are Likening Themselves to Civil Rights Activists," *Politico*, September 18, 2019, https://www.politico.com/story/2019/09/18/ california-anti-vaccine-civil-rights-1500976.

51. Jeffrey Kluger, "'They're Chipping Away.' Inside the Grassroots Effort to Fight Mandatory Vaccines," *TIME*, June 13, 2019.

52. David Taub, "Pan's Bill Would Further Restrict Vaccine Exemptions for Schoolkids," GV Wire, March 26, 2019, https://gvwire.com/2019/03/26/pans-bill-would-further- restrict-vaccine-exemptions-for-schoolkids.

53. Julie Bosman, Patricia Mazzei, and Dan Levin, "Jessica Biel Weighs in on Vaccine Fight, Drawing Fierce Pushback," *The New York Times*, June 13, 2019; Marisa Iati, "Jessica Biel Lobbied Alongside a Prominent Antivaxxer but Says She Supports Vaccines," *The Washington Post*, June 13, 2019.

54. Mays, "Anti-Vaccine Protesters."

55. Rob Schneider, tweet, September 5, 2019, Twitter.

56. Richard Pan, interview with the author, June 14, 2019.

57. Richard Pan, interview with the author, June 14, 2019.

58. Tracy Seipel and Jessica Calefati, "California's New Vaccine Law: Freshman Senator Wins Plaudits from Colleagues," *San Jose Mercury News*, July 4, 2015, https://www.mercurynews.com/2015/07/04/californias-new-vaccine-law-freshman-senator-wins-plaudits-from-colleagues.

59. Melody Gutierrez, "Anti-Vaccine Activist Assaults California Vaccine Law Author, Police Say," *Los Angeles Times*, August 21, 2019; Leilani Marie Labong, "Q&A with California Senator Richard Pan," *SacTown Magazine*, May–June 2021.

60. Gutierrez, "Anti-Vaccine Activist Assaults California Vaccine Law Author."

61. Melody Gutierrez, "Vaccine Bills Are Signed Amid Protests; Stricter Rules Will Increase Oversight of Doctors' Exemptions," *Los Angeles Times*, September 7, 2019; Marisa Iati, "California's Governor Signed a Pro-Vaccine Bill into Law This Week. Then the Protests Got Weird," *The Washington Post*, September 14, 2019; Taryn Luna, "Vaccine Bill Protester Threw Blood on California Senators, Investigation Confirms," *Los Angeles Times*, October 2, 2019; Labong, "Q&A with California Senator Richard Pan."

62. Melody Gutierrez, "Newsom Criticized the New Vaccine Bill. Anti-Vaccine Activists Are Celebrating," *Los Angeles Times*, June 4, 2019.

63. Melody Gutierrez, "California Vaccine Bill Undergoes Major Changes and Wins Support of Former Critic Newsom," *Los Angeles Times*, June 18, 2019.

64. Melody Gutierrez, "Vaccine Bill to Follow Newsom's Changes; California Governor and Senator Reach a Deal to Scale Back Parts of the Legislation and Add New Scrutiny," *Los Angeles Times*, September 7, 2019.

65. T. Rozbroj, A. Lyons, and J. Lucke, "The Mad Leading the Blind: Perceptions of the Vaccine-Refusal Movement Among Australians Who Support Vaccination," *Vaccine* 37, no. 40 (2019): 5986–93, https://doi.org/https://doi.org/10.1016/j.vaccine.2019.08.023.

66. Renée DiResta, however, remains highly skeptical of the intentions and efforts of social media platforms and writes saliently about this in her contemporary role as technical research manager at Stanford Internet Observatory, Stanford Cyber Policy Center. See "Stanford Cyber Policy Center, Renee DiResta Publications" (2022) at https://cyber.fsi.stanford.edu/publications?combine_1=%22Renee%20DiResta%22&field_pub_date_value[max][year]=2022.

67. Katie Attwell et al., "Recent Vaccine Mandates in the United States, Europe and Australia: A Comparative Study," *Vaccine* 19, no. 36 (2018): 7377–84.

68. Katie Attwell and Shevaun Drislane, "Australia's 'No Jab No Play' Policies: History, Design and Rationales," *Australian & New Zealand Journal of Public Health* 46, no. 5 (2022): 640–46; Katie Attwell and Mark C. Navin, "How Policymakers Employ Ethical Frames to Design and Introduce New Policies: The Case for Childhood Vaccine Mandates in Australia," *Policy & Politics* 50, no. 4 (2022).

69. Jay Court et al., "Labels Matter: Use and Non-Use of 'Anti-Vax' Framing in Australian Media Discourse 2008–2018," *Social Science & Medicine* 291 (2021): 114502; Katie Attwell and Adam Hannah, "Convergence on Coercion: Functional and Political Pressures as Drivers of Global Childhood Vaccine Mandates," *International Journal of Health Policy and Management* 11, no. 11 (2022): 2660–71.

70. Jeremy K. Ward, James Colgrove, and Pierre Verger. "Why France Is Making Eight New Vaccines Mandatory," *Vaccine* 36, no. 14 (2018): 1801–103.

71. Melissa Eddy, "Germany Mandates Measles Vaccine," *The New York Times*, November 14, 2019.

72. Katie Attwell et al., "Recent Vaccine Mandates in the United States, Europe and Australia."

73. "An Act to Protect Maine Children and Students from Preventable Diseases by Repealing Certain Exemptions from the Laws Governing Immunization Requirements. H.P. 586—L.D. 798," State of Maine, 2019; "Assembly Bill A2371A, Exemptions from Vaccinations Due to Religious Beliefs," State of New York, The New York State Senate, 2019; "MMR Vaccine Exemption Law Change 2019," Washington State Department of Health, accessed May 20, 2020, https://www.health ygh.org/in-the-news-1/2019/5/30/mmr-vaccine-exemption-law-change-2019.

74. Thornton and Reich's online analysis features a mother describing homeschooling her children in California due exemptions not being an option, and the scholars note an increase in homeschooling across the country, reaching 2.5 million in 2019. Courtney Thornton and Jennifer Reich, "Black Mothers and Vaccine Refusal: Gendered Racism, Healthcare, and the State," *Gender & Society* 36, no. 4 (2022): 525–51.

75. "California Health and Safety Code; 120335(h)," California Legislative Information, 2016, https://leginfo.legislature.ca.gov/faces/codes_displaySection.xhtml?section Num=120335&lawCode=HSC. For discussion of the Individualized Education Program exemption in the Nonmedical Exemptions Bill, see Ross D. Silverman and Wendy Hensel, "Squaring State Child Vaccine Policy with Individual Rights Under the Individuals With Disabilities Education Act: Questions Raised in California," *Public Health Reports* 132, no. 5 (2017): 593–96; Mohanty et al., "California's Senate Bill 277"; Dorit Reiss, "Litigating Alternative Facts: School Vaccine Mandates in the Courts," *University of Pennsylvania Journal of Constitutional Law* 21 (2018): 207.

76. Lawlor, Joe. "'No' Vote – to Keep State's New Vaccine Law – Wins by Overwhelming Margin," *Portland Press Herald*, March 4, 2020. https://www.pressherald.com/2020/03/03/no-vote-to-keep-pro-vaccine-law-leading-big-in-referendum/.

77. Peter J. Hotez, "COVID19 Meets the Antivaccine Movement," *Microbes and Infection* 22, no. 4–5 (2020): 162–64, https://doi.org/10.1016/j.micinf.2020.05.010; Liz Szabo, "Anti-Vaccine Activists Latch onto Coronavirus to Bolster Their Movement," *California Healthline*, April 24, 2020, accessed March 15, 2021, https://californiahea lthline.org/news/anti-vaccine-activists-latch-onto-coronavirus-to-bolster-their-movement; Nicholas Bogel-Burroughs, "Antivaccination Activists Are Growing Force at Virus Protests," *The New York Times*, May 2, 2020.

78. Laurel Rosenhall and Emily Hoeven, "At California Capitol, Lawmakers Want Info; Protesters Want to End Coronavirus Stay-at-Home," Cal Matters, April 20, 2020.

79. Melody Gutierrez, "Anti-Vaccine Activists, Mask Opponents Target Public Health Officials—At Their Homes," *Los Angeles Times*, June 18, 2020.

80. Meryl Kornfield, "Anti-Vaccine Protesters Temporarily Shut Down Major Coronavirus Vaccine Site at Dodger Stadium in Los Angeles," *The Washington Post*, January 31, 2021; Allyson Waller and Manny Fernandez, "Protesters Disrupt Motorists from Entering Dodger Stadium Vaccination Site," *The New York Times*, January 30, 2021.

81. "Senate Bill No. 742, Vaccination Sites: Unlawful Activities: Obstructing, Intimidating, or Harassing," California Legislative Information, 2021; Richard Pan, "Governor Signs Legislation SB 742 to Protect Our Right to Get Vaccinated," new release, October 9, 2021.

82. "California Implements First-in-the-Nation Measure to Encourage Teachers and School Staff to Get Vaccinated," Office of Governor Gavin Newsome, August 11, 2021;Brandy Zadrozny and Ben Collins, "As Vaccine Mandates Spread, Protests Follow—Some Spurred by Nurses," NBC News, August 12, 2021, accessed July 31, 2022, https://www.nbcnews.com/tech/social-media/vaccine-mandates-spread-prote sts-follow-spurred-nurses-rcna1654.

83. "California Becomes First State in Nation to Announce COVID-19 Vaccine Requirements for Schools," Office of Governor Gavin Newsom, October 1, 2021, https://www.gov.ca.gov/2021/10/01/california-becomes-first-state-in-nation-to-announce-covid-19-vaccine-requirements-for-schools. The implementation of Newsom's executive order would ultimately be delayed repeatedly, and it did not ever come into effect. Melissa Gomez and Howard Blume, "Parents in California Protest Student COVID-19 Vaccine Mandate, Keep Kids Home," *Los Angeles Times*, October 18, 2021; Noura Salahieh, Mary Beth McDade, Gene Kang, and Megan Telles, "Parents Keep Kids Home from School to Protest California COVID Vaccine Mandate," KTLA Local News, October 18, 2021, accessed September 9, 2022, https://ktla.com/news/local-news/california-parents-to-keep-kids-home-from-school-to-protest-covid-vaccine-mandate; "Statement on Timeline for COVID-19 Vaccine Requirements in Schools," California Department of Public Health, April 14, 2022, accessed August 31, 2022, https://www.cdph.ca.gov/Programs/OPA/Pages/NR22-073.aspx; Adam Beam, "California Delays Coronavirus Vaccine Mandate for Schools," ABC News, April 15, 2022, accessed August 31, 2022, https://abcnews.go.com/Health/wireStory/bill-nixing-belief-excuse-school-covid-vaccines-dies-84092599.

84. Katy Grimes, "California Lawmakers Fast-Tracking Child Health Bills to Erode Parental Rights," California Globe, February 9, 2022; "Why Us?" Capitol Resource Institute, accessed September 23, 2022, https://www.capitolresource.org/why-us.

85. Richard Pan, "State Senator Dr. Richard Pan Statement on Holding School Vaccination Requirement Legislation," press release, April 14, 2022, accessed August 31, 2022, https://sd06.senate.ca.gov/news/2022-04-14-state-senator-dr-richard-pan-statement-holding-school-vaccination-requirement.

86. Melody Gutierrez, "Bill to Allow Minors to Be Vaccinated Without Parental Consent Is Withdrawn," *Los Angeles Times*, August 31, 2022.

Chapter 6

1. Gillian Brockell, "Mandatory Immunization for the Military: As American as George Washington," *The Washington Post*, August 26, 2021; Ann M. Becker, "Smallpox in Washington's Army: Strategic Implications of the Disease During the American Revolutionary War," *Journal of Military History* 68, no. 2 (2004): 381–430; Gareth Millward, *Vaccinating Britain: Mass Vaccination and the Public Since the Second World War* (Manchester, UK: Manchester University Press, 2019).

2. *Jacobson v. Massachusetts*, U.S. Supreme Court, 197 U.S. 11 (1905); *Zucht v. King*, U.S. Supreme Court, 260 U.S. 174 (1922); *Prince v. Massachusetts*, U.S. Supreme Court, 321 U.S. 158 (1944).

3. George Rosen, *A History of Public Health* (Baltimore, MD: Johns Hopkins University Press, 1993), 68–70.

4. Rosen, *A History of Public Health*, 70.

5. Rosen, *A History of Public Health*, 70.

6. Richard Pipes, *A Concise History of the Russian Revolution* (New York: Vintage Books, 1996). Thanks to Ethan Bradley for this helpful point.

7. Richard J. Altenbaugh, *Vaccination in America: Medical Science and Children's Welfare* (London: Palgrave Macmillan 2018), 50.

8. John Duffy, *The Sanitarians: A History of American Public Health* (Urbana: University of Illinois Press, 1990), 182; citing John Duffy, "School Vaccination: The Precursor to School Medical Inspection," *Journal of the History of Medicine and Allied Sciences* 33, no. 3 (1978): 344–55.

9. Duffy, *The Sanitarians*, 55.

10. For example, quarantine regulations introduced during yellow fever epidemics in the 19th century proved difficult to maintain, and in cities such as New Orleans and Philadelphia, health boards were unable to sustain quarantines even during outbreaks. Duffy observes that "with the exception of the Boston Board of Health, the health boards appointed from the 1790s to 1830 were all temporary." Duffy, *The Sanitarians*, 62.

11. Duffy, *The Sanitarians*, 123.

12. Duffy, *The Sanitarians*, 124; but see all of Chapter 9 for the payoff.

13. Rosen, *A History of Public Health*; Jane Addams, *Twenty Years at Hull-House* (New York: Macmillan, 1910); Robert Huddleston Wiebe, *The Search for Order, 1877–1920* (New York: Hill & Wang, 1966).

14. Elena Conis, *Vaccine Nation* (Chicago: University of Chicago Press, 2015), 24.

15. Progressive Era reformers often thought of the local public school as a central locus for improving the lives of every member of the community. In addition to providing education and health care for children, they also offered evening classes for adults, entertainment for the community, and public recreation through the use of athletic facilities and fields. Altenbaugh, *Vaccination in America*, 99.

16. See, for example, the establishment of a national public health department in 1910; Daniel Sledge, *Health Divided: Public Health and Individual Medicine in the Making of the Modern American State* (Lawrence: University Press of Kansas, 2017), 38–39.

17. Duffy, *The Sanitarians*, 130.

18. Duffy, *The Sanitarians*, 182.

19. James Colgrove, *State of Immunity: The Politics of Vaccination in Twentieth-Century America* (Berkeley: University of California Press, 2006), ch. 2.

20. Colgrove, *State of Immunity*, 48.

21. Colgrove, *State of Immunity*, 51; citing Cynthia Connelly, "Prevention Through Detention: The Pediatric Tuberculosis Prevention Movement in the United States, 1909–1951" (PhD diss., University of Pennsylvania, 1999), published later as Cynthia A. Connolly, *Saving Sickly Children: The Tuberculosis Preventorium in American Life, 1909–1970* (New Brunswick, NJ: Rutgers University Press, 2008).

22. Michael Willrich, *Pox: An American History* (New York: Penguin, 2011), 271.

23. Willrich, *Pox*, 252–253.

24. Willrich, *Pox*, 14. This quotation does not appear in any known published writings of Thomas Jefferson, although it does reflect his beliefs in limited government. See "Thomas Jefferson Encyclopedia," Thomas Jefferson's Monticello, 2014, accessed July 29, 2022, https://www.monticello.org/research-education/thomas-jefferson-encyclopedia/government-best-which-governs-least-spurious-quotation. For the first record of this quote, see "Introduction to United States Magazine and Democratic Review," *United States Magazine and Democratic Review* 1, no. 1 (1837): 6. A discussion of this earlier source can be found in Paul F. Boller, Jr., *Not So! Popular Myths About America from Columbus to Clinton* (Oxford, UK: Oxford University Press, 1995), 49.

25. Willrich, *Pox*, ch. 4.

26. R. Kipling, "The White Man's Burden," 1899; Patrick Brantlinger, "Kipling's 'The White Man's Burden' and Its Afterlives," *English Literature in Transition, 1880–1920* 50, no. 2 (2007): 172–91, https://doi.org/10.1353/elt.2007.0017.

27. "The Spread of Small-Pox by Tramps," *Lancet* 163, no. 4198 (1904): 446–47; cited by Willrich, *Pox*, 6.

28. Willrich, *Pox*, 41.

29. Sledge, *Health Divided*, 1–2; Sledge cites Margaret Humphreys, *Yellow Fever and the South* (Baltimore, MD: Johns Hopkins University Press, 1994).

30. Sledge, *Health Divided*, 7; Sledge cites Werner Troesken, *Water, Race, and Disease* (Cambridge, MA: MIT Press, 2004).

31. Altenbaugh, *Vaccination in America*, 39; Willrich, *Pox*, ch. 6.

32. Willrich, *Pox*, 238. However, America's new public health laws in the late 19th and early 20th centuries sometimes exercised power even against persons who were not marginalized or powerless. William Novak argues that early public health laws sought to prevent public nuisances and regulate "noxious trades," and often targeted members of the White merchant class; William J. Novak, *The People's Welfare: Law and Regulation in Nineteenth-Century America* (Chapel Hill: The University of North Carolina Press, 1996). White, middle- or upper-class people were also subject to coercive vaccination policies. Henning Jacobson—of *Jacobson v. Massachusetts*, which we examine in the next section—was a graduate of Yale Divinity School and served as pastor of a congregation of Swedish Evangelical Lutherans in Cambridge,

Massachusetts; Colgrove, *State of Immunity*, 38–39. William H. Smith operated a delivery and hauling business in Greenpoint (a Brooklyn neighborhood) that was placed under quarantine when he refused vaccination; Colgrove, *State of Immunity*, 26–32.

33. It would have been more just to provide all Americans with healthy living conditions and to promote voluntary vaccination. But in the absence of such interventions, it would have been even worse to leave these poor and marginalized communities alone to suffer disease and spread it to other communities. We can have societies that are both healthy and free only if we are investing in robust public health infrastructure. In the absence of such provision, coercive measures can sometimes be a second-best alternative.

34. *Jacobson v. Massachusetts*, U.S. Supreme Court, 197 U.S. 11 (1905).

35. Peter S. Canellos and Joel Lau, "The Surprisingly Strong Supreme Court Precedent Supporting Vaccine Mandates," *Politico*, August 9, 2021; Darragh Roche, "What Is Jacobson V. Massachusetts? How Supreme Court Ruled on Vaccine Mandate in 1905," *Newsweek*, September 10, 2021, https://www.newsweek.com/what-jacob son-v-massachusetts-how-supreme-court-ruled-vaccine-mandate-1905-1627761; Chris Beyrer and Larry Corey, "The Conservative Supreme Court That Embraced Vaccine Mandates," Changing America, December 4, 2021, https://thehill.com/ changing-america/opinion/584236-the-conservative-supreme-court-that-embra ced-vaccine-mandates. Amanda Kaufman, "A Supreme Court Case That Originated in Mass. Could Provide a Legal Precedent for President Biden's Vaccine Mandates, Experts Say," *The Boston Globe*, September 10, 2021; Madison Hall, "How a Supreme Court Decision from 1905 Set the Stage for Vaccine Mandates," Insider, September 11, 2021, https://www.businessinsider.com/supreme-court-decision-from-1905-set-stage-for-vaccine-mandates-2021-9; "The U.S. Has a Long Precedent for Vaccine Mandates," NPR, August 29, 2021, https://www.npr.org/2021/08/29/1032169566/ the-u-s-has-a-long-precedent-for-vaccine-mandates.

36. Lawrence O. Gostin and Lindsey F. Wiley, *Public Health Law: Power, Duty, Restraint*, 3rd ed. (Oakland: University of California Press, 2016), 124, quoting from pp. 26–27 of *Jacobson*.

37. Willrich, *Pox*, 327. Perhaps it is worthwhile to reflect about the kind of "social compact" that Harlan had in mind. It has something in common with the contract that Hobbes imagines in *Leviathan* (Thomas Hobbes, *Leviathan*, edited by Richard Tuck (New York: Cambridge University Press, 1996), since it prioritizes state power to provide protection to the people, even at substantial cost to individual liberties to be left alone. It also resembles the contract Rousseau discusses in "On the Social Contract" (Jean-Jacques Rousseau, "On the Social Contract," in *The Basic Political Writings*, translated by Donald A. Cress, 141–227 [Indianapolis, IN: Hackett, 1987]), which emphasizes a democratic people's moral imperative to legislate for the common good, rather than to protect individual liberty. It is less clear how Harlan's conception of the social compact relates to more explicitly liberal conceptions of that ideal. For example, Locke argues in the *Second Treatise on Government* (John Locke, *Second Treatise of Government* [Indianapolis, IN: Hackett, 1980]) that people preserve their natural rights even after they enter civil society and that the only legitimate activity

of government is to protect those rights. Furthermore, Kant argues in *Metaphysics of Morals* (Immanuel Kant, *The Metaphysics of Morals*, edited by M. J Gregor and R. J Sullivan [Cambridge University Press, 1996]) that our freedom and moral equality with others can be realized only in a civil society under shared government, which can be understood as an object of consent of all persons. On Locke and Kant's accounts of the social contract, we could justify vaccine mandates only if we showed that they protected negative liberties or otherwise promoted human freedom. Harlan's invocation of "protection, safety, prosperity and happiness" would not be sufficient reasons to restrict liberty and freedom. We return to these questions about liberty and the idea of the social contract in Chapter 8. For now, we want only to note that there is something distinctively illiberal (or at least only weakly liberal) about Harlan's conception of the imagined social contract that binds us together under government.

38. Gostin and Wiley, *Public Health Law*, 353.
39. Gostin and Wiley, *Public Health Law*, 353; citing *Boone v. Boozman* (2002) finding that Jacobson and Zucht do not apply only in pandemics; *Boone v. Boozman*, U.S. District Court for the Eastern District of Arkansas, 217 F. Supp. 2d 938 (2002).
40. *Roman Catholic Diocese of Brooklyn, New York v. Andrew M. Cuomo, Governor of New York*, Supreme Court of the United States, 592 U.S. (2020); *National Federation of Independent Business, et al, 21A244 v. Department of Labor, Occupational Safety and Health Administration, et al.*, Supreme Court of the United States, 595 U.S. (2022).
41. *Lochner v. New York*, U.S. Supreme Court, 198 U.S. 45 (1905), no. 292.
42. Mark J. Stern, "A New Lochner Era," *Slate*, June 29, 2018, https://slate.com/news-and-politics/2018/06/the-lochner-era-is-set-for-a-comeback-at-the-supreme-court.html; Thomas B. Colby and Peter J. Smith, "The Return of Lochner," *Cornell Law Review* 100, no. 3 (2015): 527–602; Amanda Shanor, "The New Lochner," *Wisconsin Law Review* 2016, no. 1 (2016): 133–208; Gillian E. Metzger, "1930s Redux: The Administrative State Under Siege." *Harvard Law Review* 131, no. 1 (2017): 1.
43. *Jarkesy v. Securities and Exchange Commission*, United States Court of Appeals for the Fifth Circuit, no. 20–61007; (2022) ; "Federal Court Calls Unconstitutional the U.S. Securities & Exchange Commission's In-House Administrative Proceedings for Securities Fraud Cases," *National Law Review* 12, no. 231. (2022); *West Virginia et al. v. Environmental Protection Agency et al.*, Supreme Court of the United States, no. 20-1530; Charlie Savage, "E.P.A. Ruling Is Milestone in Long Pushback to Regulation of Business," *The New York Times*, June 30, 2022; Patrick Parenteau, "The Supreme Court Has Curtailed EPA'S Power to Regulate Carbon Pollution—and Sent a Warning to Other Regulators," The Conversation, July 1, 2022, https://theconversation.com/the-supreme-court-has-curtailed-epas-power-to-regulate-carbon-pollution-and-sent-a-warning-to-other-regulators-185281; Alex Guillen, "Impact of Supreme Court's Climate Ruling Spreads," *Politico*, July 20, 2022, https://www.politico.com/news/2022/07/20/chill-from-scotus-climate-ruling-hits-wide-range-of-biden-actions-00045920; Krista Mahr, "'It's a Tsunami': Legal Challenges Threatening Public Health Policy," *Politico*, May 10, 2022, https://www.politico.com/news/2022/05/10/legal-challenges-cdc-public-health-policy-00031253; "Battle over CDC's Powers Goes Far Beyond Travel Mask Mandate," NPR, April 21, 2022.

44. David Bernstein, "The Supreme Court Could Foster a New Kind of Civil War," *Politico*, June 14, 2022. Sam Baker, "The Supreme Court's Next Target Is the Executive Branch," Axios, July 25, 2022, https://www.axios.com/2022/07/05/supreme-court-conservative-climate-health-regulations.

45. Kenneth L. Garver and Bettylee Garver, "Historical Perspectives: Eugenics: Past, Present, and the Future," *American Journal of Human Genetics* 49, no. 5 (1991): 1109–18; Steven Selden, "Transforming Better Babies into Fitter Families: Archival Resources and the History of the American Eugenics Movement, 1908–1930," *Proceedings of the American Philosophical Society* 149, no. 2 (2005): 199–225.

46. *Buck v. Bell, Superintendent of State Colony Epileptics and Feeble Minded*, 274 U.S. 200; 47 S.Ct. 584; 71 L.Ed. 1000.

47. Sandra Day O'Connor, "They Often Are Half Obscure: The Rights of the Individual and the Legacy of Oliver W. Holmes," *San Diego Law Review* 29, no. 3 (1992): 385.

48. *Buck v. Bell, Superintendent of State Colony Epileptics and Feeble Minded*, United States Supreme Court, 274 U.S. 200; 47 S.Ct. 584 (emphasis added).

49. Jamal Greene, *How Rights Went Wrong: Why Our Obsession with Rights Is Tearing America Apart* (Boston: Houghton Mifflin Harcourt, 2021); Vicki C. Jackson, "Constitutional Law in an Age of Proportionality," *Yale Law Journal* 124, no. 8 (2015): 3094–196.

50. The echoes of the eugenics movement continued in some communities for decades after World War II. For example, both California and North Carolina continued their sterilization programs into the 1970s, often targeting poor women of color. On California's history of eugenics, see Alexandra Minna Stern, *Eugenic Nation: Faults and Frontiers of Better Breeding in Modern America* (Berkeley: University of California Press, 2005). For North Carolina, see Kevin Begos et al., *Against Their Will: North Carolina's Sterilization Program* (Apalachicola, FL: Gray Oak Books, 2012).

51. Ira Katznelson, "What America Taught the Nazis," *The Atlantic*, November 2017, 42–44; Alex Ross, "The Hitler Vortex: How American Racism Influenced Nazi Thought," *The New Yorker* 94, no. 11 (2018): 66, https://www.newyorker.com/magazine/2018/04/30/how-american-racism-influenced-hitler; Edwin Black, *War Against the Weak: Eugenics and America's Campaign to Create a Master Race* (New York: Dialog Press, 2003).

52. Wendy K. Mariner, George J. Annas, and Leonard H. Glantz, "Jacobson V Massachusetts: It's Not Your Great-Great-Grandfather's Public Health Law," *American Journal of Public Health* 95, no. 4 (2005): 581–90, https://doi.org/10.2105/ajph.2004.055160; Josh Blackman, "The Irrepressible Myth of Jacobson V. Massachusetts," *Buffalo Law Review* 70, no. 113 (2021); Lawrence O. Gostin, "Jacobson v Massachusetts at 100 Years: Police Power and Civil Liberties in Tension," *American Journal of Public Health* 95, no. 4 (2005): 576–81.

53. Jay Katz, "The Consent Principle of the Nuremberg Code: Its Significance Then and Now," in *The Nazi Doctors and the Nuremberg Code*, edited by George J. Annas and M. A. Grodin (New York: Oxford University Press, 1998), 236.

54. Indeed, when the Declaration of Helsinki was drafted, American representatives resisted strong protections against research on institutionalized children and prison

populations, which resulted in a weakened document on those points. Altenbaugh, *Vaccination in America*, 246–247; citing Susan E. Lederer, "Children as Guinea Pigs: Historical Perspective," *Accountability in Research* 10, no. 1 (2003): 1–16.

55. Altenbaugh, *Vaccination in America*, 138.

56. Susan E. Lederer, *Subjected to Science: Human Experimentation in America Before the Second World War* (Baltimore, MD: Johns Hopkins University Press, 1995), 74.

57. Altenbaugh, *Vaccination in America*, 139; relying on Lederer, *Subjected to Science*, and Lederer, "Children as Guinea Pigs."

58. Allan M. Brandt, "Racism and Research: The Case of the Tuskegee Syphilis Study," *Hastings Center Report* 8, no. 6 (1978): 21–29, https://doi.org/10.2307/3561468.

59. Altenbaugh, *Vaccination in America*, 146–49.

60. Altenbaugh, *Vaccination in America*, 160; citing Jonathon Moreno, *Undue Risk: Secret State Experiments on Humans* (New York: Routledge, 2001).

61. Altenbaugh, *Vaccination in America*, 185–188.

62. Altenbaugh, *Vaccination in America*, 197–98; citing David M. Oshinsky, *Polio: An American Story* (Oxford, UK: Oxford University Press, 2005), 157–58.

63. See Ruth R. Faden, Tom L. Beauchamp, and Nancy M. P. King, *A History and Theory of Informed Consent* (New York: Oxford University Press, 1986).

64. See Faden et al., *A History and Theory of Informed Consent*. Our focus here is on medical *treatment*, rather than research, which responds to a broader set of goals. See, for example, Alex John London, *For the Common Good: Philosophical Foundations of Research Ethics* (Oxford, UK: Oxford University Press, 2021).

65. Thomas Grisso and Paul S. Appelbaum, *Assessing Competence to Consent to Treatment: A Guide for Physicians and Other Health Professionals* (New York: Oxford University Press, 1998).

66. Joint Commission on Accreditation of Healthcare Organizations, *Comprehensive Accreditation Manual: CAMH for Hospitals: The Official Handbook*, 2011; Steven H. Miles et al., "Medical Ethics Education: Coming of Age," *Academic Medicine* 64, no. 12 (1989): 705–14; Ann K. Boulis and Jerry A. Jacobs, *The Changing Face of Medicine* (Ithaca, NY: Cornell University Press, 2011); Joseph A. Carrese et al., "The Essential Role of Medical Ethics Education in Achieving Professionalism: The Romanell Report," *Academic Medicine* 90, no. 6 (2015): 744–52; Alberto Giubilini, Sharyn Milnes, and Julian Savulescu, "The Medical Ethics Curriculum in Medical Schools: Present and Future," *Journal of Clinical Ethics* 27, no. 2 (2016): 129–45.

67. Dorit Reiss and Nili Karako-Eyal, "Informed Consent to Vaccination: Theoretical, Legal, and Empirical Insights," *American Journal of Law & Medicine* 45, no. 4 (2019): 357–419; Wenedy E. Parmet, "Informed Consent and Public Health: Are They Compatible When It Comes to Vaccines," *Journal of Health Care Law & Policy* 8 (2005): 71–110.

68. Aviva L. Katz and Sally A. Webb, "Informed Consent in Decision-Making in Pediatric Practice," *Pediatrics* 138, no. 2 (2016): e20161484, https://doi.org/10.1542/peds.2016-1484.

69. Colgrove, *State of Immunity*, 42.

70. George J. Annas, "Bioterrorism, Public Health, and Civil Liberties," *New England Journal of Medicine* 346 (2002): 1337–42, 1340.

71. Colgrove, *State of Immunity*, 6.

72. Ronald Bayer and Amy L. Fairchild, "The Genesis of Public Health Ethics," *Bioethics* 18, no. 6 (2004): 473–92, https://doi.org/10.1111/j.1467-8519.2004.00412.x.

73. Mark Largent, *Vaccine: The Debate in Modern* America (Baltimore, MD: Johns Hopkins University Press, 2012), 152.

74. Gostin and Wiley, *Public Health Law*, 34; citing Lawrence Wallack and Regina Lawrence, "Talking About Public Health: Developing America's 'Second Language,'" *American Journal of Public Health* 95, no. 4 (1971): 567–70.

75. The historical material in this section draws heavily on Nadja Durbach, *Bodily Matters: The Anti-Vaccination Movement in England, 1853–1907* (Durham NC: Duke University Press, 2005); but also on Robert M. Wolfe and Lisa K. Sharp, "Anti-Vaccinationists Past and Present," *BMJ* 325, no. 7361 (2002): 430–32; M. Kaufman, "The American Anti-Vaccinationists and Their Arguments," *Bulletin of the History of Medicine* 41, no. 5 (1967): 463–78.

76. The policies we discuss in this section applied to England, but they sometimes also governed one or more of the other parts of the United Kingdom (Wales, Scotland, and Northern Ireland). For simplicity, we focus only on the governance of England.

77. Wolfe and Sharp, "Anti-Vaccinationists Past and Present."

78. Wolfe and Sharp, "Anti-Vaccinationists Past and Present."

79. Government of Great Britain, *Labour Statistics: Returns of Wages Published Between 1830 and 1886* (London, printed for Her Majesty's Stationery Office by Eyre and Spottiswoode, 1887), https://babel.hathitrust.org/cgi/pt?id=nyp.33433008925 772&view=1up&seq=10&skin=2021.

80. Dorothy Porter, *Health, Civilization and the State: A History of Public Health from Ancient to Modern Times* (New York: Taylor & Francis, 2005), 128.

81. Porter, *Health, Civilization and the State*, 129.

82. Daniel A. Salmon et al., "Compulsory Vaccination and Conscientious or Philosophical Exemptions: Past, Present, and Future," *Lancet* 367, no. 9508 (2006): 436–42.

83. Quoting from Royal Commission on Vaccination, *Final Report of the Royal Commission Appointed to Inquire into the Subject of Vaccination* (London: Printed for Her Majesty's Stationery Office by Eyre and Spottiswoode,1896).

84. "The Report of the Royal Commission on Vaccination," *British Medical Journal* 2 (1896): 453–58.

85. "The Vaccination Act, 1898," *British Medical Journal* 2 (1898): 637; "Victorian Health Reform: How Did the Victorians View Compulsory Vaccination?" The National Archives, 2022, accessed August 26, 2022, https://www.nationalarchives.gov.uk/education/resources/victorian-health-reform/#:~:text=In%201898%2C%20a%20new%20Vaccination,exempting%20their%20children%20from%20vaccination; Nadja Durbach, "Class, Gender, and the Conscientious Objector to Vaccination, 1898–1907," *Journal of British Studies* 41, no. 1 (2002): 58–83.

86. Salmon et al., "Compulsory Vaccination and Conscientious or Philosophical Exemptions"; E. G. Thomas, "The Old Poor Law and Medicine," *Medical History* 24, no. 1 (1980): 1–19.

87. British Medical Association, *Childhood Immunisation: A guide for healthcare professionals*, Board of Science and Education, British Medical Association, 2003, https://www.researchgate.net/publication/281282847_Childhood_immunisationa_guide_for_healthcare_professionals_BMA_June_2003/link/55df1f7d08aeaa26af109 9ee/download.

88. "Report of the MMR Expert Group," Scottish Government, April 2002, https://www.webarchive.org.uk/wayback/archive/20160125045947/http://www.gov.scot/Publi cations/2002/04/14619/3798.

89. Although the delay was largely due to not wanting to lose staff ahead of winter, it is noteworthy that a mandate had not been implemented much earlier, when the country had full supply of vaccines. See "Delay in Making COVID Vaccine Mandatory Is 'Sensible' Ahead of Winter Pressures, Says BMA," British Medical Association, November 9, 2021.

90. "Consultation on Removing Vaccination as a Condition of Deployment for Health and Social Care Staff," Government of the United Kingdom, news release, 2022, https://www.gov.uk/government/news/consultation-on-removing-vaccination-as-a-condition-of-deployment-for-health-and-social-care-staff; "Oral Statement to Parliament on Vaccines as a Condition of Deployment," Department of Health and Social Care and The Rt Hon Sajid Javid MP, news release, January 31, 2022, https://www.gov.uk/government/speeches/oral-statement-on-vaccines-as-a-condition-of-deployment. ; Dominic Wilkinson, Alberto Giubilini, and Julina Savulescu, "Is This the End of the Road for Vaccine Mandates in Healthcare? " The Conversation, February 4, 2022, https://theconversation.com/is-this-the-end-of-the-road-for-vacc ine-mandates-in-healthcare-176310.

91. Melissa Leach and James Fairhead, *Vaccine Anxieties: Global Science, Child Health and Society* (London: Earthscan, 2007); See also Elizabeth Rough, *UK Vaccination Policy, Research Briefing* (London: House of Commons Library, 2022).

92. Salmon et al., "Compulsory Vaccination and Conscientious or Philosophical Exemptions."

93. It is also worth noting a further difference between the conditions of past and present immunization policies. Today, vaccine-preventable infectious diseases are less likely to be fatal and generally less harmful than they were in the past, largely due to improved medical care. For example, in the 18th and early 19th centuries, smallpox sometimes had a greater than 20–30% mortality rate, yet the disease would likely be much less deadly today with contemporary treatments. Furthermore, many of our newer vaccines (e.g., against varicella) protect against diseases that are much less likely to cause significant harm than were the diseases that early vaccine mandates protected against (e.g., smallpox, diphtheria, and polio). Accordingly, the health and social benefits generated by today's vaccine mandates are likely be less dramatic than those that were accomplished by mandates in previous generations. This is a further reason to believe that coercive vaccine mandates are less likely to be justified today than they were in the past. It is also another reason to think that people today will be less likely to tolerate vaccine mandates than people were in previous generations.

94. Charles Allan McCoy, "Adapting Coercion: How Three Industrialized Nations Manufacture Vaccination Compliance," *Journal of Health Politics, Policy and Law* 44, no. 6 (2019): 823–54.

Chapter 7

1. Emily Swanson and Tom Murphy, "High Trust in Doctors, Nurses in US, AP-NORC Poll Finds," AP News, August 10, 2021. https://apnews.com/article/joe-biden-busin ess-health-coronavirus-pandemic-509835fc9b663bffc83f52d248e9ef4a; Jolyon Attwooll, "Doctors Lead International 'Most Trusted' Profession Poll," The Royal Australian College of General Practitioners, news release, October 13, 2021, https:// www1.racgp.org.au/newsgp/professional/doctors-lead-international-most-trusted-profession.

2. Stephen McAndrew, "Internal Morality of Medicine and Physician Autonomy," *Journal of Medical Ethics* 45, no. 3 (2019): 198–203.

3. Alyson Sulaski Wyckoff, "Eliminate Nonmedical Immunization Exemptions for School Entry, Says AAP," American Academy of Pediatrics, August 29, 2016; Sarah Wickline Wallan, "AMA: No More Non-Medical Vaccine Exemptions—Vaccine Exemptions Should Only Be for Medical Reasons, AMA Members Say," Medpage Today, June 7, 2015; "Immunization Exemptions," American Academy of Family Physicians, 2015, https://www.aafp.org/dam/AAFP/documents/patient_care/ immunizations/vaccine-exemptions.pdf; "Statement: State Immunization Laws Should Eliminate Non-Medical Exemptions Say Internists," American College of Physicians, news release, July 29, 2015, https://www.acponline.org/acp-newsroom/ state-immunization-laws-should-eliminate-non-medical-exemptions-say-int ernists.

4. Paul A. Offit, "Vaccine Exemptions: When Do Individual Rights Trump Societal Good?" *Journal of the Pediatric Infectious Diseases Society* 4, no. 2 (2015): 89–90.

5. "AMA Supports Tighter Limitations on Immunization Opt Outs," American Medical Association, 2015, accessed December 30, 2021, http://www.ama-assn.org/ama/ pub/news/news/2015/2015-06-08-tighter-limitations-immunization-opt-outs.page; American Academy of Pediatrics Committee on Practice and Ambulatory Medicine, Committee on Infectious Diseases, Committee on State Government Affairs, Council on School Health, and Section on Administration and Practice Management, "Medical Versus Nonmedical Immunization Exemptions for Child Care and School Attendance," *Pediatrics* 138, no. 3 (2016): e20162145.

6. The authors thank University of Western Australia medical student Maddison Ayton for her research assistance in this section.

7. Ivan Waddington, "The Movement Towards the Professionalization of Medicine," *British Medical Journal* 301, no. 6754 (1990): 688–90, https://doi.org/10.1136/ bmj.301.6754.688.

8. Paul Starr, "Professionalization and Public Health: Historical Legacies, Continuing Dilemmas," *Journal of Public Health Management and Practice* (2009): S26–S30, S28.

9. David A. Johnson and Humayun J. Chaudhry, *Medical Licensing and Discipline in America: A History of the Federation of State Medical Boards* (Lanham, MD: Lexington Books, 2012); Mike Saks, *The Professions, State and the Market: Medicine in Britain, the United States and Russia* (London: Routledge, 2015); Paul Starr, *The Social Transformation of American Medicine: The Rise of a Sovereign Profession and the Making of a Vast Industry*, 2nd edition (New York: Basic Books, 2017).

10. Waddington, "The Movement Towards the Professionalization of Medicine"; Kyle Loudon and Irvine Loudon, *Medical Care and the General Practitioner, 1750–1850* (Oxford, UK: Oxford University Press, 1986); S. E. D Shortt, "Physicians, Science, and Status: Issues in the Professionalization of Anglo-American Medicine in the Nineteenth Century," *Medical History* 27, no. 1 (1983): 51–68; William G. Rothstein, *American Physicians in the Nineteenth Century: From Sects to Science* (Baltimore, MD: Johns Hopkins University Press, 1992).

11. George D. Lyman, "The Beginnings of California's Medical History," *California and Western Medicine* 23, no. 5 (1925): 561–76.

12. "Our Story," California Medical Association, 2020, accessed May 15, 2020, https://www.cmadocs.org/about.

13. Michael Willrich, *Pox: An American History* (New York: Penguin, 2011).

14. Janet Adamy and Paul Overberg, "Doctors, Once GOP Stalwarts, Now More Likely to Be Democrats," *The Wall Street Journal*, October 6, 2019, https://www.wsj.com/artic les/doctors-once-gop-stalwarts-now-more-likely-to-be-democrats-11570383523; Adam Bonica, Howard Rosenthal, and David J. Rothman, "The Political Polarization of Physicians in the United States: An Analysis of Campaign Contributions to Federal Elections, 1991 through 2012," *JAMA Internal Medicine* 174, no. 8 (2014): 1308–17, https://doi.org/10.1001/jamainternmed.2014.2105; "The Physicians Foundation 2020 Physician Survey: Part 3," The Physicians Foundation, 2020, accessed September 9, 2022, https://physiciansfoundation.org/physician-and-patient-surveys/the-phy sicians-foundation-2020-physician-survey-part-3.

15. John Duffy, *The Sanitarians: A History of American Public Health* (Urbana: University of Illinois Press, 1990), 134–135.

16. Duffy, *The Sanitarians*, 135; James E. Reeves, "The Eminent Domain of Sanitary Science, and the Usefulness of State Boards of Health in Guarding the Public Welfare," *JAMA* 1 (1883): 612–18, 612, https://doi.org/10.1001/jama.1883.02390210008001a.

17. Duffy, *The Sanitarians*, 136.

18. Duffy, *The Sanitarians*, 136.

19. James Colgrove, *State of Immunity: The Politics of Vaccination in Twentieth-Century America* (Berkeley: University of California Press, 2006), ch. 3.

20. Colgrove, *State of Immunity*, 98. For some of the relevant literature, see Allan M. Brandt and Martha Gardner, "Antagonism and Accommodation: Interpreting the Relationship Between Public Health and Medicine in the United States During the 20th Century," *American Journal of Public Health (1971)* 90, no. 5 (2000): 707–15, https://doi.org/10.2105/ajph.90.5.707; John Duffy, "The American Medical Profession and Public Health: From Support to Ambivalence," *Bulletin of the History of Medicine* 53, no. 1 (1979): 1–22.

21. Duffy, *The Sanitarians*, 217–218; citing Louis G. Stirling, "Tendencies of the Times, Medical and Otherwise," *New Orleans Medical and Surgical Journal* 72 (1920): 218–20.
22. Duffy, *The Sanitarians*, 217.
23. Colgrove, *State of Immunity*, 99–100.
24. Elena Conis, *Vaccine Nation* (Chicago: University of Chicago Press, 2015). Conis cites Sydney Halpern as arguing that this interventionist focus on children achieving "normal" development was targeted especially at the working class. See Sydney A. Halpern, *American Pediatrics: The Social Dynamics of Professionalism, 1880–1980* (Berkeley: University of California Press, 1988), cited by Conis at 24.
25. Elena Conis, *Vaccine Nation*, 36.
26. Colgrove, *State of Immunity*, 101–02.
27. Duffy, *The Sanitarians*, 276–77.
28. One of the AMA's efforts to resist Medicare involved hiring an actor named Ronald Reagan (later the governor of California and the President of the United States) to make a propaganda recording to spread the message that federal provision of health care was the first step toward socialist dictatorship. You can still find recordings of this speech online. Reagan claims that, were Medicare to be passed, "we are going to spend our sunset years telling our children and our children's children, what it once was like in America when men were free." Roger Lowenstein, "A Question of Numbers," *The New York Times*, January 16, 2005; Reagan speech: "Ronald Reagan Speaks out on Socialized Medicine," YouTube, 1961, https://www.youtube.com/watch?v=AYrl DlrLDSQ.
29. Colgrove, *State of Immunity*, 101.
30. Colgrove, *State of Immunity*, 171; Herber A. Schreier, "On the Failure to Eradicate Measles," *The New England Journal of Medicine* 290, no. 14 (1974): 803–04, https://doi.org/10.1056/nejm197404042901412. William J. Dougherty, "Community Organization for Immunization Programs," *Medical Clinics of North America* 51 (1967): 837–42.
31. Colgrove, *State of Immunity*, 123.
32. Robert Pear, "Clinton's Health Plan: A.M.A. Rebels over Health Plan in Major Challenge to President," *The New York Times*, September 30, 1993, https://www.nytimes.com/1993/09/30/us/clinton-s-health-plan-ama-rebels-over-health-plan-major-challenge-president.html. Dana Priest and Michael Weisskopf, "AMA Split on Clinton Health Plan," *The Washington Post*, December 6, 1993, https://www.washingtonpost.com/archive/politics/1993/12/06/ama-split-on-clinton-health-plan/97c9f379-c9c0-4467-b991-b7bc8f8839c5.
33. Steven Johnson, "AMA Maintains Its Opposition to Single-Payer Systems," June 11, 2019, https://www.modernhealthcare.com/physicians/ama-maintains-its-opposition-single-payer-systems.
34. "H.R. 2264—Omnibus Budget Reconciliation Act of 1993, United States Congress, 1993; "VFC Childhood Vaccine Supply Policy," Centers for Disease Control and Prevention, 2016, accessed July 29, 2022, https://www.cdc.gov/vaccines/programs/vfc/about/vac-supply-policy/index.html#:~:text=The%20Omnibus%20Budget%20Reconciliation%20Act,funding%20for%20the%20VFC%20program.

35. "The Vaccines for Children Program: At a Glance," Centers for Disease Control and Prevention, 2016, accessed July 29, 2022, https://www.cdc.gov/vaccines/programs/vfc/about/index.html.

36. Larry K. Pickering et al.,"Immunization Programs for Infants, Children, Adolescents, and Adults: Clinical Practice Guidelines by the Infectious Diseases Society of America," *Clinical Infectious Diseases* 49, no. 6 (2009): 817–40, https://doi.org/10.1086/605430; "Pediatricians Overwhelmingly Support Vaccines for Children Program," American Academy of Pediatrics, 2020, https://www.aap.org/en/news-room/news-releases/pediatrics2/2020/pediatricians-overwhelmingly-support-vaccines-for-children-program.

37. Sean T. O'Leary et al., "Pediatricians' Experiences with and Perceptions of the Vaccines for Children Program," *Pediatrics* 145, no. 3 (2020): e20191207.

38. "The Vaccines for Children Program," Centers for Disease Control & Prevention.

39. "H.R. 3590—Patient Protection and Affordable Care Act," United States Congress, 2010; "Preventive Care Benefits for Children," U.S. Department of Health & Human Services, 2022, accessed July 29, 2022, https://www.hhs.gov/healthcare/index.html.

40. Colgrove, *State of Immunity*, 62; citing American Association for Medical Progress, *Smallpox—A Preventable Disease* (New York: American Association for Medical Progress, 1924), 8–9.

41. Henry Farrell, "The Travesty of Liberalism," Crooked Timber, March 21, 2018, accessed July 29, 2022, https://crookedtimber.org/2018/03/21/liberals-against-progressives/#comment-729288.

42. Bernice L. Hausman, *Anti/Vax: Reframing the Vaccination Controversy* (Ithaca, NY: Cornell University Press, 2019), 146–147. Hausman is drawing from Nikolas Rose, *The Politics of Life Itself: Biomedicine, Power, and Subjectivity in the Twenty-First Century* (Princeton, NJ: Princeton University Press, 2007).

43. Richard Altenbaugh, *Vaccination in America: Medical Science and Children's Welfare.* (London: Palgrave Macmillan 2018), 69.

44. Altenbaugh, *Vaccination in America*, 71; see, for example, Sinclair Lewis, *Arrowsmith* (New York: Harcourt, Brace & World, 1952); Paul De Kruif, *Microbe Hunters* (San Diego, CA: Harcourt Brace Jovanovich, 1954).

45. Altenbaugh, *Vaccination in America*, 74–77.

46. John Ehrenreich, *The Cultural Crisis of Modern Medicine* (New York: Monthly Review Press, 1978); R. D. Laing, *The Divided Self: A Study of Sanity and Madness* (London: Tavistock, 1960); Thomas Szasz, *The Myth of Mental Illness* (New York: Harper & Row, 1974); Nick Crossley, "RD Laing and the British Anti-Psychiatry Movement: A Socio-Historical Analysis," *Social Science & Medicine* 47, no. 7 (1998): 877–89.

47. Kenneth Zola Irving, "Medicine as an Institution of Social Control," in *The Cultural Crisis of Modern Medicine*, edited by John Ehrenreich (New York: Monthly Review Press, 1978), 80; cited in Hausman, *Anti/Vax*, 136.

48. See, for example, Mark A. Largent, *Vaccine: The Debate in Modern America* (Baltimore, MD: Johns Hopkins University Press, 2012); Conis, *Vaccine Nation*.

49. Hausman, *Anti/Vax*, 151.

50. Alberto Giubilini, Sharyn Milnes, and Julian Savulescu, "The Medical Ethics Curriculum in Medical Schools: Present and Future," *Journal of Clinical Ethics* 27, no. 2 (2016): 129–45; Shimon M. Glick, "The Teaching of Medical Ethics to Medical Students," *Journal of Medical Ethics* 20, no. 4 (1994): 239–43; Steven Miles et al., "Medical Ethics Education: Coming of Age," *Academic Medicine* 64, no. 12 (1989): 705–14; Frederic W. Hafferty and Ronald Franks, "The Hidden Curriculum, Ethics Teaching, and the Structure of Medical Education," *Academic Medicine*, 69, no. 11 (1994): 861–71; Ellen Fox, Robert M. Arnold, and Baruch Brody, "Medical Ethics Education: Past, Present, and Future." *Academic Medicine* 70, no. 9 (1995): 761–68.

51. See note 14.

52. Robert Blendon et al., "Americans' Conflicting Views About the Public Health System, and How to Shore up Support," *Health Affairs* 29, no. 11 (2010): 2033–40, https://doi.org/10.1377/hlthaff.2010.0262.

53. Mary G. Findling, Robert J. Blendon, and John M. Benson, "Polarized Public Opinion About Public Health During the COVID-19 Pandemic: Political Divides and Future Implications," *JAMA Health Forum* 3, no. 3 (2022): e220016.

54. See, for example, Jennifer Margulis, *The Business of Baby: What Doctors Don't Tell You, What Corporations Try to Sell You, and How to Put Your Pregnancy, Childbirth, and Baby Before Their Bottom Line* (New York: Scribner, 2013); Robert Mendelsohn, *How to Raise a Healthy Child in Spite of Your Doctor: One of America's Leading Pediatricians Puts Parents Back in Control of Their Children's Health* (New York: Ballantine, 1987); Rachel Weaver, *Be Your Child's Pediatrician* (Share-A-Care Publications, 2016).

55. Douglas S. Diekema, "Rhetoric, Persuasion, Compulsion, and the Stubborn Problem of Vaccine Hesitancy," *Perspectives in Biology and Medicine* 65, no. 1 (2022): 106–23.

56. Diekema, "Rhetoric, Persuasion, Compulsion,", 119.

57. "About Paul A. Offit, MD," Children's Hospital of Philadelphia, 2022, https://www.chop.edu/doctors/offit-paul-a; Sharyl Attkisson, "How Independent Are Vaccine Defenders?" CBS News, July 25, 2008, https://www.cbsnews.com/news/how-independent-are-vaccine-defenders/; Priyanka Boghani, "Dr. Paul Offit: 'A Choice Not to Get a Vaccine Is Not a Risk-Free Choice,'" *Frontline*, March 23, 2015, https://www.pbs.org/wgbh/frontline/article/paul-offit-a-choice-not-to-get-a-vaccine-is-not-a-risk-free-choice/.

58. "About Peter Jay Hotez, M.D., Ph.D.," Baylor College of Medicine, 2022, https://www.bcm.edu/people-search/peter-hotez-23229.

59. In an amendment to a financial disclosure for a 2012 *JAMA* publication, Offit writes, "I would like to disclose that I am the co-inventor and co-patent holder of the bovine–human reassortant rotavirus vaccine Rotateq. My hospital sold the patent more than 3 years ago, but I retained interest in the vaccine through the Wistar Institute. I sold that interest 2 years ago and do not currently receive royalties from the sales of Rotateq. In addition, I have been the author of a number of books related to vaccine safety; all profits from the sales of these books are donated to charity"; Paul A. Offit, "Incomplete Financial Disclosure in a Viewpoint on Complementary and Alternative Therapies," *JAMA* 308, no. 5 (2012): 454. A 2021 *JAMA* disclosure statement by Hotez reads, "Dr. Hotez reported being a developer of vaccines against COVID-19 and other

coronaviruses as well as neglected tropical diseases, including Chagas disease, hookworm, schistotomiasis, and leishmaniasis, in which the vaccine technology is owned by Baylor College of Medicine and nonexclusively licensed to Biological E, one of India's big vaccine manufacturers, that are either in development or clinical trials, for which he has not received compensation or remuneration"; Peter J. Hotez and K. M. Venkat Narayan, "Restoring Vaccine Diplomacy," *JAMA* 325 no. 23 (2021): 2337–38.

60. Katharina T. Paul and Kathrin Loer, "Contemporary Vaccination Policy in the European Union: Tensions and Dilemmas," *Journal of Public Health Policy* 40 (2019): 166–179, https://doi.org/10.1057/s41271-019-00163-8; Charles Allan McCoy, "Adapting Coercion: How Three Industrialized Nations Manufacture Vaccination Compliance," *Journal of Health Politics, Policy and Law* 44, no. 6 (2019): 823–54, https://doi.org/10.1215/03616878-7785775; Espen Gamlund et al., "Mandatory Childhood Vaccination: Should Norway Follow?" *Etikk i praksis—Nordic Journal of Applied Ethics* 14, no. 1 (2020): 7–27, https://doi.org/10.5324/eip.v14i1.3316; Emma Cave, "Voluntary Vaccination: The Pandemic Effect," *Legal Studies* 37, no. 2 (2017): 279–304.

61. Paul and Loer, "Contemporary Vaccination Policy in the European Union"; McCoy, "Adapting Coercion"; Cave, "Voluntary Vaccination: The Pandemic Effect."

62. We note that "mixed models" operate in countries such as Italy and Australia, with publicly funded vaccines and relatively robust public health institutions underscored by increasingly coercive vaccine mandates that seek to impose consequences on vaccine refusers.

63. Centers for Disease Control and Prevention, tweet, accessed March 11, 2022, https://twitter.com/cdcgov/status/1475587321277956096?lang=en.

64. Raymond Kluender et al., "Medical Debt in the US, 2009–2020," *JAMA* 326, no. 3 (2021): 250–56, https://doi.org/10.1001/jama.2021.8694.

65. "Costs for a Hospital Stay for COVID-19," FAIR Health, April 23, 2021, http://www.fairhealth.org/article/costs-for-a-hospital-stay-for-covid-19.

66. John A. Graves, Khrysta Baig, and Melinda Buntin, "The Financial Effects and Consequences of COVID-19: A Gathering Storm," *JAMA* 326, no. 19 (2021): 1909–10, https://doi.org/10.1001/jama.2021.18863.

67. Theodore Iwashyna et al., "Continuing Cardiopulmonary Symptoms, Disability, and Financial Toxicity 1 Month After Hospitalization for Third-Wave COVID-19: Early Results from a US Nationwide Cohort," *Journal of Hospital Medicine* 16, no. 9 (2021): 531–37.

Chapter 8

1. "Senate Health Committee, Wednesday, April 8, 2015," California Senate, 2015, accessed December 31, 2021, https://www.senate.ca.gov/media/senate-health-committee-53/video.

2. Scott Anderson, "Coercion," in *The Stanford Encyclopedia of Philosophy*, edited by Edward N. Zalta (Stanford, CA: Stanford University, 2014), http://plato.stanford.edu/archives/spr2014/entries/coercion.

3. Tracey Chantler, Emilie Karafillakis, and James Wilson, "Vaccination: Is There a Place for Penalties for Non-Compliance?" *Applied Health Economics and Health Policy* 17, no. 3 (2019): 265–71, https://doi.org/10.1007/s40258-019-00460-z.

4. Lainie Friedman Ross, *Children, Families, and Health Care Decision Making* (Oxford, UK: Oxford University Press, 1998); Joel Feinberg, "The Child's Right to an Open Future," in *Whose Child? Children's Rights, Parental Authority and State Power*, edited by William Aiken and Hugh La Follette, 124–53 (Totowa, NJ: Rowman & Littlefield, 1980); Douglas S. Diekema, "Parental Refusals of Medical Treatment: The Harm Principle as Threshold for State Intervention," *Theoretical Medicine and Bioethics* 25, no. 4 (2004): 243–64.

5. Lynn Gillam, "The Zone of Parental Discretion: An Ethical Tool for Dealing with Disagreement Between Parents and Doctors About Medical Treatment for a Child," *Clinical Ethics* 11, no. 1 (2016): 1–8, https://doi.org/https://doi.org/10.1177/14777 50915622033.

6. Jeffrey Blustein, *Parents and Children: The Ethics of the Family* (Oxford, UK: Oxford University Press, 1982); David Archard, "The Obligations and Responsibilities of Parenthood," in *Procreation and Parenthood: The Ethics of Bearing and Rearing Children*, edited by David Archard and David Benatar, 103–26 (Oxford, UK: Oxford University Press, 2010). Samantha Brennan and Robert Noggle, "The Moral Status of Children: Children's Rights, Parents' Rights, and Family Justice," *Social Theory and Practice* 23, no. 1 (1997): 1–26.

7. Harry Brighouse and Adam Swift, *Family Values: The Ethics of Parent–Child Relationships* (Princeton, NJ: Princeton University Press, 2016); Daniel Groll, "Four Models of Family Interests," *Pediatrics* 134, Supplement 2 (2014): S81–S86; Paul Baines, "Family Interests and Medical Decisions for Children," *Bioethics* 31, no. 8 (2017): 599–607, https://doi.org/10.1111/bioe.12376; Lainie Friedman Ross and Alissa Hurwitz Swota, "The Best Interest Standard: Same Name but Different Roles in Pediatric Bioethics and Child Rights Frameworks," *Perspectives in Biology and Medicine* 60, no. 2 (2017): 186–97, https://doi.org/10.1353/pbm.2017.0027; Lynne Bower, "The Ethical Grounds for the Best Interest of the Child," *Cambridge Quarterly of Healthcare Ethics* 25, no. 1 (2016): 63–69, https://doi.org/10.1017/s096318011 5000298.

8. Paul A. Offit, *Bad Faith: When Religious Belief Undermines Modern Medicine* (New York: Basic Books, 2015).

9. Loretta M. Kopelman, "The Best-Interests Standard as Threshold, Ideal, and Standard of Reasonableness," *Journal of Medicine and Philosophy* 22, no. 3 (1997): 271–89, https://doi.org/10.1093/jmp/22.3.271; Erica K. Salter, "Deciding for a Child: A Comprehensive Analysis of the Best Interest Standard," *Theoretical Medicine and Bioethics* 33, no. 3 (2012): 179–98; Johan Christiaan Bester, "The Harm Principle Cannot Replace the Best Interest Standard: Problems with Using the Harm Principle for Medical Decision Making for Children," *American Journal of Bioethics* 18, no. 8

(2018): 9–19, https://doi.org/10.1080/15265161.2018.1485757; Lainie Friedman Ross, "Better Than Best (Interest Standard) in Pediatric Decision Making," *Journal Clinical Ethics* 30, no. 3 (2019): 183–95.

10. Johan Christiaan Bester, "Not a Matter of Parental Choice but of Social Justice Obligation: Children Are Owed Measles Vaccination," *Bioethics* 32, no. 9 (2018): 611–19; Douglas S. Diekema, "Physician Dismissal of Families Who Refuse Vaccination: An Ethical Assessment," *Journal of Law, Medicine & Ethics* 43, no. 3 (2015): 654–60; Douglas S. Diekema and the American Academy of Pediatrics Committee on Bioethics, "Responding to Parental Refusals of Immunization of Children," *Pediatrics* 115, no. 5 (2005): 1428–31; American Academy of Pediatrics, "Reaffirmation: Responding to Parents Who Refuse Immunization for Their Children," *Pediatrics* 131, no. 5 (2013): e1696–e96, https://doi.org/10.1542/peds.2013-0430; Kathryn M. Edwards et al., "Countering Vaccine Hesitancy," *Pediatrics* 138, no. 3 (2016): e1, https://doi.org/10.1542/peds.2016-2146.

11. The language about defeasible rights has sometimes been complicated by inaccuracy in language. Ross famously called such rights *prima facie* (W. D. Ross, *The Right and the Good*, 2nd ed., edited by Philip Stratton-Lake (Oxford, UK: Oxford University Press, 1930), but that is a misnomer because the rights Ross discussed were not merely apparent but were always rights. His point was that those rights could sometimes be infringed upon, which is to say that they are pro tanto. Hence, we describe actual rights that can be infringed upon as pro tanto rights.

12. Barbara Loe Fisher, "The Moral Right to Conscientious, Philosophical and Personal Belief Exemption to Vaccination," National Vaccine Information Center, 1997, https://www.nvic.org/vaccination-decisions/informed-consent.

13. Ruth R. Faden, Tom L. Beauchamp, and Nancy M. P. King, *A History and Theory of Informed Consent* (New York: Oxford University Press, 1986); American Society for Bioethics and Humanities, *Improving Competencies in Clinical Ethics Consultation: An Education Guide* (Chicago: American Society for Bioethics and Humanities, 2015), 25.

14. For example, the United Nations Committee on Economic, Social and Cultural Rights has found that the right to health identified in the International Convention on Economic, Social and Cultural Rights includes "the right to control one's health and body . . . such as the right to be free from torture, non-consensual medical treatment and experimentation"; UN Committee on Economic, Social and Cultural Rights, *Economic and Social Council Official Records, Report on the Twenty-Second, Twenty-Third and Twenty-Fourth Sessions, Supplement 2* (New York: United Nations, 2001), 130.

15. "Vaccine Information Statements (VISs)," Centers for Disease Control and Prevention, 2022, https://www.cdc.gov/vaccines/hcp/vis/index.html.

16. Mary Holland testimony at California Senate Judiciary Committee, April 28, 2015, https://www.senate.ca.gov/media/senate-judiciary-committee-38/video, transcribed by the authors. Holland's testimony is from 3:26:09 to 3:39:33. The transcribed portion begins at 3:29:19.

17. Dorit Reiss (law professor and parent activist) interview with the author, June 2019.

18. As Jessica Flanigan argues (following Onora O'Neill), "the principle of informed consent does not go so far as to justify harming others with one's medical choices"; Jessica Flanigan, "A Defense of Compulsory Vaccination," *HEC Forum* 26, no. 1 (2014): 5–25, 17; Onora O'Neill, "Accountability, Trust and Informed Consent in Medical Practice and Research," *Clinical Medicine* 4, no. 3 (2004): 269–76.

19. Margaret Battin, Leslie Francis, Jay Jacobson, and Charles Smith, *The Patient as Victim and Vector: Ethics and Infectious Disease* (Oxford, UK: Oxford University Press, 2009).

20. American Academy of Pediatrics Committee on Bioethics, "Policy Statement: Informed Consent in Decision Making in Pediatric Practice," *Pediatrics* 138, no. 2 (2016), e20161495; Aviva L. Katz and Sally A. Webb, "Informed Consent in Decision-Making in Pediatric Practice," *Pediatrics* 138, no. 2 (2016): e20161485, https://doi.org/10.1542/peds.2016-1485.

21. Mark C. Navin and Jason Adam Wasserman, "Reasons to Amplify the Role of Parental Permission in Pediatric Treatment," *American Journal of Bioethics* 17, no. 11 (2017): 6–14, https://doi.org/10.1080/15265161.2017.1378752.

22. "Senate Health Committee, Wednesday, April 8, 2015," California Senate, https://www.senate.ca.gov/media/senate-health-committee-53/video, accessed December 31, 2021.

23. Stefano Crenna, Antonio Osculati, and Silvia D. Visonà, "Vaccination Policy in Italy: An Update," *Journal of Public Health Research* 7, no. 3 (2018): 1523–23, https://doi.org/10.4081/jphr.2018.1523.

24. "California Health and Safety Code; 120335(h)," California Legislative Information, 2016. For discussion of the IEP exemption in the Nonmedical Exemptions Bill, see Ross D. Silverman and Wendy F. Hensel, "Squaring State Child Vaccine Policy with Individual Rights Under the Individuals with Disabilities Education Act: Questions Raised in California," *Public Health Reports* 132, no. 5 (2017): 593–96, https://www.jstor.org/stable/26374172; Salini Mohanty et al., "California's Senate Bill 277: Local Health Jurisdictions' Experiences with the Elimination of Nonmedical Vaccine Exemptions," *American Journal of Public Health* 109, no. 1 (2019): 96–101; Dorit Reiss, "Litigating Alternative Facts: School Vaccine Mandates in the Courts," *University of Pennsylvania Journal of Constitutional Law* 21, no. 1 (2018): 207.

25. Katie Attwell and Mark C. Navin, "How Policymakers Employ Ethical Frames to Design and Introduce New Policies: The Case of Childhood Vaccine Mandates in Australia," *Policy & Politics* 50, no. 4 (2022): 526–47.

26. Roland Pierik, "Mandatory Vaccination: An Unqualified Defence," *Journal of Applied Philosophy* 35, no. 2 (2018): 381–98, https://doi.org/doi:10.1111/japp.12215; see also Bester, "Not a Matter of Parental Choice."

27. Mark C. Navin, *Values and Vaccine Refusal: Hard Questions in Ethics, Epistemology and Health Care* (New York: Routledge, 2016).

28. Diane DePanfilis and Marsha K. Salus, *Child Protective Services: A Guide for Caseworkers* (Washington, DC: U.S. Department of Health and Human Services, Administration for Children and Families, Administration on Children, Youth and Families, Children's Bureau, Office on Child Abuse and Neglect, 2003;

Insoo Kim Berg and Susan Kelly, *Building Solutions in Child Protective Services* (New York: Norton, 2000).

29. Diekema, "Parental Refusals of Medical Treatment."

30. "OHCHR Dashboard," United Nations, March 9, 2022, https://indicators.ohchr.org.

31. Karen Attiah, "Why Won't the U.S. Ratify the U.N.'s Child Rights Treaty?" *The Washington Post*, November 21, 2014, https://www.washingtonpost.com/blogs/post-partisan/wp/2014/11/21/why-wont-the-u-s-ratify-the-u-n-s-child-rights-treaty; Luisa Blanchfield, "The United Nations Convention on the Rights of the Child," Congressional Research Service, July 27, 2015, https://crsreports.congress.gov/product/pdf/R/R40484/25. Recall also Capitol Resource Institute and its mobilization to prevent minors from consenting to their own COVID-19 vaccines discussed in Chapter 5. "Why Us?" Capitol Resource Institute, 2022, accessed September 23, 2022, https://www.capitolresource.org/why-us.

32. Stephanie Liou et al., "Impact of the COVID-19 Pandemic on Pediatric Firearm-Related Injuries in the USA," *Pediatrics* 147, no. 3 (2021): 103–05.

33. "Surprising Dangers of Trampolines for Kids," Cleveland Clinic, 2020, accessed March 11, 2022, https://health.clevelandclinic.org/surprising-dangers-of-trampolines-for-kids.

34. Late 19th-century advocates for children's rights wanted to protect children and families from external threats to the family from businesses and government. One of their core motivations was to protect orphans from being used in medical experiments, which often happened because doing so was "cheaper than [using] animals" (Richard J. Altenbaugh, *Vaccination in America: Medical Science and Children's Welfare* [London: Palgrave Macmillan, 2018], 59; Altenbaugh is here drawing on American Human Association, *Human Vivisection: A Statement and an Inquiry* [Chicago: American Human Association, 1899]). Early advocacy of children's rights was part of a broader rejection of nonconsensual, nontherapeutic experiments on captive populations, including children, unconscious patients in hospitals, and animals (Altenbaugh, *Vaccination in America*, 59–60; Altenbaugh is here drawing on James M. Brown, *The Reality of Human Vivisection: A Review* [1901]: folder 1, Box 16, RG 600-1 Anti-Vivisection Papers, Rockefeller University Archives). Indeed, Henry Bergh, who in 1874 co-founded the first child protective agency in the world—The Society for Prevention of Cruelty to Children—was in 1866 the founder of the world's first organization to advocate for animal rights, the American Society for Prevention of Cruelty to Animals. In both cases, he was motivated to resist unethical medical experiments. Bergh and other 19th-century children's rights advocates also objected to vaccination because they considered it to be an ongoing medical research project involving children, and they demanded that parents always retain the right to determine if their children should participate. These early advocates of children's rights rejected vaccine mandates because they believed coercive vaccination was an unjust intrusion by industry and government into family life. The cultural meaning of children's rights has clearly shifted in the past 150 years in ways that have often bypassed parental authority. But it seems likely that many Americans still think that children's rights are primarily about protecting families from outside threats, rather

than invitations for government to make decisions for children against parents' wishes.

35. Navin, *Values and Vaccine Refusal*; Flanigan, "A Defense of Compulsory Vaccination"; Jason Brennan, "A Libertarian Case for Mandatory Vaccination," *Journal of Medical Ethics* 44, no. 1 (2018): 37–43; Franklin G. Miller, "Liberty and Protection of Society During a Pandemic: Revisiting John Stuart Mill," *Perspectives in Biology and Medicine* 64, no. 2 (2021): 200–10; Tim Dare, "Mass Immunisation Programmes: Some Philosophical Issues," *Bioethics* 12, no. 2 (1998): 125–49, https://doi.org/10.1111/1467-8519.00100; Euzebiusz Jamrozik, Toby Handfield, and Michael J. Selgelid, "Victims, Vectors and Villains: Are Those Who Opt out of Vaccination Morally Responsible for the Deaths of Others?" *Journal of Medical Ethics* 42, no. 12 (2016): 762–68, https://doi.org/10.1136/medethics-2015-103327.

36. Flanigan, "A Defense of Compulsory Vaccination"; Brennan, "A Libertarian Case for Mandatory Vaccination."

37. John Stuart Mill, *On Liberty* (London: Parker & Son, 1859).

38. Lawrence O. Gostin, "When Terrorism Threatens Health: How Far Are Limitations on Personal and Economic Liberties Justified?" *Florida Law Review* 55, no. 5 (2003): 1105–70, 1148.

39. Marcel Verweij and Angus Dawson, "Ethical Principles for Collective Immunisation Programmes," *Vaccine* 22, no. 23 (2004): 3122–26, 3123.

40. Giubilini, Douglas, and Savulescu argue that causal impotence of individual acts of vaccine refusal does not undermine an individual's *moral obligation* to vaccinate (Alberto Giubilini, Thomas Douglas, and Julian Savulescu, "The Moral Obligation to Be Vaccinated: Utilitarianism, Contractualism, and Collective Easy Rescue," *Medicine, Health Care, and Philosophy* 21, no. 4 (2018): 547–60, https://doi.org/10.1007/s11019-018-9829-y. However, our focus is on the permissibility of *state coercion* to promote vaccination, and Giubilini et al. explicitly bracket that topic.

41. Alberto Giubilini, *The Ethics of Vaccination* (London: Palgrave Macmillan, 2019); Roland Pierik and Marcel Verweij, "Inducing Immunity," unpublished manuscript, 2022; Navin, *Values and Vaccine Refusal*.

42. Mariëtte Van den Hoven, "Why One Should Do One's Bit: Thinking About Free Riding in the Context of Public Health Ethics," *Public Health Ethics* 5, no. 2 (2012): 154–60, https://doi.org/10.1093/phe/phs023; Giubilini et al., "The Moral Obligation to Be Vaccinated."

43. H. L. A. Hart, "Are There Any Natural Rights?" *The Philosophical Review* 64, no. 2 (1955): 175–91, https://doi.org/10.2307/2182586; John Rawls, *A Theory of Justice*, (Cambridge, MA: Harvard University Press; 1971); George Klosko, *The Principle of Fairness and Political Obligation* (Lanham, MD: Rowman & Littlefield, 1992); Jonathon Wolff, "Political Obligation, Fairness, and Independence," *Ratio* 8 (1995): 87–99.

44. Robert Nozick, *Anarchy, the State, and Utopia* (New York: Basic Books, 1974).

45. Heidi Malm and Mark C. Navin, "Pox Parties for Grannies? Chickenpox, Exogenous Boosting, and Harmful Injustices," *American Journal of Bioethics* 20, no. 9 (2020): 45–57, https://doi.org/10.1080/15265161.2020.1795528.

46. Alberto Giubilini, "An Argument for Compulsory Vaccination: The Taxation Analogy," *Journal of Applied Philosophy* 37, no.3 (2020): 446–66.

47. For further skepticism about fairness-based enforceable obligations to vaccinate, see Ethan Bradley and Mark C. Navin, "Vaccine Refusal Is Not Free Riding," *Erasmus Journal for Philosophy and Economics* 14, no. 1 (2021): 167–81, https://doi.org/10.23941/ejpe.v14i1.555.

48. "Our View: Vaccine Mandates Serve the Greater Good in Ending a Never-Ending Pandemic," *USA Today*, October 1, 2021, accessed December 22, 2021, https://www.lohud.com/story/opinion/editorials/2021/10/01/gannett-new-york-editorial-support-covid-19-vaccination-mandates/5936672001/; Arwa Mahdawi, "Telling Anti-Vaxxers to Get the Jab Should Not Be Controversial—Even Fox News Is Doing It," *The Guardian*, October 27, 2021, https://www.theguardian.com/commentisfree/2021/oct/27/telling-anti-vaxxers-to-get-the-jab-should-not-be-controversial-even-fox-news-is-doing-it; Dennis Wagner, "The COVID Culture War: At What Point Should Personal Freedom Yield to the Common Good?" *USA Today*, August 2, 2021, accessed December 22, 2021, https://www.usatoday.com/story/news/nation/2021/08/02/covid-culture-war-masks-vaccine-pits-liberty-against-common-good/5432614001.

49. Katie Attwell and Mark C. Navin, "How Policymakers Employ Ethical Frames to Design and Introduce New Policies: The Case of Childhood Vaccine Mandates in Australia," *Policy & Politics* 50, no. 4 (2022): 526–47, https://doi.org/10.1332/030557321X16476002878591.

50. Walter Sinnott-Armstrong, "Consequentialism," in *The Stanford Encyclopedia of Philosophy*, edited by Edward N. Zalta (Stanford, CA: Metaphysics Research Lab, Stanford University, 2021), https://plato.stanford.edu/archives/fall2021/entries/consequentialism.

51. Julia Driver, "The History of Utilitarianism," In *The Stanford Encyclopedia of Philosophy*, edited by Edward N. Zalta (Stanford, CA: Metaphysics Research Lab, Stanford University, 2014), https://plato.stanford.edu/archives/win2014/entries/utilitarianism-history.

52. Stephen Holland, *Public Health Ethics* (Hoboken, NJ: Wiley, 2015).

53. Mark A. Rothstein, "Are Traditional Public Health Strategies Consistent with Contemporary American Values?" *Temple Law Review* 77, no. 2 (2004): 175–92.

54. Stuart J. Horner, "For Debate. The Virtuous Public Health Physician," *Journal of Public Health* 22, no. 1 (2000): 48–53, https://doi.org/10.1093/pubmed/22.1.48.

55. Dare, "Mass Immunisation Programmes."

56. Julian Savulescu, Igmar Persson, and Dominic Wilkinson, "Utilitarianism and the Pandemic," *Bioethics* 34, no. 6 (2020): 620–32, https://doi.org/10.1111/bioe.12771.

57. Timothy Martin Wilkinson, "Making People Be Healthy," *Journal of Primary Health Care* 1, no. 3 (2009): 244–46, 245, https://doi.org/10.1071/hc09244; Griffin Trotter, "COVID-19 and the Authority of Science," *HEC Forum* (2021): 1–28.

58. Teck Chuan Voo, Hannah Clapham, and Clarence C. Tam, "Ethical Implementation of Immunity Passports During the COVID-19 Pandemic," *Journal of Infectious Diseases* 222, no. 5 (2020): 715–18, https://doi.org/10.1093/infdis/jiaa352; Jeroen

Luyten and Philippe Beutels, "The Social Value of Vaccination Programs: Beyond Cost-Effectiveness," *Health Affairs* 35, no. 2 (2016): 212–18, https://doi.org/10.1377/hlthaff.2015.1088; Federica Angeli, Silvia Camporesi, and Giorgia Dal Fabbro, "The COVID-19 Wicked Problem in Public Health Ethics: Conflicting Evidence, or Incommensurable Values?" *Humanities & Social Sciences Communications* 8, no. 1 (2021): 1–8, https://doi.org/10.1057/s41599-021-00839-1; Douglas J. Opel, Douglas S. Diekema, and Lainie Friedman Ross, "Should We Mandate a COVID-19 Vaccine for Children?" *JAMA Pediatrics* 175, no. 2 (2021): 125–26, https://doi.org/10.1001/jamapediatrics.2020.3019.

59. Mark C. Navin and Katie Attwell, "Vaccine Mandates, Value Pluralism, and Policy Diversity," *Bioethics* 33, no. 9 (2019): 1042–49, https://doi.org/10.1111/bioe.12645.

60. Rawls, *A Theory of Justice*, sections 5, 6, 29.

61. Rawls, *A Theory of Justice*, 24

62. Tom L. Beauchamp and James F. Childress, *Principles of Biomedical Ethics* (New York: Oxford University Press, 2001).

63. Courtney Thornton and Jennifer Reich, "Black Mothers and Vaccine Refusal: Gendered Racism, Healthcare, and the State," *Gender & Society* 36, no. 4 (2022): 525–51.

64. Peter Schröder-Bäck et al., "Teaching Seven Principles for Public Health Ethics: Towards a Curriculum for a Short Course on Ethics in Public Health Programmes," *BMC Medical Ethics* 15, no. 1 (2014): 73, https://doi.org/10.1186/1472-6939-15-73; Martin White, Jean Adams, and Peter Heywood, "How and Why Do Interventions That Increase Health Overall Widen Inequalities Within Populations?" in *Health, Inequality and Public Health*, edited by S. Barbones, 65–82 (Bristol: Policy Press, 2009).

65. Madison Powers and Ruth R. Faden, *Social Justice, the Moral Foundations of Public Health and Health Policy* (Oxford, UK: Oxford University Press, 2006); Kathryn MacKay, "Utility and Justice in Public Health," *Journal of Public Health* 40, no. 3 (2018): e413–e18, https://doi.org/10.1093/pubmed/fdx169.

66. Emerson and Singer identify compelling reasons to aim for polio eradication: Failing to eradicate causes harm (or at least neglects a duty of rescue), eradication spares future generations further harms, and eradication is a means for fulfilling an existing commitment to public health. Claudia I. Emerson and Peter A. Singer, "Is There an Ethical Obligation to Complete Polio Eradication?" *Lancet* 375, no. 9723 (2010): 1340–41, https://doi.org/10.1016/s0140-6736(10)60565-x.

67. Nancy Stepan, *Eradication: Ridding the World of Diseases Forever?* (Ithaca, NY: Cornell University Press, 2011), 192; quoting Frank Fenner, *Smallpox and Its Eradication* (Geneva, Switzerland: World Health Organization, 1988), 259.

68. "Commemorating the 40th Anniversary of Smallpox Eradication," World Health Organization, May 8, 2020, https: //www.who.int/news-room/events/detail/2020/05/08/default-calendar/commemorating-the-40th-anniversary-of-smallpox-eradication.

69. *Catechism of the Catholic Church* (Vatican, Libreria Editrice Vaticana), pars. 1038–1041. Some of the relevant passages from the Christian Bible include Acts 24:15, John 5:28–29, and Matthew 25:31,32,46.

70. Robert T. Chen et al., "The Vaccine Adverse Event Reporting System (VAERS)," *Vaccine* 12, no. 6 (1994): 542–50, https://doi.org/10.1016/0264-410x(94)90315-8. In this first iteration of the figure, Chen and colleagues do not use the "natural history" terminology and instead refer to the "potential stages in the evolution of an immunisation programme." However, 10 years later, the figure is rebadged as a "natural history" in Robert T. Chen, "Evaluation of Vaccine Safety After the Events of 11 September 2001: Role of Cohort and Case–Control Studies," *Vaccine* 22, no. 15 (2004): 2047–53.

71. Frank Fenner et al., "What Is Eradication?" in *The Eradication of Infectious Diseases*, edited by W. R. Dowdle and Donald R. Hopkins (New York: Wiley, 1998), 11; this is the definition of the International Task Force for Disease Eradication: Centers for Disease Control, "International Task Force for Disease Eradication," *MMWR Morbidity and Mortality Weekly Review* 39 (1990): 209–217.

72. Stepan, *Eradication*, 7.

73. Stepan, *Eradication*, 15.

74. Elena Conis, *Vaccine Nation* (Chicago: University of Chicago Press, 2015), 6.

75. James Colgrove, *State of Immunity: The Politics of Vaccination in Twentieth-Century America* (Berkeley: University of California Press, 2006), 155.

76. Colgrove, *State of Immunity*, 155.

77. Colgrove, *State of Immunity*, ch. 5.

78. Colgrove, *State of Immunity*, 150.

79. Sharon Otterman, "Why Polio, Once Eliminated, Is Testing N.Y. Health Officials," *The New York Times*, October 3, 2022, https://www.nytimes.com/2022/10/03/nyregion/polio-new-york-eradication.html.

80. "Measles Cases and Outbreaks," Centers for Disease Control and Prevention, August 9, 2022, https://www.cdc.gov/measles/cases-outbreaks.html.

81. Stepan, *Eradication*, 97. Here, Stepan is quoting Malcolm Gladwell, "The Mosquito Killer," *The New Yorker*, July 2, 2001, 49.

82. Karin Brulliard, "CIA Vaccine Program Used in Bin Laden Hunt in Pakistan Sparks Criticism," *The Washington Post*, July 22, 2011, sec. World, http://www.washingtonpost.com/world/asia-pacific/pakistan-fights-polio-in-shadow-of-cia-ruse/2011/07/21/gIQAQqmcSI_story.html.

83. Katie Moisse, "The Lasting Fallout of Fake Vaccination Programs," ABC News, May 21, 2014, https://abcnews.go.com/Health/lasting-fallout-fake-vaccination-programs/story?id=23795483; "How the CIA's Fake Vaccination Campaign Endangers Us All," *Scientific American*, May 1, 2013, https://www.scientificamerican.com/article/how-cia-fake-vaccination-campaign-endangers-us-all.

84. Stepan, *Eradication*, 8.

Chapter 9

1. Olivia M. Vaz et al., "Mandatory Vaccination in Europe," *Pediatrics* 145, no. 2 (2020): e20190620, https://doi.org/10.1542/peds.2019-0620; Brynley Hull et al., "'No

Jab, No Pay': Catch-Up Vaccination Activity During Its First Two Years," *Medical Journal of Australia* 213, no. 8 (2020): 364–69, https://doi.org/10.5694/mja2.50780.

2. Paul Delamater et al., "Elimination of Nonmedical Immunization Exemptions in California and School-Entry Vaccine Status," *Pediatrics* 143 (2019): 1–9; Jacqueline K. Olive et al., "The State of the Antivaccine Movement in the United States: A Focused Examination of Nonmedical Exemptions in States and Counties," *PLoS Medicine* 15, no. 6 (2018): e1002578, https://doi.org/10.1371/journal.pmed.1002578.

3. Hillel Y Levin et al., "Stopping the Resurgence of Vaccine-Preventable Childhood Diseases: Policy, Politics, and Law," *University of Illinois Law Review* 2020, no. 1 (2020): 233–72.

4. For discussion of some of these costs, see Hillel Y. Levin, Allan J. Jacobs, and Kavita Shah Arora, "To Accommodate or Not to Accommodate: (When) Should the State Regulate Religion to Protect the Rights of Children and Third Parties," *Washington and Lee Law Review* 73 (2016): 915. Notably, it is uncommon for discussions about vaccine mandates to consider these burdens and costs. See, for example, Chris Feudtner and Edgar K. Marcuse, "Ethics and Immunization Policy: Promoting Dialogue to Sustain Consensus," *Pediatrics* 107, no. 5 (2001): 1158–64, https://doi.org/10.1542/peds.107.5.1158. Feudtner and Marcuse aim to identify the costs and benefits associated with vaccine mandates, but they never consider children who are denied access to formal education. Such an obvious oversight may result from the assumption that threats of school exclusion are effective. More generally, there is good reason to think that the cost–benefit analyses of health policy experts routinely ignore or undervalue ethics issues implicated by the policies they advocate. See Paul Menzel et al., "Toward a Broader View of Values in Cost-Effectiveness Analysis of Health," *Hastings Center Report* 29, no. 3 (1999): 7–15, https://doi.org/10.2307/3528187.

5. These are implementation issues that could be "fixed" if there were sufficient political will to do so, but private schools (and their alumni and families) tend to have substantial political power. Implementation issues are also exacerbated by a mission discrepancy between educational institutions (including their frontline workers) and public health goals. For more on this, see Mark C. Navin, Andrea T. Kozak, and Katie Attwell, "School Staff and Immunization Governance: Missed Opportunities for Public Health Promotion," *Vaccine* 40, no. 51 (2022): 7433–39, https://doi.org/https://doi.org/10.1016/j.vaccine.2021.07.061.

6. Delamater et al., "Elimination of Nonmedical Immunization Exemptions."

7. Sharon Otterman, "Why Polio, Once Eliminated, Is Testing N.Y. Health Officials," *The New York Times*, October 3, 2022, https://www.nytimes.com/2022/10/03/nyregion/polio-new-york-eradication.html.

8. Navin et al., "School Staff and Immunization Governance"; Mark C. Navin et al., "School Staff as Vaccine Advocates: Perspectives on Vaccine Mandates and the Student Registration Process," *Vaccine*, 41, no. 5 (2022): 1169–75.

9. Otterman, "Why Polio, Once Eliminated, Is Testing N.Y. Health Officials."

10. Consider, for example, that mask mandates often had very high approval ratings, especially in early stages of the COVID-19 pandemic, but that school districts across the country often rescinded or chose not to enforce mask mandates after angry

parents appeared at school board meetings. Marlene Lenthang, "How School Board Meetings Have Become Emotional Battlegrounds for Debating Mask Mandates," ABC News, August 29, 2021, https://abcnews.go.com/US/school-board-meeti ngs-emotional-battlegrounds-debating-mask-mandates/story?id=79657733;   Katie Reilly, "School Masking Mandates Are Going to Court. Here's Why the Issue Is So Complicated," *TIME*, August 30, 2021, https://time.com/6103134/parents-fight-sch ool-mask-mandates/.; Julia C. Wong, "Masks Off: How US School Boards Became 'Perfect Battlegrounds' for Vicious Culture Wars," *The Guardian*, August 24, 2021, https://www.theguardian.com/us-news/2021/aug/24/mask-mandates-covid-school-boards; Tori Powell, "Virginia Parent Charged After She Threatens to 'Bring Every Single Gun Loaded' over School's Mask Dispute," CBS News, 2022, https://www.cbsn ews.com/news/virginia-school-board-gun-threat-face-mask-dispute.

11. See, for example, Mark Largent's description of why he requested NMEs for his children. Largent supports vaccines, but his children's illnesses and his family's tragedies had kept them from completing required vaccines in time for daycare and school enrollment. Mark A. Largent, *Vaccine: The Debate in Modern America* (Baltimore, MD: Johns Hopkins University Press, 2012), 4.

12. Kavin M. Patel et al., "Evaluation of Trends in Homeschooling Rates After Elimination of Nonmedical Exemptions to Childhood Immunizations in California, 2012–2020," *JAMA Open* 5, no. 2 (2022): e2146467, https://doi.org/10.1001/jama networkopen.2021.46467; Pamela McDonald et al., "Exploring California's New Law Eliminating Personal Belief Exemptions to Childhood Vaccines and Vaccine Decision-Making Among Homeschooling Mothers in California," *Vaccine* 37, no. 5 (2019): 742–50.

13. "California Health and Safety Code; 120335(h)," California Legislative Information. For discussion of the IEP exemption in the Nonmedical Exemptions Bill, see Ross D. Silverman and Wendy F. Hensel, "Squaring State Child Vaccine Policy with Individual Rights Under the Individuals With Disabilities Education Act: Questions Raised in California," *Public Health Reports* 132, no. 5 (2017): 593–96; Salini Mohanty et al., "California's Senate Bill 277: Local Health Jurisdictions' Experiences with the Elimination of Nonmedical Vaccine Exemptions," *American Journal of Public Health* 109, no. 1 (2019): 96–101; Doris Reiss, "Litigating Alternative Facts: School Vaccine Mandates in the Courts," *University of Pennsylvania Journal of Constitutional Law* 21 (2018): 207.

14. *Whitlow v. California*, 203 F. Supp. 4d 1079 (C.D. California 2016).

15. Delamater et al., "Elimination of Nonmedical Immunization Exemptions."

16. Courtney Thornton and Jennifer Reich, "Black Mothers and Vaccine Refusal: Gendered Racism, Healthcare, and the State," *Gender & Society* 36, no. 4 (2022): 525–51, 539.

17. Julie Leask and Margie Danchin, "Imposing Penalties for Vaccine Rejection Requires Strong Scrutiny," *Journal of Paediatrics and Child Health* 53 (2007): 439–44.

18. Charles Bruner, Anne Discher, and Hedy Chang, "Chronic Elementary Absenteeism: A Problem Hidden in Plain Sight," Attendance Works and Child & Family Policy Center,

2011, https://ies.ed.gov/ncee/edlabs/regions/west/relwestFiles/pdf/508_Chronic_ Elementary_Absence_AW_C_FPC_2011.pdf; Catherine E. Ross and Chia-ling Wu, "The Links Between Education and Health," *American Sociological Review* 60, no. 5 (1995): 719–45, https://doi.org/10.2307/2096319; Richard G. Rogers, Robert A. Hummer, and Bethany G. Everett, "Educational Differentials in U.S. Adult Mortality: An Examination of Mediating Factors," *Social Science Research* 42, no. 2 (2013): 465–81, https://doi.org/10.1016/j.ssresearch.2012.09.003; Elizabeth M. Lawrence, Richard G. Rogers, and Anna Zajacova, "Educational Attainment and Mortality in the United States: Effects of Degrees, Years of Schooling, and Certification," *Population Research and Policy Review* 35, no. 4 (2016): 501–25, https://doi.org/10.1007/s11113-016-9394-0.

19. Katie Attwell and Mark C. Navin, "How Policymakers Employ Ethical Frames to Design and Introduce New Policies: The Case of Childhood Vaccine Mandates in Australia," *Policy & Politics* 50, no. 4 (2022), https://doi.org/10.1332/030557321 X16476002878591; Katie Attwell and David T. Smith, "Parenting as Politics: Social Identity Theory and Vaccine Hesitant Communities," *International Journal of Health Governance* 22, no. 3 (2017): 183–98; Leask and Danchin, "Imposing Penalties for Vaccine Rejection"; Mark C. Navin and Mark A. Largent, "Improving Nonmedical Vaccine Exemption Policies: Three Case Studies," *Public Health Ethics* 10, no. 3 (2017): 225–34, http://dx.doi.org/10.1093/phe/phw047.

20. Heidi Larson, *Stuck: How Vaccine Rumors Start and Why They Don't Go Away* (Oxford, UK: Oxford University Press, 2020), 39–41. See also "Thimerosal in Vaccines: A Joint Statement of the American Academy of Pediatrics and the Public Health Service," Centers for Disease Control and Prevention, July 9, 1999, www. cdc.gov/mmwr/preview/mmwrhtml/mm4826a3.htm; Centers for Disease Control and Prevention, "Impact of the 1999 AAP/USPHS Joint Statement on Thimerosal in Vaccines on Infant Hepatitis B Vaccination Practices," *MMWR Morbidity and Mortality Weekly Report* 50, no. 6 (2001): 94–97. Also, Jason Schwartz discusses how a similar commitment to maintain public trust in vaccines led to the withdrawal of the RotaShield rotavirus vaccine in 1999 and also how that decision resulted in similar kinds of unintended consequences as did CDC and AAP advocacy for the removal of thimerosal: Jason L. Schwartz, "The First Rotavirus Vaccine and the Politics of Acceptable Risk," *Milbank Quarterly* 90, no. 2 (2012): 278–310, https://doi.org/ 10.1111/j.1468-0009.2012.00664.x.

21. Nayanah Siva, "Thiomersal Vaccines Debate Continues Ahead of UN Meeting," *Lancet* 379, no. 9834 (2012): 2328–28, https://doi.org/10.1016/s0140-6736(12)61002-2; Walter A. Orenstein et al., "Global Vaccination Recommendations and Thimerosal," *Pediatrics* 131, no. 1 (2013): 149–51, https://doi.org/10.1542/peds.2012-1760.

22. Schwartz, "The First Rotavirus Vaccine."

23. Hannah Henry (advocate, Vaccinate California), interview with the author, June 12, 2019.

24. "Understanding Qanon's Connection to American Politics, Religion, and Media Consumption," Public Religion Research Institute, May 27, 2021, accessed February 10, 2022, https://www.prri.org/research/qanon-conspiracy-american-politics-report.

25. Claire Gecewicz, "'New Age' Beliefs Common Among Both Religious and Nonreligious Americans," Pew Research Center, October 1, 2018, accessed February 10, 2022, https://www.pewresearch.org/fact-tank/2018/10/01/new-age-beliefs-common-among-both-religious-and-nonreligious-americans.

26. Scott Neuman, "1 in 4 Americans Thinks the Sun Goes Around the Earth, Survey Says," NPR, February 14, 2014, sec: America, accessed February 10, 2022, https://www.npr.org/sections/thetwo-way/2014/02/14/277058739/1-in-4-americans-think-the-sun-goes-around-the-earth-survey-says.

27. James Colgrove and Abigail Lowin, "A Tale of Two States: Mississippi, West Virginia, and Exemptions to Compulsory School Vaccination Laws," *Health Affairs* 35, no. 2 (2016): 348–55, https://doi.org/10.1377/hlthaff.2015.1172.

28. There were many conflicts about the introduction of new vaccines, both to the recommended schedule and to the lists of vaccines mandated for school enrollment. However, these conflicts were often short-lived, and they were not aligned with Democrat versus Republican political conflicts. See, for example, Elena Conis, *Vaccine Nation* (Chicago; University of Chicago Press, 2021).

29. Neal D. Goldstein, Joanna S. Suder, and Jonathan Purtle, "Trends and Characteristics of Proposed and Enacted State Legislation on Childhood Vaccination Exemption, 2011–2017," *American Journal of Public Health* 109, no. 1 (2019): 102.

30. Jennifer S. Rota et al., "Processes for Obtaining Nonmedical Exemptions to State Immunization Laws," *American Journal of Public Health* 91, no. 4 (2001): 645–48; Saad B. Omer et al., "Legislative Challenges to School Immunization Mandates, 2009–2012," *JAMA* 311, no. 6 (2014): 620–21, https://doi.org/10.1001/jama.2013.282 869; Nina R. Blank, Arthur L. Caplan, and Catherine Constable, "Exempting Schoolchildren from Immunizations: States with Few Barriers Had Highest Rates of Nonmedical Exemptions," *Health Affairs* 32, no. 7 (2013): 1282–90, https://search.proquest.com/docview/1412283879?accountid=14681.

31. Goldstein et al., "Trends and Characteristics of Proposed and Enacted State Legislation."

32. Dan M. Kahan, "Vaccine Risk Perceptions and Ad Hoc Risk Communication: An Empirical Assessment," Yale Law & Economics Research Paper 491, January 27, 2014, 12.

33. Catherine Flores Martin (Executive Director of California Immunization Coalition), interview with the author, June 2019. In 2015, the American Medical Association and the American College of Physicians called on their state chapters to work to eliminate NMEs. In 2016, the American Academy of Pediatrics did likewise. "Elimination of Non-Medical Exemptions from State Immunization Laws," American College of Physicians, 2015, https://www.acponline.org/acp_policy/policies/non_medical_exe mptions_policy_2015.pdf; "AMA Supports Tighter Limitations on Immunization Opt-Outs," American Medical Association, 2015, https://www.ama-assn.org/cont ent/ama-supportstighter-limitations-immunization-opt-outs; American Academy of Pediatrics, "Medical Versus Nonmedical Immunization Exemptions for Child Care and School Attendance," *Pediatrics* 138, no. 3 (2016): e20162145.

34. Nicholas Riccardi, "Vaccine Skeptics Find Unexpected Allies in Conservative GOP," PBS NewsHour, February 6, 2015, https://www.pbs.org/newshour/health/vaccine-skeptics-find-unexpected-allies-conservative-gop.

35. Email follow-up to June 2019 interview with Leah Russin (advocate, Vaccinate California), sent February 8, 2022.

36. "SB-277 Public Health: Vaccinations (2015–2016)—Votes," California Legislative Information, accessed August 26, 2022, https://leginfo.legislature.ca.gov/faces/bill VotesClient.xhtml?bill_id=201520160SB277; "SB-276 Immunizations: Medical Exemptions (2019–2020)—Votes," California Legislative Information, accessed August 26, 2022, https://leginfo.legislature.ca.gov/faces/billVotesClient.xhtml?bill_ id=201920200SB276.

37. "States with Religious and Philosophical Exemptions from School Immunization Requirements," National Conference of State Legislatures, 2022, accessed March 16, 2022, https://www.ncsl.org/research/health/school-immunization-exemption-state-laws.aspx.

38. "A02371 (A Bill to Repeal Section 2164 of the Public Health Law)—Legislative Summary," New York State Assembly, 2019, https://nyassembly.gov/leg/?defa ult_fld=&leg_video=&bn=A02371&term=2019&Summary=Y&Actions= Y&Committee%26nbspVotes=Y&Floor%26nbspVotes=Y&Memo=Y; Associated Press, "New York Ends Religious Exemption to Vaccine Mandate for Schoolchildren," *The Guardian*, June 14, 2019, https://www.theguardian.com/us-news/2019/jun/ 13/new-york-vaccines-measles-religion; "Title 20-A: Education, Chapter 223, Sub-chapter 2: Immunization, §6355. Enrollment in school," Maine Legislature, 2019; Evan Simko-Bednarski, "Maine Bars Residents from Opting out of Immunizations for Religious or Philosophical Reasons," CNN Health, May 27, 2019; "Legislative Summary HB 1638–2019-20, Promoting Immunity Against Vaccine Preventable Diseases," Washington State Legislature, 2019, https://app.leg.wa.gov/billsumm ary?BillNumber=1638&Initiative=false&Year=2019#rollCallPopup; Rachel La Corte, "Washington House Passes Bill Limiting Vaccine Exemptions," *The Seattle Times*, March 7, 2019, https://www.seattletimes.com/seattle-news/bill-limiting-vacc ine-exemptions-passes-washington-state-house; "Connecticut House Bill 6423 an Act Concerning Immunizations. Legislative summary," LegiScan, 2021, https://legis can.com/CT/bill/HB06423/2021; Jodi Latina, Bob Wilson, and Kent Pierce, "Gov. Lamont Signs Bill to Repeal Religious Exemption for Childhood Vaccinations After Large Protest During Senate Vote," WTNH News, April 27, 2021, https://www.wtnh. com/news/connecticut/hartford/state-senate-takes-up-repeal-of-religious-exempt ion-for-childhood-vaccinations-after-passing-in-house.

39. "State Partisan Composition," National Conference of State Legislatures, February 1, 2022, https://www.ncsl.org/research/about-state-legislatures/partisan-composition. aspx#.

40. Breanna Fernandes et al., "US State-Level Legal Interventions Related to COVID-19 Vaccine Mandates," *JAMA* 327, no. 2 (2021): 178–79, https://doi.org/10.1001/ jama.2021.22122; Neal D. Goldstein, Joanna S. Suder, and Jonathan Purtle, "Trends and Characteristics of Proposed and Enacted State Legislation on Childhood

Vaccination Exemption, 2011–2017," *American Journal of Public Health* 109, no. 1 (2019): 102; Jonathon Purtle et al., "Who Votes for Public Health? U.S. Senator Characteristics Associated with Voting in Concordance with Public Health Policy Recommendations (1998–2013)," *SSM—Population Health* 3 (2017): 136–40, https://doi.org/10.1016/j.ssmph.2016.12.011.

41. Owen Dyer, "Republican Candidates Cast Doubt on Vaccines in US Presidential Debate," *BMJ* 351 (2015).

42. Jan-Werner Müller, "The American Right Is Pushing 'Freedom over Fear.' It Won't Stop the Virus," *The Guardian*, July 16, 2020, sec. Opinion, https://www.theguardian.com/commentisfree/2020/jul/16/coronavirus-american-right-freedom-over-fear; Charlie Warzel, "Protesting for the Freedom to Catch the Coronavirus," *The New York Times*, April 19, 2020, sec. Opinion, https://www.nytimes.com/2020/04/19/opinion/coronavirus-trump-protests.html; "Coronavirus: Donald Trump Vows Not to Order Americans to Wear Masks," BBC News, July 18, 2020, sec. US & Canada, https://www.bbc.com/news/world-us-canada-53453468; Terry Tang, Ken Moritsugu, and Lisa Marie Pane, "US Virus Outbreaks Stir Clash over Masks, Personal Freedom," AP News, April 20, 2021, https://apnews.com/article/health-us-news-ap-top-news-international-news-virus-outbreak-54374ff841dfd84323a1fb86d1e93180;   Sara Burnett and Brian Slodysko, "'LIBERATE MICHIGAN!' Right-Wing Coalitions Rallying to End Lockdown Restrictions amid Coronavirus Pandemic," *Chicago Tribune*, April 17, 2020, https://www.chicagotribune.com/coronavirus/ct-nw-coronavirus-protesters-trump-20200417-7oad7qhicrc53eywjaclwdsaoq-story.html.

43. "States That Issued Lockdown and Stay-at-Home Orders in Response to the Coronavirus (COVID-19) Pandemic, 2020," BallotPedia, https://ballotpedia.org/States_that_issued_lockdown_and_stay-at-home_orders_in_response_to_the_coronavirus_(COVID-19)_pandemic,_2020; Lauren Aratani, "How Did Face Masks Become a Political Issue in America?" *The Guardian*, June 29, 2020, https://www.theguardian.com/world/2020/jun/29/face-masks-us-politics-coronavirus; Maria Milosh et al., "Biden's Plea for Masks Will Fail. Blame Political Polarization," *The Washington Post*, April 13, 2021, https://www.washingtonpost.com/outlook/2021/04/13/masks-mandate-partisanship-politics/; Thomas B. Edsall, "When the Mask You're Wearing 'Tastes Like Socialism," *The New York Times*, May 20, 2020, https://www.nytimes.com/2020/05/20/opinion/coronavirus-trump-partisanship.html; Griffin Connolly, "Democrats Want to Fine Colleagues $1,000 for Not Wearing Masks in Congress, as Three Fall Ill with COVID," Independent, January 12, 2021, https://www.independent.co.uk/news/world/americas/us-politics/covid-masks-house-democrats-1000-fine-b1786276.html; Jonathon Mahler, "A Governor on Her Own, with Everything at Stake," *The New York Times*, April 25, 2021, https://www.nytimes.com/2020/06/25/magazine/gretchen-whitmer-coronavirus-michigan.html.

44. "As Cases Spread Across U.S. Last Year, Pattern Emerged Suggesting Link Between Governors' Party Affiliation and COVID-19 Case and Death Numbers," Johns Hopkins Bloomberg School of Public Health, March 10, 2021, accessed March 16, 2021, https://publichealth.jhu.edu/2021/as-cases-spread-across-us-last-year-pattern-emerged-suggesting-link-between-governors-party-affiliat

ion-and-covid-19-case-and-death-numbers; Brain Neelon et al., "Associations Between Governor Political Affiliation and COVID-19 Cases, Deaths, and Testing in the U.S.," *American Journal of Preventive Medicine* 61, no. 1 (2021): 115–19, https://doi.org/10.1016/j.amepre.2021.01.034.

45. "Coronavirus: Armed Protesters Enter Michigan Statehouse," BBC News, May 1, 2020, https://www.bbc.com/news/world-us-canada-52496514; Pete Williams, "Man Who Plotted to Kidnap Michigan Gov. Gretchen Whitmer Sentenced to over 6 Years," NBC News Digital, August 26, 2021, https://www.nbcnews.com/politics/justice-dep artment/man-who-plotted-kidnap-michigan-gov-gretchen-whitmer-sentenced-over-n1277582.

46. Nicholas Riccardi, "Vaccine Skeptics Find Unexpected Allies."

47. Stuart Anderson, "Trump Takes Credit for Vaccine Created by Others, Including Immigrants," *Forbes*, December 1, 2020, https://www.forbes.com/sites/stuartander son/2020/12/01/trump-takes-credit-for-vaccine-created-by-others-including-imm igrants; Olafimihan Oshin, "Trump Takes Credit for Vaccine Rollout: 'One of the Greatest Miracles of the Ages,'" *The Hill*, May 25, 2021, https://thehill.com/homen ews/administration/555247-trump-takes-credit-for-vaccine-rollout-one-of-the-greatest-miracles.

48. Maggie Haberman, "Trump and His Wife Received Coronavirus Vaccine Before Leaving the White House," *The New York Times*, March 1, 2021, https://www.nytimes.com/2021/03/01/us/politics/donald-trump-melania-coronavirus-vaccine.html.

49. Brian Stetler, "Tucker Carlson's Fox News Colleagues Call out His Dangerous Anti-Vaccination Rhetoric," CNN, May 6, 2021, https://www.cnn.com/2021/05/06/media/tucker-carlson-anti-vaccination-monologue/index.html; Aaron Blake, "Tucker Carlson's Worst Vaccine Segment Yet," *The Washington Post*, May 6, 2021, https://www.washingtonpost.com/politics/2021/05/06/tucker-carlsons-worst-vaccine-segment-yet.

50. Lisa Lerer, "How Republican Vaccine Opposition Got to This Point: News Analysis," *The New York Times*, July 17, 2021, https://www.nytimes.com/2021/07/17/us/polit ics/coronavirus-vaccines-republicans.html.

51. See, for example, Elliot Davis, "States Are Banning COVID-19 Vaccine Requirements," *U.S. News & World Report*, April 30, 2021, https://www.usnews.com/news/best-sta tes/articles/2021-04-30/these-states-are-banning-covid-19-vaccine-requirements; Mitchell Hannah, "12 States Banning COVID-19 Vaccine Mandates & How They Affect Healthcare Workers," *Becker's Hospital Review*, October 12, 2021, https://www.beckershospitalreview.com/workforce/11-states-banning-covid-19-vaccine-mandates-how-it-affects-healthcare-workers.html; Rachel Treisman, "Some States Are Working to Prevent COVID-19 Vaccine Mandates," NPR News, August 2, 2021, https://www.npr.org/2021/08/02/1023809875/states-ban-covid-vaccine-mandates.

52. Amy Goldstein, "Anger over Mask Mandates, Other COVID Rules, Spurs States to Curb Power of Public Health Officials," *The Washington Post*, December 25, 2021, https://www.washingtonpost.com/health/2021/12/25/covid-public-health-laws-res tricted/.

53. We reported some of this data in a research letter led by Katie's student Breanna Fernandes: Fernandes et al., "US State-Level Legal Interventions."

54. Jonathon Ambarian, "Montana House Passes One Bill on Vaccine Exemptions, Two Others Fall Short," KTVH News, February 25, 2021, https://www.ktvh.com/news/montana-house-passes-one-bill-on-vaccine-exemptions-two-others-fall-short.

55. Aaron Blake, "The GOP's Vaccine Skeptic Wing Has a Breakthrough in Tennessee," *The Washington Post*, July 13, 2022, https://www.washingtonpost.com/politics/2021/07/13/gops-vaccine-skeptic-wing-has-breakthrough-tennessee.

56. Aaron Blake, "The Question Republicans Still Can't Really Answer on Vaccine Mandates," *The Washington Post*, September 25, 2021, https://www.washingtonpost.com/politics/2021/09/25/question-republicans-still-cant-really-answer-vaccine-mandates.

57. Patricia Mazzei, "As G.O.P. Fights Mask and Vaccine Mandates, Florida Takes the Lead," *The New York Times*, November 17, 2021; Lori Rozsa, "Desantis Brings Back Florida Lawmakers to Crack Down on Pandemic Mandates," *The Washington Post*, November 14, 2021, https://www.washingtonpost.com/nation/2021/11/14/desantis-brings-back-florida-lawmakers-crack-down-pandemic-mandates; Alison Durkee, "Florida Gov. Desantis Signs Sweeping Legislation Restricting Vaccine Mandates, School Mask Rules and More," *Forbes*, November 18, 2021, https://www.forbes.com/sites/alisondurkee/2021/11/18/florida-gov-desantis-signs-sweeping-legislation-restricting-vaccine-mandates-school-mask-rules-and-more/?sh=52150a7c77cf.

58. Fernandes et al., "US State-Level Legal Interventions."

59. Hsueh-Fen Chen and Saleema A. Karim, "Relationship Between Political Partisanship and COVID-19 Deaths: Future Implications for Public Health," *Journal of Public Health* 44, no. 3 (2022): 716–23, https://doi.org/10.1093/pubmed/fdab 136; Ariel Fridman, Rachel Gershon, and Ayelet Gneezy, "COVID-19 and Vaccine Hesitancy: A Longitudinal Study," *PLoS One* 16, no. 4 (2021): e0250123–e23, https://doi.org/10.1371/journal.pone.0250123; Wolfgang Stroebe et al., "Politicization of COVID-19 Health-Protective Behaviors in the United States: Longitudinal and Cross-National Evidence," *PLoS One* 16, no. 10 (2021): e0256740–e40, https://doi.org/10.1371/journal.pone.0256740.

60. Ashley Kirzinger et al., "KFF COVID-19 Vaccine Monitor: The Increasing Importance of Partisanship in Predicting COVID-19 Vaccination Status," KFF, November 16, 2021, https://www.kff.org/coronavirus-covid-19/poll-finding/importance-of-parti sanship-predicting-vaccination-status.

61. "Do You Think Parents Should Be Required to Have Their Children Vaccinated Against Measles, Mumps, and Rubella?" YouGovAmerica, 2022, accessed August 5, 2022, https://today.yougov.com/topics/health/survey-results/daily/2022/02/17/1ad4c/2.

62. "Do You Think Parents Should Be Required to Have Their Children Vaccinated Against Infectious Diseases?" YouGovAmerica, 2022, accessed August 5, 2022, https://today.yougov.com/topics/health/survey-results/daily/2022/02/17/1ad4c/1.

63. "Do You Think Parents Should Be Required to Have Their Children Vaccinated Against COVID-19 if They Are Eligible for the Vaccine?" YouGovAmerica, 2022, accessed August 5, 2022, https://today.yougov.com/topics/health/survey-results/daily/2022/02/17/1ad4c/3.

64. Ashley Gambrell, Maria Sundaram, and Robert A. Bednarczyk, "Estimating the Number of US Children Susceptible to Measles Resulting from COVID-19-Related Vaccination Coverage Declines," *Vaccine* 40, no. 32 (2022): 4574–79, https://doi.org/10.1016/j.vaccine.2022.06.033.

65. Ranee Seither et al., "Vaccination Coverage with Selected Vaccines and Exemption Rates Among Children in Kindergarten—United States, 2020–21 School Year," *MMWR Morbidity and Mortality Weekly Report* 71, no. 16 (2022): 561–68, https://doi.org/10.15585/mmwr.mm7116a1.

66. Benjamin Mueller and Jan Hoffman, "Routine Childhood Vaccinations in the U.S. Slipped During the Pandemic," *The New York Times*, April 21, 2022.

67. "COVID-19 Pandemic Fuels Largest Continued Backslide in Vaccinations in Three Decades," UNICEF, 2022, accessed August 5, 2022, https://www.unicef.org/cuba/en/press-releases/covid-19-pandemic-fuels-largest-continued-backslide-vaccinations-three-decades; Stephanie Nolen, "Sharp Drop in Global Childhood Vaccinations Imperils Millions of Lives," *The New York Times*, July 14, 2022, https://www.nytimes.com/2022/07/14/health/childhood-vaccination-rates-decline.html.

Chapter 10

1. Yascha Mounk, *The Great Experiment: Why Diverse Democracies Fall Apart and How They Can Endure* (London: Penguin, 2022).

2. Sindiso Nyathi et al., "The 2016 California Policy to Eliminate Nonmedical Vaccine Exemptions and Changes in Vaccine Coverage: An Empirical Policy Analysis," *PLoS Medicine* 16, no. 12 (2019): e1002994, https://doi.org/10.1371/journal.pmed.1002994.

3. Samantha Stein, "Are Today's Youth Even More Self-Absorbed (and Less Caring) Than Generations Before?" *Psychology Today*, June 5, 2010, https://www.psychologytoday.com/us/blog/what-the-wild-things-are/201006/are-today-s-youth-even-more-self-absorbed-and-less-caring; Joel Stein, "Millennials: The Me Me Me Generation," *TIME*, May 20, 2013, https://time.com/247/millennials-the-me-me-me-generation; Jamil Zaki, "What, Me Care? Young Are Less Empathetic," *Scientific American*, January 1, 2011, https://www.scientificamerican.com/article/what-me-care; Lynne Malcom, "Research Says Young People Today Are More Narcissistic Than Ever," ABC Australia, 2014, https://www.abc.net.au/radionational/programs/allinthemind/young-people-today-are-more-narcissistic-than-ever/5457236.

4. Marcia Meldrum, "'A Calculated Risk': The Salk Polio Vaccine Field Trials of 1954," *BMJ* 317, no. 7167 (1998): 1233–36; The Takeaway, "Could You Patent the Sun?" WNYC Studios, podcast, December 12, 2016, https://www.wnycstudios.org/podcasts/takeaway/segments/retro-report-patenting-sun.

5. Dorothy Porter, *Health, Civilization and the State a History of Public Health from Ancient to Modern Times* (New York: Taylor & Francis, 2005), 298.

6. Michael Willrich, *Pox: An American History* (New York: Penguin, 2011), 9; Willrich is here citing both "Smallpox Epidemic," *The New York Times*, December 4, 1900, and "Topics of the Times," *The New York Times*, December 12, 1900.

7. James Colgrove, *State of Immunity: The Politics of Vaccination in Twentieth-Century America* (Berkeley: University of California Press, 2006), quoting a Connecticut board of health officer; C. Bell, "Compulsory Vaccination: Should It Be Enforced by Law?" *JAMA* 28 (1897): 49–53.

8. John Duffy, *The Rudolph Matas History of Medicine in Louisiana Volume I* (Baton Rouge: Louisiana State University Press, 1985).

9. David M. Oshinsky, *Polio: An American Story* (Oxford, UK: Oxford University Press, 2006).

10. Gareth Williams, *Paralysed with Fear: The Story of Polio* (Basingstoke, UK: Palgrave Macmillan, 2013).

11. Jane S. Smith, *Patenting the Sun: Polio and the Salk Vaccine* (New York: Morrow, 1990); Stephan Kinsella, "Patent and Penicillin," Mises Institute, 2006, https://mises. org/wire/patent-and-penicillin.

12. Michael Lipka, "Mainline Protestants Make up Shrinking Number of U.S. Adults," Pew Research Center, 2015, https://www.pewresearch.org/fact-tank/2015/05/18/ mainline-protestants-make-up-shrinking-number-of-u-s-adults; George Marsden, *The Twilight of the American Enlightenment: The 1950s and the Crisis of Liberal Belief* (New York: Basic Books, 2014); "White Evangelicals See Trump as Fighting for Their Beliefs, Though Many Have Mixed Feelings About His Personal Conduct," Pew Research Center, 2020, https://www.pewforum.org/2020/03/12/white-evangelicals-see-trump-as-fighting-for-their-beliefs-though-many-have-mixed-feelings-about-his-personal-conduct.

13. Juliana Menasce Horowitz, Ruth Igielnik, and Rakesh Kochhar, "Trends in Income and Wealth Inequality," Pew Research Center, 2020, https://www.pewresearch. org/social-trends/2020/01/09/trends-in-income-and-wealth-inequality; "Current Decade Rates as Worst in 50 Years," Pew Research Center, 2009, https://www.pewr esearch.org/politics/2009/12/21/current-decade-rates-as-worst-in-50-years; Kim Parker, Rich Morin, and Juliana Menasce Horowitz. "Looking to the Future, Public Sees an America in Decline on Many Fronts," Pew Research Center, 2019, https:// www.pewresearch.org/social-trends/2019/03/21/public-sees-an-america-in-decl ine-on-many-fronts; "Confidence in Institutions," Gallup, 2021, https://news.gallup. com/poll/1597/confidence-institutions.aspx; Lee Rainie and Andrew Perrin, "Key Findings About Americans' Declining Trust in Government and Each Other," Pew Research Center, 2019, https://www.pewresearch.org/fact-tank/2019/07/22/key-findings-about-americans-declining-trust-in-government-and-each-other.

14. Charles T. Clotfelter, *After Brown: The Rise and Retreat of School Desegregation* (Princeton, NJ: Princeton University Press, 2006); Richard Rothstein, *The Color of Law: A Forgotten History of How Our Government Segregated America Paperback* (New York: Liveright, 2018); Judith Hellerstein and David Neumark, *Workplace Segregation in the United States: Race, Ethnicity, and Skill* (Cambridge, MA: National Bureau of Economic Research, 2005), https://www.nber.org/system/files/working_ papers/w11599/w11599.pdf.

15. Robert D. Putnam, *Bowling Alone : The Collapse and Revival of American Community* (New York: Simon & Schuster, 2001).

16. Richard J. Altenbaugh, *Vaccination in America: Medical Science and Children's Welfare* (New York: Palgrave Macmillan, 2018), 61–62.

17. See, for example, Gilbert H. Welch, Steven Woloshin, and Lisa Schwartz, *Overdiagnosed: Making People Sick in the Pursuit of Health* (Boston: Beacon, 2018); Jeremy A. Greene, *Prescribing by Numbers: Drugs and the Definition of Disease* (Baltimore, MD: Johns Hopkins University Press, 2008); Joseph Durmit, *Drugs for Life. How Pharmaceutical Companies Define Our Health* (Durham, NC: Duke University Press, 2012); John Abramson, *Overdo$Ed America: The Broken Promise of American Medicine* (New York: Harper Perennial, 2008); Marcia Angell, *The Truth About the Drug Companies: How They Deceive Us and What to Do About It* (London: Random House, 2005); Peter Gotzsche, *Deadly Medicines and Organised Crime. How Big Pharma Has Corrupted Healthcare* (London: Routledge, 2013). See also Beth Macy, *Dopesick: Dealers, Doctors, and the Drug Company That Addicted America* (Boston: Little, Brown, 2018); Associated Press, "Book About Sackler Family and Opioid Crisis Wins UK Prize," *U.S. News & World Report*, November 16, 2021, https://www.usnews.com/news/entertainment/articles/2021-11-16/book-about-sackler-family-and-opioid-crisis-wins-uk-prize#:~:text=%E2%80%9CEmpire%20of%20Pain%E2%80%9D%20traces%20the,of%20lawsuits%20and%20bankruptcy%20proceedings.

18. Bernice L. Hausman, *Anti/Vax: Reframing the Vaccination Controversy* (Ithaca, NY: Cornell University Press, 2019), 170–72.

19. Hausman, *Anti/Vax*, 217.

20. Hausman, *Anti/Vax*, 217.

21. "Public Trust in Government: 1958–2021," Pew Research Center, 2021, https://www.pewresearch.org/politics/2021/05/17/public-trust-in-government-1958-2021.

22. Sean Westwood et al., "The Tie That Divides: Cross-National Evidence of the Primacy of Partyism," *European Journal of Political Research* 57, no. 2 (2018): 333–54, https://doi.org/10.1111/1475-6765.12228.

23. John Gramlich, "Young Americans Are Less Trusting of Other People—and Key Institutions—Than Their Elders," Pew Research Center, 2019, https://www.pewresearch.org/fact-tank/2019/08/06/young-americans-are-less-trusting-of-other-people-and-key-institutions-than-their-elders.

24. Kevin Vallier, *Trust in a Polarized Age* (Oxford, UK: Oxford University Press, 2020), 2.

25. Jonathon Berman, *Anti-Vaxxers: How to Challenge a Misinformed Movement* (Cambridge, MA: MIT Press, 2020); Maya Goldenberg, *Vaccine Hesitancy: Public Trust, Expertise and the War on Science* (Pittsburgh, PA: University of Pittsburgh, 2021).

26. Stephanie Coontz, *The Way We Never Were: American Families and the Nostalgia Trap* (New York: Basic Books, 2016).

27. Lilliana Mason, *Uncivil Agreement: How Politics Became Our Identity* (Chicago: University of Chicago Press, 2018).

28. Mason, *Uncivil Agreement*, 14.

29. Ezra Klein, *Why We're Polarized* (New York: Avid Reader Press/Simon & Schuster, 2021), 60–64.

30. Henri Tajfel, "Social Identity and Intergroup Behaviour," *Social Science Information* 13, no. 2 (1974): 65–93, https://doi.org/10.1177/053901847401300204. See also Katie Attwell and David T. Smith, "Parenting as Politics: Social Identity Theory and Vaccine Hesitant Communities," *International Journal of Health Governance* 22, no. 3 (2017): 183–98.

31. Patrick R. Miller and Pamela Johnston Conover, "Red and Blue States of Mind: Partisan Hostility and Voting in the United States," *Political Research Quarterly* 68, no. 2 (2015): 225–39, https://doi.org/10.1177/1065912915577208.

32. Julie Leask, "Target the fence-sitters," *Nature*. 473, no. 7348 (2011): 443–45.

33. Miller and Johnston Conover, "Red and Blue States of Mind"; Saad B. Omer, Cornelia Betsch, and Julie Leask, "Mandate Vaccination with Care," *Nature* 571, no. 7766 (2019): 469, https://doi.org/10.1038/d41586-019-02232-0; Julie Leask and Margie Danchin, "Imposing Penalties for Vaccine Rejection Requires Strong Scrutiny," *Journal of Paediatrics and Child Health* 53, no. 5 (2017): 439–44; Brendan Nyhan, "Why California's Approach to Tightening Vaccine Rules Has Potential to Backfire," *The New York Times*, April 14, 2015; Mark C. Navin and Mark A. Largent, "Improving Nonmedical Vaccine Exemption Policies: Three Case Studies," *Public Health Ethics* 10, no. 3 (2017): 225–34, http://dx.doi.org/10.1093/phe/phw047.

34. Douglas J. Opel et al., "Childhood Vaccine Exemption Policy: The Case for a Less Restrictive Alternative," *Pediatrics* 137, no. 4 (2016): e20154230.

35. Katie Attwell and Mark C. Navin, "Childhood Vaccination Mandates: Scope, Sanctions, Severity, Selectivity, and Salience," *Milbank Quarterly* 97, no. 4 (2019): 978–1014, https://doi.org/10.1111/1468-0009.12417. See also Katie Attwell, Jeremy K. Ward, and Sian Tomkinson, "Manufacturing Consent for Vaccine Mandates: A Comparative Case Study of Communication Campaigns in France and Australia," *Frontiers in Communication* 6 (2020): 598602, https://doi.org/10.3389/fcomm.2021.598602.

36. New York state has been testing sewage for polio, helping authorities to identify the risks of further infections. Sharon Otterman, "Why Polio, Once Eliminated, Is Testing N.Y. Health Officials," *The New York Times*, October 3, 2022, https://www.nyti mes.com/2022/10/03/nyregion/polio-new-york-eradication.html.

37. See, e.g., Samuel Freeman, "Illiberal Libertarians: Why Libertarianism Is Not a Liberal View," *Philosophy & Public Affairs* 30, no. 2 (2001): 105–51.

38. Katie Attwell and Mark C. Navin, "Bosses Shouldn't Demand That You Be Vaccinated," *The New York Times*, February 26, 2021.

39. Liam Drew, "Did COVID Vaccine Mandates Work? What the Data Say," *Nature*, News Feature, July 6, 2022; Mark A. Rothstein, Wendy E. Parmet, and Dorit Reiss, "Employer-Mandated Vaccination for COVID-19," *American Journal of Public Health* 111, no. 6 (2021): 1061–64, https://doi.org/10.2105/ajph.2020.306166; Karen Mulligan and Jeffrey E. Harris, "COVID-19 Vaccination Mandates for School and Work Are Sound Public Policy," USC, Schaeffer Center for Health Policy & Economics, 2021, https://healthpolicy.usc.edu/wp-content/uploads/2022/07/USC_Schaeffer_Covid19-VaccineMandates_WhitePaper.pdf; Stella Talic et al.,

"Effectiveness of Public Health Measures in Reducing the Incidence of COVID-19, SARS-COV-2 Transmission, and COVID-19 Mortality: Systematic Review and Meta-Analysis," *BMJ* 375 (2021): e068302; Katie Attwell et al, "COVID-19 Vaccine Mandates: An Australian Attitudinal Study," *Vaccine* 40, no. 51 (2022): 7360–69, https://doi.org/https://doi.org/10.1016/j.vaccine.2021.11.056.

40. Sophie Aubrey, "They're Trying to Protect Their Babies, Yet They're Accused of Being 'Over the Top'," *Sydney Morning Herald*, July 21, 2019; Samantha Carlson, Kerrie Wiley, and Peter McIntyre, "'No Vax, No Visit'? If Mum Was Vaccinated Baby Is Already Protected Against Whooping Cough," The Conversation, May 30, 2016, https://theconversation.com/no-vax-no-visit-if-mum-was-vaccinated-baby-is-alre ady-protected-against-whooping-cough-59374; Christen Johnson, "Should You Ask Family and Friends to Get Vaccinated Before Visiting Your Newborn?" *Chicago Tribune*, April 4, 2019, https://www.chicagotribune.com/lifestyles/sc-fam-social-gra ces-parents-asking-relatives-vaccine-pre-baby-0416-story.html; Katie McPherson, "How to Ask Loved Ones to Get Vaccinated Before Meeting Your Newborn," Romper, November 20, 2020, https://www.romper.com/pregnancy/how-to-ask-family-to-get-the-flu-shot-other-vaccines-before-baby-is-born.

41. Bill Bishop, *The Big Sort: Why the Clustering of Like-Minded America Is Tearing Us Apart* (Boston: Mariner Books, 2009); Amy Chua, *Political Tribes: Group Instinct and the Fate of Nations* (London: Penguin, 2018); Klein, *Why We're Polarized*; David French, *Divided We Fall: America's Secession Threat and How to Restore Our Nation* (New York: St. Martin's, 2020); Mason, *Uncivil Agreement*.

42. Steven R. Daniels, "The Evolution of Attitudes on Same-Sex Marriage in the United States, 1988–2014," *Social Science Quarterly* 100, no. 5 (2019): 1651–63, https://doi.org/10.1111/ssqu.12673; Hye-Yon Lee and Diana C. Mutz, "Changing Attitudes Toward Same-Sex Marriage: A Three-Wave Panel Study," *Political Behavior* 41, no. 3 (2018): 701–22, https://doi.org/10.1007/s11109-018-9463-7.

43. Ginny E. Garcia, Richard Lewis, and Joanne Ford-Robertson, "Attitudes Regarding Laws Limiting Black–White Marriage: A Longitudinal Analysis of Perceptions and Related Behaviors," *Journal of Black Studies* 46, no. 2 (2015): 199–217; Zhenchao Qian and Daniel T. Lichter, "Changing Patterns of Interracial Marriage in a Multiracial Society," *Journal of Marriage and Family* 73, no. 5 (2011): 1065–84, https://doi.org/10.1111/j.1741-3737.2011.00866.x; Zhenchao Qian, "Breaking the Last Taboo: Interracial Marriage in America," *Contexts* 4, no. 4 (2005): 33–37, https://doi.org/10.1525/ctx.2005.4.4.33.

Bibliography

6abc Action News—WPVI Philadelphia. "1991: The Philly Measles Outbreak That Killed 9 Children." February 6, 2015. https://6abc.com/1991-outbreak-faith-tabernacle-first-century-gospel-measles/504818.

A Voice for Choice. "About A Voice for Choice." 2020. Accessed December 29, 2021. http://avoiceforchoice.org/about-avfc.

Abi-Jaoude, Elia, Karline Treurnicht Naylor, and Antonio Pignatiello. "Smartphones, Social Media Use and Youth Mental Health." *Canadian Medical Association Journal* 192, no. 6 (2020): E136–E41. https://doi.org/10.1503/cmaj.190434.

Abramson, John. *Overdo$Ed America: The Broken Promise of American Medicine.* New York: Harper Perennial, 2008.

Adamy, Janet, and Paul Overberg. "Doctors, Once GOP Stalwarts, Now More Likely to Be Democrats." *The Wall Street Journal*, October 6, 2019. https://www.wsj.com/articles/doctors-once-gop-stalwarts-now-more-likely-to-be-democrats-11570383523.

Addams, Jane. *Twenty Years at Hull-House.* New York: Macmillan, 1910.

Age of Autism. "Tea Party Joins Canary Party in Opposing Vaccine Mandates in California." 2012. Accessed February 16, 2021. https://www.ageofautism.com/2012/06/tea-party-joins-canary-party-in-opposing-vaccine-mandates-in-california.html.

Allen, Arthur. *Vaccine: The Controversial Story of Medicine's Greatest Lifesaver.* New York: Norton, 2008.

Allen, Ben. "Senators Richard Pan and Ben Allen to Introduce Legislation to End California's Vaccine Exemption Loophole." News release. April 5, 2023. https://sd24.senate.ca.gov/news/press-release/senators-richard-pan-and-ben-allen-introduce-legislation-end-californias-vaccine.

Altenbaugh, Richard J. *Vaccination in America: Medical Science and Children's Welfare.* London: Palgrave Macmillan, 2018.

Ambarian, Jonathon. "Montana House Passes One Bill on Vaccine Exemptions, Two Others Fall Short." KTVH News. February 25, 2021. https://www.ktvh.com/news/montana-house-passes-one-bill-on-vaccine-exemptions-two-others-fall-short.

American Academy of Family Physicians. "Immunization Exemptions." 2015. https://www.aafp.org/dam/AAFP/documents/patient_care/immunizations/vaccine-exemptions.pdf.

American Academy of Pediatrics. "Care of the Young Athlete Patient Education Handout—Football." 2010. https://www.aap.org/globalassets/publications/coya/football.1.0.pdf.

American Academy of Pediatrics. "Care of the Young Athlete Patient Education Handout—Ice Hockey." 2011. https://www.aap.org/globalassets/publications/coya/icehockey.1.0.pdf.

American Academy of Pediatrics. "Reaffirmation: Responding to Parents Who Refuse Immunization for Their Children." *Pediatrics* 131, no. 5 (2013): e1696–e96. https://doi.org/10.1542/peds.2013-0430.

American Academy of Pediatrics. "Tackling in Youth Football." *Pediatrics* 136, no. 5 (2015): e1419–e30. https://doi.org/10.1542/peds.2015-3282.

American Academy of Pediatrics. "Incidence of Concussion in Youth Ice Hockey Players." *Pediatrics* 137, no. 2 (2016): e20151633–e33. https://doi.org/10.1542/peds.2015-1633.

American Academy of Pediatrics. "Pediatricians Overwhelmingly Support Vaccines for Children Program." News release. 2020. https://www.aap.org/en/news-room/news-releases/pediatrics2/2020/pediatricians-overwhelmingly-support-vaccines-for-child ren-program.

American Academy of Pediatrics California. "SB 277 (Pan & Allen) Elimination of CA Personal Belief Exemption for School-Entry Vaccines." News release. March 30, 2015. http://aap-ca.org/letter/sb-277-pan-allen-elimination-of-ca-personal-belief-exempt ion-for-school-entry-vaccines.

American Academy of Pediatrics Committee on Bioethics. "Policy Statement: Informed Consent in Decision-Making in Pediatric Practice." *Pediatrics* 138, no. 2 (2016): e20161495.

American Academy of Pediatrics Committee on Practice and Ambulatory Medicine, Committee on Infectious Diseases, Committee on State Government Affairs, Council on School Health, and Section on Administration and Practice Management. "Medical Versus Nonmedical Immunization Exemptions for Child Care and School Attendance." *Pediatrics* 138, no. 3 (2016): e20162145. https://doi.org/10.1542/peds.2016-2145.

American Association for Medical Progress. *Smallpox—A Preventable Disease.* New York: American Association for Medical Progress, 1924.

American College of Physicians. "Elimination of Non-Medical Exemptions from State Immunization Laws." 2015. Accessed February 10, 2022. https://www.acponline.org/acp_policy/policies/non_medical_exemptions_policy_2015.pdf.

American College of Physicians. "Statement: State Immunization Laws Should Eliminate Non-Medical Exemptions Say Internists." News release. 2015. https://www.acponline.org/acp-newsroom/state-immunization-laws-should-eliminate-non-medical-exe mptions-say-internists.

American Human Association. *Human Vivisection: A Statement and an Inquiry.* Chicago: American Human Association, 1899.

American Medical Association. "AMA Supports Tighter Limitations on Immunization Opt Outs." June 8, 2015. Accessed March 31, 2016. https://www.ama-assn.org/press-center/press-releases/ama-supports-tighter-limitations-immunization-opt-outs.

American Medical Association. "AMA Policy Advocates to Eliminate Non-Medical Vaccine Exemptions." June 13, 2019. Accessed July 5, 2022. https://www.ama-assn.org/press-center/press-releases/ama-policy-advocates-eliminate-non-medical-vaccine-exemptions.

American Society for Bioethics and Humanities. *Improving Competencies in Clinical Ethics Consultation: An Education Guide.* Chicago: American Society for Bioethics and Humanities, 2015.

Amin, Avnika, Robert Bednarczyk, Cara Ray, Kala Melchiori, Jesse Graham, Jeffrey Huntsinger, and Saad Omer. "Association of Moral Values with Vaccine Hesitancy." *Nature Human Behaviour* 1, no. 12 (2017): 873–80. https://doi.org/10.1038/s41 562-017-0256-5.

Amit Aharon, Anat, Haim Nehama, Shmuel Rishpon, and Orna Baron-Epel. "Parents with High Levels of Communicative and Critical Health Literacy Are Less Likely to Vaccinate Their Children." *Patient Education and Counseling* 100, no. 4 (2016): 768–75. https://doi.org/10.1016/j.pec.2016.11.016.

Anderson, Scott. "Coercion." In *Stanford Encyclopedia of Philosophy*, edited by E. N. Zalta. Stanford, CA: Stanford University, 2021.

Anderson, Stuart. "Trump Takes Credit for Vaccine Created by Others, Including Immigrants." *Forbes*, December 1, 2020. https://www.forbes.com/sites/stuartander son/2020/12/01/trump-takes-credit-for-vaccine-created-by-others-including-imm igrants.

Angeli, Federica, Silvia Camporesi, and Giorgia Dal Fabbro. "The COVID-19 Wicked Problem in Public Health Ethics: Conflicting Evidence, or Incommensurable Values?" *Humanities & Social Sciences Communications* 8, no. 1 (2021): 1–8. https://doi.org/ 10.1057/s41599-021-00839-1.

Angell, Marcia. *The Truth About the Drug Companies. How They Deceive Us and What to Do About It*. London: Random House, 2005.

Annas, George J. "Bioterrorism, Public Health, and Civil Liberties." *New England Journal of Medicine* 346, no. 17 (2002): 1337–42. https://doi.org/10.1056/nejm20020425 3461722.

Anthony, N., M. Reed, A. M. Leff, J. Huffer, and B. Stephens. "Immunization: Public Health Programming Through Law Enforcement." *American Journal of Public Health (1971)* 67, no. 8 (1977): 763–64. https://doi.org/10.2105/ajph.67.8.763.

Aratani, Lauren. "How Did Face Masks Become a Political Issue in America?" *The Guardian*, June 29, 2020. https://www.theguardian.com/world/2020/jun/29/face-masks-us-politics-coronavirus.

Archard, David. "The Obligations and Responsibilities of Parenthood." In *Procreation and Parenthood: The Ethics of Bearing and Rearing Children*, edited by D. Archard and D. Benatar, 103–26. Oxford, UK: Oxford University Press, 2010.

Ariely, Dan. *Predictably Irrational, Revised and Expanded Edition: The Hidden Forces That Shape Our Decisions* (rev. ed.). New York: Harper Perennial, 2010.

Associated Press. "New York Ends Religious Exemption to Vaccine Mandate for Schoolchildren" *The Guardian*, June 14, 2019. https://www.theguardian.com/us-news/ 2019/jun/13/new-york-vaccines-measles-religion.

Associated Press. "Book About Sackler Family and Opioid Crisis Wins Uk Prize." U.S. *News & World Report*, November 16, 2021. https://www.usnews.com/news/entert ainment/articles/2021-11-16/book-about-sackler-family-and-opioid-crisis-wins-uk-prize#:~:text=%E2%80%9CEmpire%20of%20Pain%E2%80%9D%20traces%20 the,of%20lawsuits%20and%20bankruptcy%20proceedings.

Attiah, Karen. "Why Won't the U.S. Ratify the U.N.'S Child Rights Treaty?" *The Washington Post*, November 21, 2014. https://www.washingtonpost.com/blogs/post-partisan/wp/ 2014/11/21/why-wont-the-u-s-ratify-the-u-n-s-child-rights-treaty.

Attkisson, Sharyl. "How Independent Are Vaccine Defenders?" CBS News. July 25, 2008. https://www.cbsnews.com/news/how-independent-are-vaccine-defenders.

Attwell, Katie. "The Politics of Picking: Selective Vaccinators and Population-Level Policy." *SSM—Population Health* 7 (2019): 100342. https://doi.org/10.1016/ j.ssmph.2018.100342.

Attwell, Katie, and Shevaun Drislane. "Australia's 'No Jab No Play' Policies: History, Design and Rationales." *Australian & New Zealand Journal of Public Health* 46, no. 5 (2022): 640–46. https://doi.org/10.1111/1753-6405.13289

Attwell, Katie, and Melanie Freeman. "I Immunise: An Evaluation of a Values-Based Campaign to Change Attitudes and Beliefs." *Vaccine* 33, no. 46 (2015): 6235–40. https:// doi.org/http://dx.doi.org/10.1016/j.vaccine.2015.09.092.

Attwell, Katie, and Adam Hannah. "Convergence on Coercion: Functional and Political Pressures as Drivers of Global Childhood Vaccine Mandates." *International Journal of Health Policy and Management* 11, no. 11 (2022): 2660–71. https://doi.org/10.34172/ijhpm.2022.6518.

Attwell, Katie, Adam Hannah, and Julie Leask. "COVID-19: Talk of 'Vaccine Hesitancy' Lets Governments Off the Hook." *Nature* 602 (2022): 574–77.

Attwell, Katie, Julie Leask, S. B. Meyer, P. Rokkas, and P. R. Ward. "Vaccine Rejecting Parents' Engagement with Expert Systems That Inform Vaccination Programs." *Journal of Bioethical Inquiry* 14, no. 1 (2017): 65–76.

Attwell, Katie, Samantha Meyer, and Paul Ward. "The Social Basis of Vaccine Questioning and Refusal: A Qualitative Study Employing Bourdieu's Concepts of 'Capitals' and 'Habitus.'" *International Journal of Environmental Research and Public Health* 15, no. 5 (2018): 1044. http://www.mdpi.com/1660-4601/15/5/1044.

Attwell, Katie, and Mark Navin. "Bosses Shouldn't Demand That You Be Vaccinated." *The New York Times*, February 26, 2021.

Attwell, Katie, and Mark Navin. "How Policymakers Employ Ethical Frames to Design and Introduce New Policies: The Case of Childhood Vaccine Mandates in Australia." *Policy & Politics* 50, no. 4 (2022). https://doi.org/10.1332/030557321X16476002878591.

Attwell, Katie, and Mark C. Navin. "Childhood Vaccination Mandates: Scope, Sanctions, Severity, Selectivity, and Salience." *Milbank Quarterly* 97, no. 4 (2019): 978–1014. https://doi.org/10.1111/1468-0009.12417.

Attwell, Katie, Mark C. Navin, Pierluigi Lopalco, Christine Jestin, Sabine Reiter, and Saad B. Omer. "Recent Vaccine Mandates in the United States, Europe and Australia: A Comparative Study." *Vaccine* 19, no. 36 (2018): 7377–84. https://doi.org/10.1016/j.vaccine.2018.10.019.

Attwell, Katie, Marco Rizzi, Lara McKenzie, Samantha J. Carlson, Leah Roberts, Sian Tomkinson, and Chris Blyth. "COVID-19 Vaccine Mandates: An Australian Attitudinal Study." *Vaccine* 40, no. 51 (2022): 7360–69. https://doi.org/https://doi.org/10.1016/j.vaccine.2021.11.056.

Attwell, Katie, and David T. Smith. "Parenting as Politics: Social Identity Theory and Vaccine Hesitant Communities." *International Journal of Health Governance* 22, no. 3 (2017): 183–98.

Attwell, Katie, and David T. Smith. "Hearts, Minds, Nudges and Shoves: (How) Can We Mobilise Communities for Vaccination in a Marketised Society?" *Vaccine* 36, no. 44 (2018): 6506–08. https://doi.org/10.1016/j.vaccine.2017.08.005.

Attwell, Katie, Jeremy K. Ward, and Sian Tomkinson. "Manufacturing Consent for Vaccine Mandates: A Comparative Case Study of Communication Campaigns in France and Australia." *Frontiers in Communication* 6 (2020): 598602. https://doi.org/10.3389/fcomm.2021.598602.

Attwell, Katie, Paul R. Ward, Samantha Meyer, Philippa Rokkas, and Julie Leask. "'Do-It-Yourself': Vaccine Rejection and Complementary and Alternative Medicine (CAM)." *Social Science and Medicine* 196 (2018): 106–14.

Attwooll, Jolyon. "Doctors Lead International 'Most Trusted' Profession Poll." News release. October 13, 2021. https://www1.racgp.org.au/newsgp/professional/doctors-lead-international-most-trusted-profession.

Atwell, Jessica E., Josh Van Otterloo, Jennifer Zipprich, Kathleen Winter, Kathleen Harriman, Daniel A. Salmon, Neal A. Halsey, and Saad B. Omer. "Nonmedical Vaccine Exemptions and Pertussis in California, 2010." *Pediatrics* 132, no. 4 (2013): 624–30. https://doi.org/10.1542/peds.2013-0878.

Aubrey, Sophie. "They're Trying to Protect Their Babies, Yet They're Accused of Being 'Over the Top.'" *Sydney Morning Herald*, July 21, 2019.

Australian Government. "COVID-19 Vaccine Rollout Update." 2022. Accessed July 5, 2022. https://www.health.gov.au/sites/default/files/documents/2022/07/covid-19-vaccine-rollout-update-5-july-2022.pdf.

Bailey, Beth L. *America's Army Making the All-Volunteer Force.* Cambridge, MA: Harvard University Press, 2009. doi:10.4159/9780674053526.

Baines, Paul. "Family Interests and Medical Decisions for Children." *Bioethics* 31, no. 8 (2017): 599–607. https://doi.org/10.1111/bioe.12376.

Baird, G., A. Pickles, E. Simonoff, T. Charman, P. Sullivan, S. Chandler, T. Loucas, et al. "Measles Vaccination and Antibody Response in Autism Spectrum Disorders." *Archives of Disease in Childhood* 93, no. 10 (2008): 832–37. https://doi.org/10.1136/adc.2007.122937.

Baker, Sam. "The Supreme Court's Next Target Is the Executive Branch." Axios, July 5, 2022. https://www.axios.com/2022/07/05/supreme-court-conservative-climate-health-regulations.

Baker, Stephanie Alice, and Michael James Walsh. "'A Mother's Intuition: It's Real and We Have to Believe in It': How the Maternal Is Used to Promote Vaccine Refusal on Instagram." *Information, Communication & Society*, online January 23, 2022. https://doi.org/10.1080/1369118x.2021.2021269.

BallotPedia. "States That Issued Lockdown and Stay-at-Home Orders in Response to the Coronavirus (COVID-19) Pandemic, 2020." 2022. https://ballotpedia.org/States_that_issued_lockdown_and_stay-at-home_orders_in_response_to_the_coronavirus_(COVID-19)_pandemic,_2020.

Balmer, Randall. *Bad Faith: Race and the Rise of the Religious Right.* Grand Rapids, MI: Eerdmans, 2021.

Battin, Margaret, Leslie Francis, Jay Jacobson, and Charles Smith. *The Patient as Victim and Vector: Ethics and Infectious Disease.* Oxford, UK: Oxford University Press, 2009.

Baumgartner, Frank R., and Bryan D. Jones. *Agendas and Instability in American Politics.* Chicago: University of Chicago Press, 1993.

Bay Area Rapid Transit. "About," 2022. https://www.bart.gov/about.

Bayer, Ronald, and Amy L. Fairchild. "The Genesis of Public Health Ethics." *Bioethics* 18, no. 6 (2004): 473–92. https://doi.org/10.1111/j.1467-8519.2004.00412.x.

Baylor College of Medicine. "About Peter Jay Hotez, M.D., Ph.D." 2022. https://www.bcm.edu/people-search/peter-hotez-23229.

BBC News. "Coronavirus: Armed Protesters Enter Michigan Statehouse." May 1, 2020. https://www.bbc.com/news/world-us-canada-52496514.

BBC News "Coronavirus: Donald Trump Vows Not to Order Americans to Wear Masks." July 18, 2020. https://www.bbc.com/news/world-us-canada-53453468.

Beam, Adam. "California Delays Coronavirus Vaccine Mandate for Schools." April 14, 2022. https://apnews.com/article/covid-health-gavin-newsom-california-sacramento-7e58e25b2d5194979995c864533772ab.

Beauchamp, Tom L., and James F. Childress. *Principles of Biomedical Ethics.* New York: Oxford University Press, 2001.

Becker, Ann M. "Smallpox in Washington's Army: Strategic Implications of the Disease During the American Revolutionary War." *Journal of Military History* 68, no. 2 (2004): 381–430. https://doi.org/10.1353/jmh.2004.0012.

Bedford, Helen, Katie Attwell, Margie Danchin, Helen Marshall, Paul Corben, and Julie Leask. "Vaccine Hesitancy, Refusal and Access Barriers: The Need for Clarity in Terminology." *Vaccine* 36, no. 44 (2018): 6556–58. https://doi.org/10.1016/j.vacc ine.2017.08.004.

Bednarczyk, Robert A., Adrian R. King, Ariana Lahijani, and Saad B. Omer. "Current Landscape of Nonmedical Vaccination Exemptions in the United States: Impact of Policy Changes." *Expert Review of Vaccines* 18, no. 2 (2019): 175–90. https://doi.org/ 10.1080/14760584.2019.1562344.

Begos, Kevin, Danielle Deaver, John Railey, Scott Sexton, and Paul Lombardo. *Against Their Will: North Carolina's Sterilization Program.* Apalachicola, FL: Gray Oak Books, 2012.

Bell, C. "Compulsory Vaccination: Should It Be Enforced by Law?" *JAMA* 28 (1897): 49–53.

Bell, Stephen, Andrew Hindmoor, and Frank Mols. "Persuasion as Governance: A State-Centric Relational Perspective." *Public Administration* 88, no. 3 (2010): 851–70. https:// doi.org/10.1111/j.1467-9299.2010.01838.x.

Berezin, Mabel, and Alicia Eads. "Risk Is for the Rich? Childhood Vaccination Resistance and a Culture of Health." *Social Science & Medicine* 165 (2016): 233–45. https://doi.org/ 10.1016/j.socscimed.2016.07.009.

Berg, Insoo Kim, and Susan Kelly. *Building Solutions in Child Protective Services.* New York: Norton, 2000.

Berman, Jonathan M. *Anti-Vaxxers: How to Challenge a Misinformed Movement.* Cambridge, MA: MIT Press, 2020.

Bernstein, David. "The Supreme Court Could Foster a New Kind of Civil War." *Politico*, June 14, 2022. https://www.politico.com/news/magazine/2022/06/14/supreme-court-civil-war-00039543.

Bernstein, J., and M. Navin. "Reciprocity, Vulnerability, and the Moral Significance of Herd Immunity," *Journal of Applied Philosophy* (forthcoming).

Bester, Johan C. "Not a Matter of Parental Choice but of Social Justice Obligation: Children Are Owed Measles Vaccination." *Bioethics* 32, no. 9 (2018): 611–19. https://doi.org/ 10.1111/bioe.12511.

Bester, Johan Christiaan. "The Harm Principle Cannot Replace the Best Interest Standard: Problems with Using the Harm Principle for Medical Decision Making for Children." *American Journal of Bioethics* 18, no. 8 (2018): 9–19. https://doi.org/ 10.1080/15265161.2018.1485757.

Beyrer, Chris, and Larry Corey. "The Conservative Supreme Court That Embraced Vaccine Mandates." Changing America. December 4, 2021. https://thehill.com/chang ing-america/opinion/584236-the-conservative-supreme-court-that-embraced-vacc ine-mandates.

Bialik, Mayim. *Beyond the Sling: A Real-Life Guide to Raising Confident, Loving Children the Attachment Parenting Way.* New York: Gallery Books, 2012.

Birkland, Thomas. "Focusing Events, Mobilization, and Agenda Setting." *Journal of Public Policy* 18, no. 1 (1998): 53–74.

Bishop, Bill. *The Big Sort: Why the Clustering of Like-Minded America Is Tearing Us Apart.* Boston: Mariner Books, 2009.

Biss, Eula. *On Immunity: An Inoculation.* Minneapolis, MN: Greywolf, 2014.

Black, Edwin. *War Against the Weak: Eugenics and America's Campaign to Create a Master Race.* New York: Dialog Press, 2003.

Blackman, Josh. "The Irrepressible Myth of Jacobson V. Massachusetts." *Buffalo Law Review* 70, no. 113 (2021).

Blake, Aaron. "The GOP's Vaccine Skeptic Wing Has a Breakthrough in Tennessee." *The Washington Post*, July 13, 2021. https://www.washingtonpost.com/politics/2021/07/13/gops-vaccine-skeptic-wing-has-breakthrough-tennessee.

Blake, Aaron. "The Question Republicans Still Can't Really Answer on Vaccine Mandates." *The Washington Post*, September 25, 2021. https://www.washingtonpost.com/politics/2021/09/25/question-republicans-still-cant-really-answer-vaccine-mandates.

Blake, Aaron. "Tucker Carlson's Worst Vaccine Segment Yet." *The Washington Post*, May 6, 2021.

Blanchfield, Luisa. "The United Nations Convention on the Rights of the Child." Congressional Research Service. 2015. https://crsreports.congress.gov/product/pdf/R/R40484/25.

Blank, Nina R., Arthur L. Caplan, and Catherine Constable. "Exempting Schoolchildren from Immunizations: States with Few Barriers Had Highest Rates of Nonmedical Exemptions." *Health Affairs* 32, no. 7 (2013): 1282–90. https://doi.org/10.1377/hlthaff.2013.0239.

Blendon, Robert, John Benson, Gillian SteelFisher, and John Connolly. "Americans' Conflicting Views About the Public Health System, and How to Shore up Support." *Health Affairs* 29, no. 11 (2010): 2033–40. https://doi.org/10.1377/hlthaff.2010.0262.

Blume, Stuart. "Anti-Vaccination Movements and Their Interpretations." *Social Science & Medicine* 62, no. 3 (2006): 628–42. http://dx.doi.org/10.1016/j.socscimed.2005.06.020.

Blustein, Jeffrey. *Parents and Children: The Ethics of the Family.* Oxford, UK: Oxford University Press, 1982.

Bogel-Burroughs, Nicholas. "Antivaccination Activists Are Growing Force at Virus Protests." *The New York Times*, May 2, 2020.

Boghani, Priyanka. "Dr. Paul Offit: 'A Choice Not to Get a Vaccine Is Not a Risk-Free Choice.'" *Frontline*, March 23, 2015. https://www.pbs.org/wgbh/frontline/article/paul-offit-a-choice-not-to-get-a-vaccine-is-not-a-risk-free-choice.

Boller, Paul F., Jr. *Not So! Popular Myths About America from Columbus to Clinton.* Oxford, UK: Oxford University Press, 1995.

Bonica, Adam, Howard Rosenthal, and David J. Rothman. "The Political Polarization of Physicians in the United States: An Analysis of Campaign Contributions to Federal Elections, 1991 Through 2012." *JAMA Internal Medicine* 174, no. 8 (2014): 1308–17. https://doi.org/10.1001/jamainternmed.2014.2105.

Bonn, Dorothy. "Texas Law Allows Conscientious Immunisation Exemptions." *Lancet Infectious Diseases* 3, no. 9 (2003): 525. https://doi.org/10.1016/s1473-3099(03)00751-5.

Bosman, Julie, Patricia Mazzei, and Dan Levin. "Jessica Biel Weighs in on Vaccine Fight, Drawing Fierce Pushback." *The New York Times*, June 13, 2019.

Boulis, Ann K., and Jerry A. Jacobs. *The Changing Face of Medicine.* New York: Cornell University Press, 2011.

Bowyer, Lynne. "The Ethical Grounds for the Best Interest of the Child." *Cambridge Quarterly of Healthcare Ethics* 25, no. 1 (2016): 63–69. https://doi.org/10.1017/s0963180115000298.

Bozzola, Elena, Giulia Spina, Rino Agostiniani, Sarah Barni, Rocco Russo, Elena Scarpato, Antonio Di Mauro, et al. "The Use of Social Media in Children and Adolescents: Scoping Review on the Potential Risks." *International Journal of Environmental Research and Public Health* 19, no. 16 (2022): 9960. https://www.mdpi.com/1660-4601/19/16/9960.

Bradford, W. D., and A. Mandich. "Some State Vaccination Laws Contribute to Greater Exemption Rates and Disease Outbreaks in the United States." *Health Affairs* 34, no. 8 (2015): 1383–90. https://doi.org/10.1377/hlthaff.2014.1428.

Bradley, Ethan, and Mark Navin. "Vaccine Refusal Is Not Free Riding." *Erasmus Journal for Philosophy and Economics* 14, no. 1 (2021): 167–81. https://doi.org/10.23941/ejpe. v14i1.555.

Brady, Jonann, and Stephanie Dahle. "Celeb Couple to Lead 'Green Vaccine' Rally—Jenny McCarthy and Jim Carrey Talk About Autism March on 'GMA.'" ABC News. June 4, 2008. https://abcnews.go.com/GMA/OnCall/story?id=4987758.

Brandt, Allan M. "Racism and Research: The Case of the Tuskegee Syphilis Study." *The Hastings Center Report* 8, no. 6 (1978): 21–29. https://doi.org/10.2307/3561468.

Brandt, Allan M., and Martha Gardner. "Antagonism and Accommodation: Interpreting the Relationship Between Public Health and Medicine in the United States During the 20th Century." *American Journal of Public Health (1971)* 90, no. 5 (2000): 707–15. https://doi.org/10.2105/ajph.90.5.707.

Brantlinger, Patrick. "Kipling's 'The White Man's Burden' and Its Afterlives." *English Literature in Transition, 1880–1920* 50, no. 2 (2007): 172–91. https://doi.org/10.1353/ elt.2007.0017.

Brennan, Jason. "A Libertarian Case for Mandatory Vaccination." *Journal of Medical Ethics* 44, no. 1 (2018): 37–43. https://doi.org/10.1136/medethics-2016-103486.

Brennan, Julia M., Robert A. Bednarczyk, Jennifer L. Richards, Kristen E. Allen, Gohar J. Warraich, and Saad B. Omer. "Trends in Personal Belief Exemption Rates Among Alternative Private Schools: Waldorf, Montessori, and Holistic Kindergartens in California, 2000–2014." *American Journal of Public Health* 107, no. 1 (2017): 108–12. https://doi.org/10.2105/ajph.2016.303498.

Brennan, Samantha, and Robert Noggle. "The Moral Status of Children: Children's Rights, Parents' Rights, and Family Justice." *Social Theory and Practice* 23, no. 1 (1997): 1–26. https://doi.org/10.5840/soctheorpract19972311.

Brennan, Zachary. "FDA Survey Finds Americans Don't Really Understand Drug Approvals." Regulatory Affairs Professionals Society. December 17, 2019. Accessed August 31, 2022. https://www.raps.org/news-and-articles/news-articles/2019/12/fda-survey-finds-americans-dont-really-understand.

Breslow, Jason, and Chris Amico. "What Are the Vaccine Exemption Laws in Your State? United States." *Frontline*, March 24, 2015. https://www.pbs.org/wgbh/frontline/article/ what-are-the-vaccine-exemption-laws-in-your-state.

Brewer, Noel T., Megan E. Hall, Teri L. Malo, Melissa B. Gilkey, Beth Quinn, and Christine Lathren. "Announcements Versus Conversations to Improve HPV Vaccination Coverage: A Randomized Trial." *Pediatrics* 139, no. 1 (2017): e20161764. https://doi. org/10.1542/peds.2016-1764.

Brighouse, Harry, and Adam Swift. *Family Values: The Ethics of Parent–Child Relationships.* Princeton, NJ: Princeton University Press, 2016.

British Medical Association. "Childhood Immunisation: A Guide for Healthcare Professionals." Board of Science and Education, British Medical Association, June 2003.

British Medical Association. "Delay in Making COVID Vaccine Mandatory Is 'Sensible' Ahead of Winter Pressures, Says BMA." News release. November 9, 2021. https://www. bma.org.uk/bma-media-centre/delay-in-making-covid-vaccine-mandatory-is-sensi ble-ahead-of-winter-pressures-says-bma.

Brockell, Gillian. "Mandatory Immunization for the Military: As American as George Washington." *The Washington Post*, August 26, 2021.

Brown, James. *The Reality of Human Vivisection: A Review*. 1901.

Brown, Phil, Brian Mayer, Stephen Zavestoski, Theo Luebke, Joshua Mandelbaum, and Sabrina McCormick. "The Health Politics of Asthma: Environmental Justice and Collective Illness Experience in the United States." *Social Science & Medicine* 57, no. 3 (2003): 453–64. https://doi.org/10.1016/s0277-9536(02)00375-1.

Brulliard, Karin. "CIA Vaccine Program Used in Bin Laden Hunt in Pakistan Sparks Criticism." *The Washington Post*, July 22, 2011. https://www.washingtonpost.com/world/asia-pacific/pakistan-fights-polio-in-shadow-of-cia-ruse/2011/07/21/gIQAQqmcSI_story.html.

Bruner, Charles, Discher Anne, and Hedy Chang. "Chronic Elementary Absenteeism: A Problem Hidden in Plain Sight." Attendance Works and Child & Family Policy Center. 2011. https://ies.ed.gov/ncee/edlabs/regions/west/relwestFiles/pdf/508_Chronic_Elementary_Absence_AW_C_FPC_2011.pdf.

Brunson, Emily K. "The Impact of Social Networks on Parents' Vaccination Decisions." *Pediatrics* 131, no. 5 (2013): e1397–e404. https://doi.org/10.1542/peds.2012-2452.

Bump, Philip. "Nearly Half of Republicans Agree with 'Great Replacement Theory.'" *The Washington Post*, May 9, 2022. https://www.washingtonpost.com/politics/2022/05/09/nearly-half-republicans-agree-with-great-replacement-theory.

Burnett, Sara, and Brian Slodyysko. "'Liberate Michigan!' Right-Wing Coalitions Rallying to End Lockdown Restrictions Amid Coronavirus Pandemic." *Chicago Tribune*, April 17, 2020. https://www.chicagotribune.com/coronavirus/ct-nw-coronavirus-protesters-trump-20200417-7oad7qhicrc53eywjaclwdsaoq-story.html.

Butrica, Barbara, A., and S. Karamcheva Nadia. "The Relationship Between Automatic Enrollment and DC Plan Contributions: Evidence from a National Survey of Older Workers." Proceedings. Annual Conference on Taxation and Minutes of the Annual Meeting of the National Tax Association 109 (2016): 1–33.

Buttenheim, Alison M., and David A. Asch. "Making Vaccine Refusal Less of a Free Ride." *Human Vaccines & Immunotherapeutics* 9, no. 12 (2013): 2674–75. https://doi.org/10.4161/hv.26676.

Buttenheim, Alison M., Malia Jones, and Yelena Baras. "Exposure and Vulnerability of California Kindergarteners to Intentionally Unvaccinated Children." *American Journal of Public Health* 102, 8, no. 1 (2012): 59–67. https://doi.org/10.2105/AJPH.2012.300821.

Buttenheim, Alison M., Malia Jones, Caitlin McKown, Daniel A. Salmon, and Saad B. Omer. "Conditional Admission, Religious Exemption Type, and Nonmedical Vaccine Exemptions in California Before and After a State Policy Change." *Vaccine* 36, no. 26 (2018): 3789–93. https://doi.org/https://doi.org/10.1016/j.vaccine.2018.05.050.

Calfas, J. "U.S. COVID-19 Deaths Top 800,000." *The Wall Street Journal*, December 14, 2021.

California Department of Public Health. "Statement on Timeline for COVID-19 Vaccine Requirements in Schools." April 14, 2022. Accessed August 31, 2022. https://www.cdph.ca.gov/Programs/OPA/Pages/NR22-073.aspx.

California Legislative Information. "SB-277 Public Health: Vaccinations (2015–2016)— Votes." 2015. Accessed August 26, 2022. https://leginfo.legislature.ca.gov/faces/billVotesClient.xhtml?bill_id=201520160SB277.

California Legislative Information. "SB-277 Public Health: Vaccinations, Bill History." 2015. https://leginfo.legislature.ca.gov/faces/billHistoryClient.xhtml?bill_id=2015201 60SB277.

California Legislative Information. "California Health and Safety Code; 120335(H)." 2016. https://leginfo.legislature.ca.gov/faces/codes_displaySection.xhtml?section Num=120335&lawCode=HSC.

California Legislative Information. "SB-276 Immunizations: Medical Exemptions (2019– 2020)—Votes." 2019. Accessed August 26, 2022. https://leginfo.legislature.ca.gov/ faces/billVotesClient.xhtml?bill_id=201920200SB276.

California Legislative Information. "Senate Bill No. 742, Vaccination Sites: Unlawful Activities: Obstructing, Intimidating, or Harassing." 2021. https://leginfo.legislature. ca.gov/faces/billNavClient.xhtml?bill_id=202120220SB742.

California Medical Association. "Senate Bill 277 Clears Senate Judiciary Committee." April 29, 2015. Accessed August 7, 2020. https://www.cmadocs.org/newsroom/news/ view/ArticleId/32506/Senate-Bill-277-clears-Senate-Judiciary-Committee.

California Medical Association. "Our Story." 2020. Accessed May 15, 2020. https://www. cmadocs.org/about.

California Senate. "Senate Health Committee, Wednesday, April 8, 2015." 2015. Accessed December 31, 2021. https://www.senate.ca.gov/media/senate-health-committee-53/ video.

California Senate. "Senate Judiciary Committee, Tuesday, April 28, 2015." 2015.

California.com. "California State Facts: The First Time the Golden State . . ." 2022. Accessed October 26, 2022. https://www.california.com/california-state-facts-first- time-golden-state.

Calvin, Kris. "SB 277 (Pan & Allen) Elimination of CA Personal Belief Exemption for School-Entry Vaccines." https://shorturl.at/gjxGY.

Canellos, Peter S., and Joel Lau. "The Surprisingly Strong Supreme Court Precedent Supporting Vaccine Mandates." *Politico*, 2021. https://www.politico.com/news/magaz ine/2021/09/08/vaccine-mandate-strong-supreme-court-precedent-510280.

Canon, Gabrielle. "'California Is America, Only Sooner': How the Progressive State Could Shape Biden's Policies." *The Guardian*, January 22, 2021. https://www.theguardian. com/us-news/2021/jan/22/biden-administration-california-kamala-harris-gavin- newsom.

Capitol Resource Institute. "Why Us?" 2022. Accessed September 23, 2022. https://www. capitolresource.org/why-us.

Caplan, Arthur L. "Is Disease Eradication Ethical?" *Lancet* 373, no. 9682 (2009): 2192–93. https://doi.org/10.1016/s0140-6736(09)61179-x.

Capozzola, Christopher. *Uncle Sam Wants You: World War I and the Making of the Modern American Citizen.* Oxford, UK: Oxford University Press, 2008.

Carless, W. "Month Before Buffalo Shooting, Poll Finds, 7 in 10 Republicans Believed in 'Great Replacement' Ideas." *USA Today*, May 5, 2022. https://www.usatoday.com/story/ news/nation/2022/06/01/great-replacement-theory-poll-republicans-democrats/746 1913001/?gnt-cfr=1.

Carlson, Samantha J., Gracie Edwards, Christopher C. Blyth, Barbara Nattabi, and Katie Attwell. "'Corona Is Coming': COVID-19 Vaccination Perspectives and Experiences Amongst Culturally and Linguistically Diverse West Australians." *Health Expectations* 25, no. 6 (2022): 3062–72. https://doi.org/10.1111/hex.13613.

Carlson, Samantha, Kerrie Wiley, and Peter McIntyre. "No Vax, No Visit'? If Mum Was Vaccinated Baby Is Already Protected Against Whooping Cough." The Conversation. May 30, 2016. https://theconversation.com/no-vax-no-visit-if-mum-was-vaccinated-baby-is-already-protected-against-whooping-cough-59374.

Carrese, Joseph A., Janet Malek, Katie Watson, Lisa Soleymani Lehmann, Michael J. Green, Laurence B. McCullough, Gail Geller, Clarence H. Braddock, and David J. Doukas. "The Essential Role of Medical Ethics Education in Achieving Professionalism: The Romanell Report." *Academic Medicine* 90, no. 6 (2015): 744–52. https://doi.org/10.1097/acm.0000000000000715.

Catechism of the Catholic Church. Vatican: Libreria Editrice Vaticana. 1993. https://www.vatican.va/archive/ENG0015/_INDEX.HTM.

Cave, Emma. "Voluntary Vaccination: The Pandemic Effect." *Legal Studies* 37, no. 2 (2017): 279–304.

CBS Sacramento. "Actor Rob Schneider Joins in Protest Against Anti-Vaccination Bill." CBS News. September 5, 2012. https://sacramento.cbslocal.com/2012/09/05/actor-rob-schneider-joins-in-protest-against-anti-vaccination-bill.

CBS San Francisco. "Lawmaker Against California Vaccine Bill SB277 Makes Comparison to Internment Camps." CBS News. June 9, 2015. https://sanfrancisco.cbslocal.com/2015/06/09/sb277-vaccine-bill-assemblyman-jim-patterson-internment-camps.

Centers for Disease Control and Prevention. "International Task Force for Disease Eradication." *MMWR Morbidity and Mortality Weekly Review* 39 (1990): 209–217.

Centers for Disease Control and Prevention. "Thimerosal in Vaccines: A Joint Statement of the American Academy of Pediatrics and the Public Health Service." *MMWR Morbidity and Mortality Weekly Report* 48, no. 26 (1999): 563–56.

Centers for Disease Control and Prevention. "Impact of the 1999 AAP/USPHS Joint Statement on Thimerosal in Vaccines on Infant Hepatitis B Vaccination Practices." *MMWR Morbidity and Mortality Weekly Report* 50, no. 6 (2001): 94–97.

Centers for Disease Control and Prevention. "National and State Vaccination Coverage Among Children Aged 19–35 Months—United States, 2010." *MMWR Morbidity and Mortality Weekly Report* 60, no. 34 (2011): 1157–63. Accessed September 23, 2022. https://www.cdc.gov/mmwr/preview/mmwrhtml/mm6034a2.htm.

Centers for Disease Control and Prevention. "Measles—United States, January 1–May 23, 2014." *MMWR Morbidity and Mortality Weekly Report* 63, no. 22 (2014): 496–99. https://www.cdc.gov/mmwr/preview/mmwrhtml/mm6322a4.htm.

Centers for Disease Control and Prevention. "Measles Outbreak—California, December 2014–February 2015." *MMWR Morbidity and Mortality Weekly Report* 64, no. 6 (2015): 153–54. https://www.cdc.gov/mmwr/preview/mmwrhtml/mm6406a5.htm.

Centers for Disease Control and Prevention. "The Vaccines for Children Program: At a Glance." 2016. Accessed July 29, 2022. https://www.cdc.gov/vaccines/programs/vfc/about/index.html.

Centers for Disease Control and Prevention. "VFC Childhood Vaccine Supply Policy." 2016. Accessed July 29, 2022. https://www.cdc.gov/vaccines/programs/vfc/about/vac-supply-policy/index.html#:~:text=The%20Omnibus%20Budget%20Reconciliation%20Act,funding%20for%20the%20VFC%20program.

Centers for Disease Control and Prevention. "Measles Cases and Outbreaks." 2021. Accessed December 21, 2021. https://www.cdc.gov/measles/cases-outbreaks.html.

Centers for Disease Control and Prevention. "Twitter, December 27, 2021," tweet, accessed March 11, 2022, https://twitter.com/cdcgov/status/1475587321277956096?lang=en.

Centers for Disease Control and Prevention. "Measles Cases and Outbreaks by Year (2010–2022)." 2022. Accessed August 19, 2022. https://www.cdc.gov/measles/cases-outbreaks.html.

Centers for Disease Control and Prevention. "Measles, Mumps, and Rubella (MMR) Vaccination: What Everyone Should Know." 2021. Accessed August 26, 2022. https://www.cdc.gov/vaccines/vpd/mmr/public/index.html.

Centers for Disease Control and Prevention. "COVID Data Tracker: COVID-19 Vaccinations in the United States." 2022. Accessed April 5, 2023. https://covid.cdc.gov/covid-data-tracker/#vaccinations_vacc-people-fully-percent-pop12.

Centers for Disease Control and Prevention. "Vaccine Information Statements (VISs)." 2022. https://www.cdc.gov/vaccines/hcp/vis/index.html.

Chabria, Anita. "California Mandatory Vaccination Bill Breezes Through Senate Committee." *The Guardian*, April 30, 2015.

Change.org. "Petition: Concerned Americans for Parental Rights & Vaccine Exemptions Oppose CA AB2109. 2012. https://www.change.org/p/concerned-americans-for-parental-rights-vaccine-exemptions-oppose-ca-ab2109.

Chantler, Tracey, Emilie Karafillakis, and James Wilson. "Vaccination: Is There a Place for Penalties for Non-Compliance?" *Applied Health Economics and Health Policy* 17, no. 3 (2019): 265–71. https://doi.org/10.1007/s40258-019-00460-z.

Chassiakos, Yolanda Reid, Jenny Radesky, Dimitri Christakis, Megan A. Moreno, Corinn Cross, David Hill, Nusheen Ameenuddin, et al. "Children and Adolescents and Digital Media." *Pediatrics* 138, no. 5 (2016): e20162593. https://doi.org/10.1542/peds.2016-2593.

Chemical and Engineering News. "San Francisco Bans Phthalates, Bisphenol A." June 12, 2006. Accessed October 26, 2022. https://cen.acs.org/articles/84/i24/San-Francisco-bans-phthalates-bisphenol.html.

Chen, Feifan, Yalin He, and Yuan Shi. "Parents' and Guardians' Willingness to Vaccinate Their Children Against COVID-19: A Systematic Review and Meta-Analysis." *Vaccines* 10, no. 2 (2022): 179.

Chen, Hsueh-Fen, and Saleema A. Karim. "Relationship Between Political Partisanship and COVID-19 Deaths: Future Implications for Public Health." *Journal of Public Health* 44, no. 3 (2022): 716–23. https://doi.org/10.1093/pubmed/fdab136.

Chen, Robert T. "Evaluation of Vaccine Safety After the Events of 11 September 2001: Role of Cohort and Case–Control Studies." *Vaccine* 22, no. 15 (2004): 2047–53. https://doi.org/10.1016/j.vaccine.2004.01.023.

Chen, Robert T., Suresh C. Rastogi, John R. Mullen, Scott W. Hayes, Stephen L. Cochi, Jerome A. Donlon, and Steven G. Wassilak. "The Vaccine Adverse Event Reporting System (VAERS)." *Vaccine* 12, no. 6 (1994): 542–50. https://doi.org/10.1016/0264-410x(94)90315-8.

Children's Hospital of Philadelphia. "About Paul A. Offit, Md." 2022. https://www.chop.edu/doctors/offit-paul-a.

Choi, James J., David Laibson, Brigitte C. Madrian, and Andrew Metrick. "Optimal Defaults." *American Economic Review* 93, no. 2 (2003): 180–85. https://doi.org/10.1257/000282803321947010.

Chua, Amy. *Political Tribes: Group Instinct and the Fate of Nations.* London: Penguin, 2018.

Circle of Mamas. "About Us." 2020. Accessed December 29, 2021. https://circleofmamas.com.

Cirillo, Vincent J. "Two Faces of Death: Fatalities from Disease and Combat in America's Principal Wars, 1775 to Present." *Perspectives in Biology and Medicine* 51, no. 1 (2008): 121–33. https://doi.org/10.1353/pbm.2008.0005.

Cleveland Clinic. "Surprising Dangers of Trampolines for Kids." 2020. Accessed March 11, 2022. https://health.clevelandclinic.org/surprising-dangers-of-trampolines-for-kids.

Clotfelter, Charles T. *After Brown: The Rise and Retreat of School Desegregation.* Princeton, NJ: Princeton University Press, 2006.

"Clueless: Celebrities Make Us Sick." *The Economist*, 2014. https://www.economist.com/united-states/2014/06/28/clueless.

Cohen, Shlomo. "Nudging and Informed Consent." *American Journal of Bioethics* 13, no. 6 (2013): 3–11. https://doi.org/10.1080/15265161.2013.781704.

Colby, Thomas B., and Peter J. Smith. "The Return of Lochner." *Cornell Law Review* 100, no. 3 (2015): 527–602.

Colgrove, James. *State of Immunity: The Politics of Vaccination in Twentieth-Century America.* Berkeley: University of California Press, 2006.

Colgrove, James. "Parents Were Fine with Sweeping School Vaccination Mandates Five Decades Ago—But COVID-19 May Be a Different Story." The Conversation, October 22, 2021. Accessed August 31, 2022. https://theconversation.com/parents-were-fine-with-sweeping-school-vaccination-mandates-five-decades-ago-but-covid-19-may-be-a-different-story-168899.

Colgrove, James, and Abigail Lowin. "A Tale of Two States: Mississippi, West Virginia, and Exemptions to Compulsory School Vaccination Laws." *Health Affairs* 35, no. 2 (2016): 348–55. https://doi.org/10.1377/hlthaff.2015.1172.

Conis, Elena. *Vaccine Nation.* Chicago: University of Chicago Press, 2015. http://press.uchicago.edu/ucp/books/book/chicago/V/bo14237741.html.

Conis, Elena. "The History of the Personal Belief Exemption." *Pediatrics* 145, no. 4. (2020): e20192551. https://doi.org/10.1542/peds.2019-2551.

Conis, Elena, and Jonathan Kuo. "Historical Origins of the Personal Belief Exemption to Vaccination Mandates: The View from California." *Journal of the History of Medicine and Allied Sciences* 76, no. 2 (2021): 167–90. https://doi.org/10.1093/jhmas/jrab003.

Connolly, Cynthia Anne. "Prevention Through Detention: The Pediatric Tuberculosis Preventorium Movement in the United States, 1909–1951." ProQuest Dissertations, 1999.

Connolly, Griffin. "Democrats Want to Fine Colleagues $1,000 for Not Wearing Masks in Congress, as Three Fall Ill with COVID." Independent, January 12, 2021. https://www.independent.co.uk/news/world/americas/us-politics/covid-masks-house-democrats-1000-fine-b1786276.html.

Connolly, Máire A., and David L. Heymann. "Deadly Comrades: War and Infectious Diseases." *Lancet* 360, no. 1 (2002): s23–s24. https://doi.org/10.1016/s0140-6736(02)11807-1.

"Contact Political Representatives to Express Opposition to AB2109." Facebook. 2012. Accessed February 16, 2021.

Cook, Chris. "Why Are Steiner Schools So Controversial?." BBC Newsnight. August 4, 2014.

Cookingham, Lisa M., and Ginny L. Ryan. "The Impact of Social Media on the Sexual and Social Wellness of Adolescents." *Journal of Pediatric & Adolescent Gynecology* 28, no. 1 (2015): 2–5. https://doi.org/10.1016/j.jpag.2014.03.001.

Coontz, Stephanie. *The Way We Never Were: American Families and the Nostalgia Trap.* New York: Basic Books, 2016.

Court, Jay, Stacy M. Carter, Katie Attwell, Julie Leask, and Kerrie Wiley. "Labels Matter: Use and Non-Use of 'Anti-Vax' Framing in Australian Media Discourse 2008–2018." *Social Science & Medicine* 291 (2021): 114502. https://doi.org/https://doi.org/10.1016/j.socscimed.2021.114502.

Crawshaw, Alison F., Yasmin Farah, Anna Deal, Kieran Rustage, Sally E. Hayward, Jessica Carter, Felicity Knights, et al. "Defining the Determinants of Vaccine Uptake and Undervaccination in Migrant Populations in Europe to Improve Routine and COVID-19 Vaccine Uptake: A Systematic Review." *Lancet Infectious Diseases* 22, no. 9 (2022): E254–E66. https://doi.org/10.1016/s1473-3099(22)00066-4.

Cremin, Lawrence. *American Education: The National Experience, 1783–1876.* New York: HarperCollins, 1980.

Crenna, Stefano, Antonio Osculati, and Silvia D. Visonà. "Vaccination Policy in Italy: An Update." *Journal of Public Health Research* 7, no. 3 (2018): 1523–23. https://doi.org/10.4081/jphr.2018.1523.

Crossley, Nick. "RD Laing and the British Anti-Psychiatry Movement: A Socio-Historical Analysis." *Social Science & Medicine* 47, no. 7 (1998): 877–89.

Cutler, Alison. "Doctor Wrote Bogus COVID Vaccine Exemptions for Patients, Washington Officials Say." *The News Tribune*, December 30, 2021. https://www.thenewstribune.com/news/nation-world/national/article256945317.html.

Daniels, R. Steven. "The Evolution of Attitudes on Same-Sex Marriage in the United States, 1988–2014." *Social Science Quarterly* 100, no. 5 (2019): 1651–63. https://doi.org/10.1111/ssqu.12673.

Dare, Tim. "Mass Immunisation Programmes: Some Philosophical Issues." *Bioethics* 12, no. 2 (1998): 125–49. https://doi.org/10.1111/1467-8519.00100.

Davis, Elliott. "States Are Banning COVID-19 Vaccine Requirements." *U.S. News & World Report*, April 30, 2021. https://www.usnews.com/news/best-states/articles/2021-04-30/these-states-are-banning-covid-19-vaccine-requirements.

Davis, Robert L. "Measles–Mumps–Rubella and Other Measles-Containing Vaccines Do Not Increase the Risk for Inflammatory Bowel Disease: A Case–Control Study from the Vaccine Safety Datalink Project." *JAMA* 285, no. 24 (2001): 3073.

de Freytas-Tamura, Kimiko. "Bastion of Anti-Vaccine Fervor: Progressive Waldorf Schools." *The New York Times*, June 13 2019.

De Kruif, Paul *Microbe Hunters*. San Diego, CA: Harcourt Brace Jovanovich, 1954.

de Tocqueville, Alexis. *Democracy in America: A New Translation by Arthur Goldhammer*, edited and translated by A. Goldhammer. New York: Library of America, 2004.

Dean, C. "Letter from Santa Barbara County Health Officer to County School and Childcare Administrators." 2016. https://www.avoiceforchoiceadvocacy.org/wp-content/uploads/2016/06/SBCPHD-MEPP-Letter.pdf.

Dekker, Sidney. *Drift into Failure*. Burlington, VT: Ashgate, 2012.

Dee, Deborah L., Ruowei Li, Li-Ching Lee, and Laurence Grummer-Strawn. "Associations Between Breastfeeding Practices and Young Children's Language and Motor Skill Development." *Pediatrics* 119, Supplement 1 (2007): S92–S98. https://doi.org/10.1542/peds.2006-2089N.

Delamater, Paul L., Timothy F. Leslie, and Y. Tony Yang. "Changes in Medical Exemptions from Immunization in California After Elimination of Personal Belief Exemptions." *JAMA* 318, no. 9 (2017): 863–64. https://doi.org/10.1001/jama.2017.9242.

Delamater, Paul L., S. Cassandra Pingali, Alison M. Buttenheim, Daniel A. Salmon, Nicola P. Klein, and Saad B. Omer. "Elimination of Nonmedical Immunization Exemptions in California and School-Entry Vaccine Status." *Pediatrics* 143, no. 6 (2019): 1–9. https://doi.org/10.1542/peds.2018-3301.

Dempsey, Amanda, Bethany M. Kwan, Nicole M. Wagner, Jennifer Pyrzanowski, Sarah E. Brewer, Carter Sevick, Komal Narwaney, Kenneth Resnicow, and Jason Glanz. "A Values-Tailored Web-Based Intervention for New Mothers to Increase Infant Vaccine Uptake: Development and Qualitative Study." *Journal of Medical Internet Research* 22, no. 3 (2020): e15800. https://doi.org/10.2196/15800.

Denemark, David, Tauel Harper, and Katie Attwell. "Vaccine Hesitancy and Trust in Government: A Cross-National Analysis." *Australian Journal of Political Science* 57, no. 2 (2022): 145–63. https://doi.org/10.1080/10361146.2022.2037511.

DePanfilis, Diane, and Marsha K. Salus. *Child Protective Services: A Guide for Caseworkers.* Washington, DC: U.S. Department of Health and Human Services, Administration for Children and Families, Administration on Children, Youth and Families, Children's Bureau, Office on Child Abuse and Neglect, 2003.

Department of Health and Social Care and The Rt Hon Sajid Javid MP. "Oral Statement to Parliament on Vaccines as a Condition of Deployment, 31 January 2022." News release. 2022. https://www.gov.uk/government/speeches/oral-statement-on-vaccines-as-a-condition-of-deployment.

"Details of Rally to Oppose Ab2109." Facebook. 2012. Accessed February 16, 2021.

Dettlaff, Alan J., Stephanie L. Rivaux, Donald J. Baumann, John D. Fluke, Joan R. Rycraft, and Joyce James. "Disentangling Substantiation: The Influence of Race, Income, and Risk on the Substantiation Decision in Child Welfare." *Children and Youth Services Review* 33, no. 9 (2011): 1630–37. https://doi.org/10.1016/j.childyouth.2011.04.005.

Dickson, E. J. "A Guide to 17 Anti-Vaccination Celebrities." *Rolling Stone*, 2019. https://www.rollingstone.com/culture/culture-features/celebrities-anti-vaxxers-jessica-biel-847779.

Diekema, Douglas S. "Parental Refusals of Medical Treatment: The Harm Principle as Threshold for State Intervention." *Theoretical Medicine and Bioethics* 25, no. 4 (2004): 243–64. https://doi.org/10.1007/s11017-004-3146-6.

Diekema, Douglas S. "Personal Belief Exemptions from School Vaccination Requirements." *Annual Review of Public Health* 35 (2014): 275–92. https://doi.org/10.1146/annurev-publhealth-032013-182452.

Diekema, Douglas S. "Physician Dismissal of Families Who Refuse Vaccination: An Ethical Assessment." *Journal of Law, Medicine & Ethics* 43, no. 3 (2015): 654–60. https://doi.org/10.1111/jlme.12307.

Diekema, Douglas S. "Rhetoric, Persuasion, Compulsion, and the Stubborn Problem of Vaccine Hesitancy." *Perspectives in Biology and Medicine* 65, no. 1 (2022): 106–23. https://doi.org/10.1353/pbm.2022.0006.

Diekema, Douglas S., and the American Academy of Pediatrics Committee on Bioethics. "Responding to Parental Refusals of Immunization of Children." *Pediatrics* 115, no. 5 (2005): 1428–31.

Dimala, Christian Akem, Benjamin Momo Kadia, Miriam Aiwokeh Mbong Nji, and Ndemazie Nkafu Bechem. "Factors Associated with Measles Resurgence in the United States in the Post-Elimination Era." *Scientific Reports* 11, no. 1 (2021): 51. https://doi.org/10.1038/s41598-020-80214-3.

DiResta, Renée. "How California's Terrible Vaccination Policy Puts Kids at Risk." November 19, 2014. https://blog.noupsi.de/post/103050754027/vaxviz.

DiResta, Renée. "Personal Exemptions from Reason." *Slate*, no. 8, April 2015. http://www.slate.com/articles/health_and_science/medical_examiner/2015/04/california_anti_vaccine_movement_politics_wealth_bob_sears_and_robert_f.html.

DiResta, Renée. "A Win for Evidence-Based Policy, and California Kids." July 3, 2015. https://blog.noupsi.de/post/123134500352/sb277.

DiResta, Renée. "Renée Diresta." 2021. http://www.reneediresta.com.

Dougherty, William J. "Community Organization for Immunization Programs." *Medical Clinics of North America* 51 (1967): 837–42.

Downs, Jim. "Reconstructing an Epidemic. Smallpox Among Former Slaves, 1862–1868." In *Sick from Freedom: African-American Illness and Suffering During the Civil War and Reconstruction*, 65–93. Oxford, UK: Oxford University Press, 2012.

Drew, Liam. "Did COVID Vaccine Mandates Work? What the Data Say." *Nature*, News Feature, July 6, 2022. https://www.nature.com/articles/d41586-022-01827-4.

Driver, Julia. "The History of Utilitarianism." In *The Stanford Encyclopedia of Philosophy*, edited by Edward N. Zalta. Stanford, CT: Metaphysics Research Lab, Stanford University, 2014.

Dubé, Eve, Dominique Gagnon, and Noni E. MacDonald. "Strategies Intended to Address Vaccine Hesitancy: Review of Published Reviews." *Vaccine* 33, no. 34 (2015): 4191–203.

Dubé, Eve, Caroline Laberge, Maryse Guay, Paul Bramadat, Real Roy, and Julie Bettinger. "Vaccine Hesitancy: An Overview." *Human Vaccines and Immunotherapeutics* 9, no. 8 (2013): 1763–73.

Dubé, Eve, M. Vivion, and Noni E. MacDonald. "Vaccine Hesitancy, Vaccine Refusal and the Anti-Vaccine Movement: Influence, Impact, and Implications." *Expert Review of Vaccines* 14, no. 1 (2015): 99–117.

Duffy, John. "School Vaccination: The Precursor to School Medical Inspection." *Journal of the History of Medicine and Allied Sciences* 33, no. 3 (1978): 344–55. https://doi.org/10.1093/jhmas/XXXIII.3.344.

Duffy, John. "The American Medical Profession and Public Health: From Support to Ambivalence." *Bulletin of the History of Medicine* 53, no. 1 (1979): 1–22.

Duffy, John. *The Rudolph Matas History of Medicine in Louisiana Volume I.* Baton Rouge: Louisiana State University Press, 1985.

Duffy, John. *The Sanitarians: A History of American Public Health.* Urbana: University of Illinois Press, 1990.

Dumit, Joseph. *Drugs for Life: How Pharmaceutical Companies Define Our Health.* Durham, NC: Duke University Press, 2012.

Dunn, Lauren, and Linda Carroll. "Some Doctors Helping Anti-Vaccine Parents Get Medical Exemptions." NBC News. 2019. https://www.nbcnews.com/health/kids-health/some-doctors-helping-anti-vaccine-parents-get-medical-exemptions-n963011.

Durbach, Nadja. "Class, Gender, and the Conscientious Objector to Vaccination, 1898–1907." *Journal of British Studies* 41, no. 1 (2002): 58–83. https://doi.org/10.1086/386254.

Durbach, Nadja. *Bodily Matters: The Anti-Vaccination Movement in England, 1853–1907.* Durham, NC: Duke University Press, 2005.

Durkee, Alison. "Florida Gov. Desantis Signs Sweeping Legislation Restricting Vaccine Mandates, School Mask Rules and More." *Forbes*, 2021. https://www.forbes.com/sites/alisondurkee/2021/11/18/florida-gov-desantis-signs-sweeping-legislation-restricting-vaccine-mandates-school-mask-rules-and-more/?sh=52150a7c77cf.

Dyer, Owen. "Republican Candidates Cast Doubt on Vaccines in US Presidential Debate." *BMJ* 351 (2015). https://doi.org/http://dx.doi.org/10.1136/bmj.h5006.

Dyer, Owen. "Philippines Measles Outbreak Is Deadliest Yet as Vaccine Scepticism Spurs Disease Comeback." *BMJ* 364 (2019): 1739. https://doi.org/https://doi.org/10.1136/bmj.l739

Eddy, Melissa. "Germany Mandates Measles Vaccine." *The New York Times*, November 14, 2019. https://www.nytimes.com/2019/11/14/world/europe/germany-measles-vaccine.html.

"Editorial: Anti-Vaxxers Have Found a Way Around California's Strict New Immunization Law. They Need to Be Stopped." *Los Angeles Times*, November 8, 2017. http://www.latimes.com/opinion/editorials/la-ed-vaccine-exemption-crackdown-20171108-story.html.

Edsall, Thomas B. "When the Mask You're Wearing 'Tastes Like Socialism.'" *The New York Times*, 2020. https://www.nytimes.com/2020/05/20/opinion/coronavirus-trump-partisanship.html.

Edwards, Kathryn M., Jesse M. Hackell, Carrie L. Byington, Yvonne A. Maldonado, Elizabeth D. Barnett, H. Dele Davies, Kathryn M. Edwards, et al. "Countering Vaccine Hesitancy." *Pediatrics* 138, no. 3 (2016): e1. https://doi.org/10.1542/peds.2016-2146.

Ehrenreich, John. *The Cultural Crisis of Modern Medicine*, edited by John Ehrenreich. New York: Monthly Review Press, 1978.

Emerson, Claudia I., and Peter A. Singer. "Is There an Ethical Obligation to Complete Polio Eradication?" *Lancet* 375, no. 9723 (2010): 1340–41. https://doi.org/10.1016/s0140-6736(10)60565-x.

Environmental Law Institute. "Mary D. Nichols and State of California Receive Environmental Achievement Award—Special Award Presentation by Rep. Henry Waxman." 2014. Accessed October 26, 2022. https://www.eli.org/mary-d-nichols-and-state-california-receive-environmental-achievement-award-special-award.

Evans, John H. *The History and Future of Bioethics: A Sociological View*. Oxford, UK: Oxford University Press, 2014.

Ewert, Cody D. *Making Schools American: Nationalism and the Origin of Modern Educational Politics*. Baltimore, MD: Johns Hopkins University Press, 2022.

Faden, Ruth R., Tom L. Beauchamp, and Nancy M. P. King. *A History and Theory of Informed Consent*. New York: Oxford University Press, 1986.

FAIR Health. "Costs for a Hospital Stay for COVID-19." 2021. http://www.fairhealth.org/article/costs-for-a-hospital-stay-for-covid-19.

Farrell, Henry. "The Travesty of Liberalism." Crooked Timber. March 21, 2018. https://crookedtimber.org/2018/03/21/liberals-against-progressives/#comment-729288.

Feemster, Kristen A. "Overview: Special Focus Vaccine Acceptance." *Human Vaccines & Immunotherapeutics* 9, no. 8 (2013): 1752–54. https://doi.org/10.4161/hv.26217.

Feinberg, Joel. "The Child's Right to an Open Future." In *Whose Child? Children's Rights, Parental Authority and State Power*, edited by William Aiken and Hugh La Follett, 124–53. Totowa, NJ: Rowman & Littlefield, 1980.

Feinberg, Joel. *Harm to Self*. New York: Oxford University Press, 1986.

Fenner, Frank. *Smallpox and Its Eradication*. Geneva, Switzerland: World Health Organization, 1988.

Fenner, Frank, A. J. Hall, and Walter R. Dowdle. "What Is Eradication?" In *The Eradication of Infectious Disease*, edited by Walter R. Dowdle and Donald R. Hopkins, 3–17. New York: Wiley, 1998.

Fernandes, Breanna, Mark C. Navin, Dorit Reiss, Saad B. Omer, and Katie Attwell. "US State-Level Legal Interventions Related to COVID-19 Vaccine Mandates." *JAMA* 327, no. 2 (2021): 178–79. https://doi.org/10.1001/jama.2021.22122.

Fetters, Ashley, and Gerrit De Vynck. "How Wellness Influencers Are Fueling the Anti-Vaccine Movement." *The Washington Post*, September 12, 2021.

Feudtner, Chris, and Edgar K. Marcuse. "Ethics and Immunization Policy: Promoting Dialogue to Sustain Consensus." *Pediatrics* 107, no. 5 (2001): 1158–64. https://doi.org/10.1542/peds.107.5.1158.

Findling, Mary G., Robert J. Blendon, and John M. Benson. "Polarized Public Opinion About Public Health During the COVID-19 Pandemic: Political Divides and Future Implications." *JAMA Health Forum* 3, no. 3 (2022): e220016. https://doi.org/10.1001/jamahealthforum.2022.0016.

Fine Maron, Dina "Improved Vaccination Rates Would Fall Victim to Senate Health Cuts." Scientific American., July 10, 2017. https://www.scientificamerican.com/article/improved-vaccination-rates-would-fall-victim-to-senate-health-cuts.

Fisher, Barbara Loe. "The Moral Right to Conscientious, Philosophical and Personal Belief Exemption to Vaccination." National Vaccine Information Center. 1997. https://www.nvic.org/vaccination-decisions/informed-consent.

Flanigan, Jessica. "A Defense of Compulsory Vaccination." *HEC Forum* 26, no. 1 (2014): 5–25. https://doi.org/10.1007/s10730-013-9221-5.

Flusberg, Stephen J., Teenie Matlock, and Paul H. Thibodeau. "War Metaphors in Public Discourse." *Metaphor and Symbol* 33, no. 1 (2018): 1–8. https://doi.org/https://doi.org/10.1080/10926488.2018.1407992.

Fombonne, Eric, Rita Zakarian, Andrew Bennett, Linyan Meng, and Diane McLean-Heywood. "Pervasive Developmental Disorders in Montreal, Quebec, Canada: Prevalence and Links with Immunizations." *Pediatrics* 118, no. 1 (2006): e139–e50. https://doi.org/10.1542/peds.2005-2993.

Foulkes, Mathieu. "How the Spiritual 'Waldorf' Movement Is Connected to German Vaccine Scepticism." The Local de. 2021. https://www.thelocal.de/20211123/how-a-spiritual-movement-is-connected-to-german-vaccine-scepticism/.

Fox, Ellen, Robert M. Arnold, and Baruch Brody. "Medical Ethics Education: Past, Present, and Future." *Academic Medicine* 70, no. 9 (1995): 761–68.

Freeman, Samuel. "Illiberal Libertarians: Why Libertarianism Is Not a Liberal View." *Philosophy & Public Affairs* 30, no. 2 (2001): 105–51.

French, David. *Divided We Fall: America's Secession Threat and How to Restore Our Nation.* New York: St. Martin's, 2020.

Fridman, Ariel, Rachel Gershon, and Ayelet Gneezy. "COVID-19 and Vaccine Hesitancy: A Longitudinal Study." *PLoS One* 16, no. 4 (2021): e0250123–e23. https://doi.org/10.1371/journal.pone.0250123.

Gagneur, Arnaud, Julie Bergeron, Virginie Gosselin, Anne Farrands, and Geneviève Baron. "A Complementary Approach to the Vaccination Promotion Continuum: An Immunization-Specific Motivational-Interview Training for Nurses." *Vaccine* 37, no. 20 (2019): 2748–56. https://doi.org/https://doi.org/10.1016/j.vaccine.2019.03.076.

Gagneur, Arnaud, Thomas Lemaître, Virginie Gosselin, Anne Farrands, Nathalie Carrier, Geneviève Petit, Louis Valiquette, and Philippe De Wals. "A Postpartum Vaccination Promotion Intervention Using Motivational Interviewing Techniques Improves Short-Term Vaccine Coverage: Promovac Study." *BMC Public Health* 18, no. 1 (2018): 811. https://doi.org/10.1186/s12889-018-5724-y.

Gallup. "Confidence in Institutions." 2021. https://news.gallup.com/poll/1597/confide nce-institutions.aspx.

Gambrell, Ashley, Maria Sundaram, and Robert A. Bednarczyk. "Estimating the Number of US Children Susceptible to Measles Resulting from COVID-19-Related Vaccination Coverage Declines." *Vaccine* 40, no. 32 (2022): 4574–79. https://doi.org/10.1016/j.vacc ine.2022.06.033.

Gamlund, Espen, Karl Erik Müller, Kathrine Knarvik Paquet, and Carl Tollef Solberg. "Mandatory Childhood Vaccination: Should Norway Follow?" *Etikk i praksis—Nordic Journal of Applied Ethics* 14, no. 1 (2020): 7–27. https://doi.org/10.5324/eip.v14i1.3316.

Garcia, Ginny E., Richard Lewis, and Joanne Ford-Robertson. "Attitudes Regarding Laws Limiting Black–White Marriage: A Longitudinal Analysis of Perceptions and Related Behaviors." *Journal of Black Studies* 46, no. 2 (2015): 199–217. https://doi.org/10.1177/ 0021934714568017.

Garver, Kenneth L., and Bettylee Garver. "Historical Perspectives: Eugenics: Past, Present, and the Future." *American Journal of Human Genetics* 49, no. 5 (1991): 1109–18.

Gawande, Atul. *The Checklist Manifesto: How to Get Things Right.* London: Profile Books, 2010.

Gecewicz, Claire. "'New Age' Beliefs Common Among Both Religious and Nonreligious Americans." Pew Research Center. October 1, 2018. https://www.pewresearch.org/ fact-tank/2018/10/01/new-age-beliefs-common-among-both-religious-and-nonreligi ous-americans.

George, Daniel R., Erin R. Whitehouse, and Peter J. Whitehouse. "Asking More of Our Metaphors: Narrative Strategies to End the 'War on Alzheimer's' and Humanize Cognitive Aging." *American Journal of Bioethics* 16, no. 10 (2016): 22–24. https://doi. org/10.1080/15265161.2016.1214307.

Gidengil, Courtney, Christine Chen, Andrew M. Parker, Sarah Nowak, and Luke Matthews. "Beliefs Around Childhood Vaccines in the United States: A Systematic Review." *Vaccine* 37, no. 45 (2019): 6793–802. https://doi.org/http://dx.doi.org/ 10.1016/j.vaccine.2019.08.068.

Gillam, Lynn. "The Zone of Parental Discretion: An Ethical Tool for Dealing with Disagreement Between Parents and Doctors About Medical Treatment for a Child." *Clinical Ethics* 11, no. 1 (2016): 1–8. https://doi.org/https://doi.org/10.1177/147775091 5622033.

Giubilini, Alberto. *The Ethics of Vaccination.* Cham, Switzerland: Palgrave Pivot, 2019.

Giubilini, Alberto. "An Argument for Compulsory Vaccination: The Taxation Analogy." *Journal of Applied Philosophy* 37, no. 3 (2020): 446–66. https://doi.org/10.1111/ japp.12400.

Giubilini, Alberto, Thomas Douglas, and Julian Savulescu. "The Moral Obligation to Be Vaccinated: Utilitarianism, Contractualism, and Collective Easy Rescue." *Medicine, Health Care, and Philosophy* 21, no. 4 (2018): 547–60. https://doi.org/10.1007/s11 019-018-9829-y.

Giubilini, Alberto, Sharyn Milnes, and Julian Savulescu. "The Medical Ethics Curriculum in Medical Schools: Present and Future." *J Clinical Ethics* 27, no. 2 (2016): 129–45.

Gladwell, Malcolm "The Mosquito Killer." *The New Yorker*, July 2, 2001. https://www. newyorker.com/magazine/2001/07/02/the-mosquito-killer.

Glick, Shimon M. "The Teaching of Medical Ethics to Medical Students." *Journal of Medical Ethics* 20, no. 4 (1994): 239–43. https://doi.org/10.1136/jme.20.4.239.

Goldenberg, Maya J. *Vaccine Hesitancy: Public Trust, Expertise and the War on Science.* Pittsburgh, PA: University of Pittsburgh Press, 2021.

Goldstein, Amy. "Anger over Mask Mandates, Other COVID Rules, Spurs States to Curb Power of Public Health Officials." *The Washington Post*, December 25, 2021.

Goldstein, Neal D., and Joanna S. Suder. "Towards Eliminating Nonmedical Vaccination Exemptions Among School-Age Children." *Delaware Journal of Public Health* 8, no. 1 (2022): 84–88. https://doi.org/10.32481/djph.2022.03.014.

Goldstein, Neal D., Joanna S. Suder, and Jonathan Purtle. "Trends and Characteristics of Proposed and Enacted State Legislation on Childhood Vaccination Exemption, 2011–2017." *American Journal of Public Health* 109, no. 1 (2019): 102. https://doi.org/http://dx.doi.org/10.2105/AJPH.2018.304765.

Gomez, Melissa, and Howard Blume ———. "Parents in California Protest Student COVID-19 Vaccine Mandate, Keep Kids Home." *Los Angeles Times*, October 18, 2021.

Gorski, David. "California Bill AB 2109: The Antivaccine Movement Attacks School Vaccine Mandates Again." Science-Based Medicine, March 26, 2012. https://sciencebasedmedicine.org/antivaccine-activists-attack-vaccine-mandates.

Gostin, Lawrence O. "When Terrorism Threatens Health: How Far Are Limitations on Personal and Economic Liberties Justified?" *Florida Law Review* 55, no. 5 (2003): 1105–70.

Gostin, Lawrence O. "Jacobson V Massachusetts at 100 Years: Police Power and Civil Liberties in Tension." *American Journal of Public Health* 95, no. 4 (2005): 576–81. https://doi.org/10.2105/ajph.2004.055152.

Gostin, Lawrence O., and Lindsey Wiley. *Public Health Law: Power, Duty, Restraint.* 3rd ed. Oakland: University of California Press, 2016.

Gotzsche, Peter. *Deadly Medicines and Organised Crime: How Big Pharma Has Corrupted Healthcare.* London: Routledge, 2013.

Government of Canada. "COVID-19 Vaccination in Canada—Vaccination Coverage." 2022. Accessed April 15, 2023. https://health-infobase.canada.ca/covid-19/vaccination-coverage/.

Government of Great Britain. *Labour Statistics. Returns of Wages Published Between 1830 and 1886.* London: Printed for her Majesty's Stationery Office by Eyre and Spottiswoode, 1887.

Government of the United Kingdom. "Consultation on Removing Vaccination as a Condition of Deployment for Health and Social Care Staff." News release. January 31, 2022. https://www.gov.uk/government/news/consultation-on-removing-vaccination-as-a-condition-of-deployment-for-health-and-social-care-staff.

Government of the United Kingdom. "Coronavirus (COVID-19) in the UK—Vaccination in the United Kingdom." 2022. Accessed July 5, 2022, https://coronavirus.data.gov.uk/details/vaccinations.

Gramlich, John. "Young Americans Are Less Trusting of Other People—and Key Institutions—Than Their Elders." Pew Research Center. 2019. https://www.pewresearch.org/fact-tank/2019/08/06/young-americans-are-less-trusting-of-other-people-and-key-institutions-than-their-elders.

Graves, John A., Khrysta Baig, and Melinda Buntin. "The Financial Effects and Consequences of COVID-19: A Gathering Storm." *JAMA* 326, no. 19 (2021): 1909–10. https://doi.org/10.1001/jama.2021.18863.

Greene, Jamal. *How Rights Went Wrong: Why Our Obsession with Rights Is Tearing America Apart.* Boston, MA: Houghton Mifflin Harcourt, 2021.

Greene, Jeremy A. *Prescribing by Numbers: Drugs and the Definition of Disease*. Baltimore, MD: Johns Hopkins University Press, 2008.

Grimes, Katy. "California Lawmakers Fast-Tracking Child Health Bills to Erode Parental Rights." California Globe. February 9, 2022. https://californiaglobe.com/articles/california-lawmakers-fast-tracking-child-health-bills-to-erode-parental-rights.

Grisso, Thomas, and Paul S. Appelbaum. *Assessing Competence to Consent to Treatment : A Guide for Physicians and Other Health Professionals*. New York: Oxford University Press, 1998.

Groll, Daniel. "Four Models of Family Interests." *Pediatrics* 134, Supplement 2 (2014): S81–S86. https://doi.org/10.1542/peds.2014-1394C.

Gromis, Ashley, and Ka-Yuet Liu. "Spatial Clustering of Vaccine Exemptions on the Risk of a Measles Outbreak." *Pediatrics* 149, no. 1 (2022): e2021050971. https://doi.org/10.1542/peds.2021-050971.

Guillen, Alex. "Impact of Supreme Court's Climate Ruling Spreads." *Politico*, July 20, 2022. https://www.politico.com/news/2022/07/20/chill-from-scotus-climate-ruling-hits-wide-range-of-biden-actions-00045920.

Gumbel, Andrew. "US States Face Fierce Protests from Anti-Vaccine Activists." *The Guardian*, April 10, 2015. https://www.theguardian.com/us-news/2015/apr/10/anti-vaccine-protest-california-facts.

Gutierrez, Melody. "California's School Immunizations Bill Passes Another Committee." SFGate, April 28, 2015. https://www.sfgate.com/news/article/California-s-school-immunizations-bill-passes-6229962.php.

Gutierrez, Melody. "Gov. Newsom Criticized the New Vaccine Bill. Anti-Vaccine Activists Are Celebrating." *Los Angeles Times*, June 4, 2019.

Gutierrez, Melody. "California Vaccine Bill Undergoes Major Changes and Wins Support of Former Critic Newsom." *Los Angeles Times*, June 18, 2019.

Gutierrez, Melody. "Anti-Vaccine Activist Assaults California Vaccine Law Author, Police Say." *Los Angeles Times*, August 21, 2019.

Gutierrez, Melody. "Vaccine Bill to Follow Newsom's Changes; California Governor and Senator Reach a Deal to Scale Back Parts of the Legislation and Add New Scrutiny." *Los Angeles Times*, September 7, 2019.

Gutierrez, Melody. "Vaccine Bills Are Signed Amid Protests; Stricter Rules Will Increase Oversight of Doctors' Exemptions." *Los Angeles Times*, September 7, 2019.

Gutierrez, Melody. "Anti-Vaccine Activists, Mask Opponents Target Public Health Officials—at Their Homes." *Los Angeles Times*, June 18, 2020.

Gutierrez, Melody. "Bill to Allow Minors to Be Vaccinated without Parental Consent Is Withdrawn." *Los Angeles Times, 31 August 2022* (Los Angeles), 2021.

Haberman, Maggie. "Trump and His Wife Received Coronavirus Vaccine Before Leaving the White House." *The New York Times*, March 1, 2021. https://www.nytimes.com/2021/03/01/us/politics/donald-trump-melania-coronavirus-vaccine.html.

Hacker, Jacob S. "Privatizing Risk Without Privatizing the Welfare State: The Hidden Politics of Social Policy Retrenchment in the United States." *American Political Science Review* 98, no. 2 (2004): 243–60. https://doi.org/10.1017/s0003055404001121.

Hafferty, Frederic W., and Ronald Franks. "The Hidden Curriculum, Ethics Teaching, and the Structure of Medical Education." *Academic Medicine* 69, no. 11 (1994): 861–71. https://doi.org/10.1097/00001888-199411000-00001.

Haidt, Jonathon. *The Righteous Mind: Why Good People Are Divided by Politics and Religion*. New York: Penguin, 2012.

Hall, Madison. "How a Supreme Court Decision from 1905 Set the Stage for Vaccine Mandates." Insider. September 11, 2021. https://www.businessinsider.com/supreme-court-decision-from-1905-set-stage-for-vaccine-mandates-2021-9.

Halpern, Scott D., Peter A. Ubel, and David A. Asch. "Harnessing the Power of Default Options to Improve Health Care." *New England Journal of Medicine* 357, no. 13 (2007): 1340–44. https://doi.org/10.1056/NEJMsb071595.

Halpern, Sydney Ann. *American Pediatrics: The Social Dynamics of Professionalism, 1880–1980*. Berkeley: University of California Press, 1988.

Hannah, Mitchell. "12 States Banning COVID-19 Vaccine Mandates & How They Affect Healthcare Workers." *Becker's Hospital Review*, October 12, 2021. https://www.becker shospitalreview.com/workforce/11-states-banning-covid-19-vaccine-mandates-how-it-affects-healthcare-workers.html.

Hart, H. L. A. "Are There Any Natural Rights?" *The Philosophical Review* 64, no. 2 (1955): 175–91. https://doi.org/10.2307/2182586.

Hausman, Bernice L. *Anti/Vax: Reframing the Vaccination Controversy*. Ithaca, NY: Cornell University Press, 2019. doi:10.7591/j.ctvdtphd8.

Hawkins, Summer Sherburne, Krisztina Horvath, Jessica Cohen, Lydia E. Pace, and Christopher F. Baum. "Associations Between Insurance-Related Affordable Care Act Policy Changes with HPV Vaccine Completion." *BMC Public Health* 21, no. 1 (2021): 304. https://doi.org/10.1186/s12889-021-10328-4.

Hellerstein, Judith, and David Neumark. "Workplace Segregation in the United States: Race, Ethnicity, and Skill." National Bureau of Economic Research, 2005. https://www.nber.org/system/files/working_papers/w11599/w11599.pdf.

Heron, Jon, Jean Golding, and the Alspac Study Team. "Thimerosal Exposure in Infants and Developmental Disorders: A Prospective Cohort Study in the United Kingdom Does Not Support a Causal Association." *Pediatrics* 114, no. 3 (2004): 577–83. https://doi.org/10.1542/peds.2003-1176-L.

Hinman, Alan. "Position Paper." *Pediatric Research* 13, no. 5 (1979): 689–96. https://doi.org/https://doi.org/10.1203/00006450-197905001-00007.

Hobbes, Thomas. *Leviathan*, edited by Richard Tuck. New York: Cambridge University Press, 1996.

Hodge, James G., and L. Gostin. "School Vaccination Requirements: Historical, Social and Legal Perspectives." *Kentucky Law Journal* 90 (2002): 831.

Holland, Stephen. *Public Health Ethics*. Hoboken, NJ: Wiley, 2015.

Horner, J. Stuart. "For Debate. The Virtuous Public Health Physician." *Journal of Public Health* 22, no. 1 (2000): 48–53. https://doi.org/10.1093/pubmed/22.1.48.

Hotez, Peter J. "COVID19 Meets the Antivaccine Movement." *Microbes and Infection* 22, no. 4–5 (2020): 162–64. https://doi.org/10.1016/j.micinf.2020.05.010.

Hotez, Peter J., and K. M. Venkat Narayan. "Restoring Vaccine Diplomacy." *JAMA* 325, no. 23 (2021): 2337–38. https://doi.org/10.1001/jama.2021.7439.

"How the CIA's Fake Vaccination Campaign Endangers Us All." *Scientific American*, May 1, 2013. https://www.scientificamerican.com/article/how-cia-fake-vaccination-campa ign-endangers-us-all.

Howe, Lexington. "Vaccine Bill SB 276 Is Not 'California for All,' Opposition Says." *The Village News*, September 17, 2019. https://www.villagenews.com/story/2019/09/12/news/vaccine-bill-sb-276-is-not-california-for-all-opposition-says/57425.html.

Hull, Brynley P., Frank H. Beard, Alexandra J. Hendry, Aditi Dey, and Kristine Macartney. "'No Jab, No Pay': Catch-Up Vaccination Activity During Its First Two Years." *Medical Journal of Australia* 213, no. 8 (2020): 364–69. https://doi.org/10.5694/mja2.50780.

Humphreys, Margaret. *Yellow Fever and the South.* Baltimore, MD: Johns Hopkins University Press, 1994.

Humphreys, Margaret. *Yellow Fever and the South.* Rev. ed. Baltimore, MD: Johns Hopkins University Press, 1999.

Huntsberry, Will. "Medical Board Charges San Diego Doctor Who's Doled out Dozens of Vaccine Exemptions." Voice of San Diego, October 24, 2019.

Huntsberry, Will. "One Doctor Is Responsible For a Third of All Medical Vaccine Exemptions in San Diego." Voice of San Diego, March 18, 2019.

Hviid, Anders, Michael Stellfeld, Jan Wohlfahrt, and Mads Melbye. "Association Between Thimerosal-Containing Vaccine and Autism." *JAMA* 290, no. 13 (2003): 1763–66. https://doi.org/10.1001/jama.290.13.1763.

Iati, Marisa. "California's Governor Signed a Pro-Vaccine Bill into Law This Week. Then the Protests Got Weird." *The Washington Post*, September 14, 2019.

Iati, Marisa. "Jessica Biel Lobbied Alongside a Prominent Antivaxxer but Says She Supports Vaccines." *The Washington Post*, June 13, 2019.

Ibrahim, Baffa Sule, Rabi Usman, Yahaya Mohammed, Zainab Datti, Oyeladun Okunromade, Aisha Ahmed Abubakar, and Patrick Mboya Nguku. "Burden of Measles in Nigeria: A Five-Year Review of Case-Based Surveillance Data, 2012–2016." *Pan African Medical Journal* 32, 1 (2019): 5. https://doi.org/10.11604/pamj.supp.2019.32.1.13564.

Imdad, Aamer, Boldtsetseg Tserenpuntsag, Debra S. Blog, Neal A. Halsey, Delia E. Easton, and Jana Shaw. "Religious Exemptions for Immunization and Risk of Pertussis in New York State, 2000–2011." *Pediatrics* 132, no. 1 (2013): 37–43. https://doi.org/10.1542/peds.2012-3449.

"Introduction to United States Magazine and Democratic Review." *United States Magazine and Democratic Review* 1, no. 1 (1837).

Iwashyna, Theodore J., Lee A. Kamphuis, Stephanie J. Gundel, Aluko A. Hope, Sarah Jolley, Andrew J. Admon, Ellen Caldwell, et al. "Continuing Cardiopulmonary Symptoms, Disability, and Financial Toxicity 1 Month After Hospitalization for Third-Wave COVID-19: Early Results from a US Nationwide Cohort." *Journal of Hospital Medicine* 16, no. 9 (2021): 531–37. https://doi.org/10.12788/jhm.3660.

Jackson, Vicki C. "Constitutional Law in an Age of Proportionality." *The Yale Law Journal* 124, no. 8 (2015): 3094–196.

Jamison, Amelia M., Sandra Crouse Quinn, and Vicki S. Freimuth. "'You Don't Trust a Government Vaccine': Narratives of Institutional Trust and Influenza Vaccination Among African American and White Adults." *Social Science & Medicine* 221 (2019): 87–94. https://doi.org/10.1016/j.socscimed.2018.12.020.

Jamrozik, Euzebiusz, Toby Handfield, and Michael J. Selgelid. "Victims, Vectors and Villains: Are Those Who Opt out of Vaccination Morally Responsible for the Deaths of Others?" *Journal of Medical Ethics* 42, no. 12 (2016): 762–68. https://doi.org/10.1136/medethics-2015-103327.

Jarrett, Caitlin, Rose Wilson, Maureen O'Leary, Elisabeth Eckersberger, and Heidi J. Larson. "Strategies for Addressing Vaccine Hesitancy—A Systematic Review." *Vaccine* 33, no. 34 (2015): 4180–90. https://doi.org/10.1016/j.vaccine.2015.04.040.

Jelleyman, Tim, and Andrew Ure. "Attitudes to Immunisation: A Survey of Health Professionals in the Rotorua District." *New Zealand Medical Journal* 117, no. 1189 (2004): U769.

Jenkins, Marina C., and Megan A. Moreno. "Vaccination Discussion Among Parents on Social Media: A Content Analysis of Comments on Parenting Blogs." *Journal of Health Communication* 25, no. 3 (2020): 232–42. https://doi.org/10.1080/10810 730.2020.1737761.

Johns Hopkins Bloomberg School of Public Health. "As Cases Spread Across U.S. Last Year, Pattern Emerged Suggesting Link Between Governors' Party Affiliation and COVID-19 Case and Death Numbers." March 10, 2021. Accessed March 16, 2021. https://publichealth.jhu.edu/2021/as-cases-spread-across-us-last-year-pattern-emer ged-suggesting-link-between-governors-party-affiliation-and-covid-19-case-and-death-numbers.

Johnson, Christen. "Should You Ask Family and Friends to Get Vaccinated Before Visiting Your Newborn?" *Chicago Tribune*, April 4, 2019. https://www.chicagotribune. com/lifestyles/sc-fam-social-graces-parents-asking-relatives-vaccine-pre-baby-0416-story.html.

Johnson, David A., and Humayun J. Chaudhry. *Medical Licensing and Discipline in America: A History of the Federation of State Medical Boards*. Lanham, MD: Lexington Books, 2012.

Johnson, Steven. "AMA Maintains Its Opposition to Single-Payer Systems." Modern Healthcare. June 11, 2019. https://www.modernhealthcare.com/physicians/ama-maintains-its-opposition-single-payer-systems.

Joint Commission on Accreditation of Healthcare Organizations. *Comprehensive Accreditation Manual: CAMH for Hospitals: The Official Handbook*. 2011.

Jones, Bryan D., and Frank R. Baumgartner. *The Politics of Attention: How Government Prioritizes Problems*. Chicago: University of Chicago Press, 2005.

Jones, Malia, Alison M. Buttenheim, Daniel Salmon, and Saad B. Omer. "Mandatory Health Care Provider Counseling for Parents Led to a Decline in Vaccine Exemptions in California." *Health Affairs* 37, no. 9 (2018): 1494–502. https://doi.org/10.1377/hlth aff.2018.0437.

Jonsen, Albert R.. *The Birth of Bioethics*. New York: Oxford University Press, 1998.

Joseph, Lauren J., and Stephen Cranney. "Self-Esteem Among Lesbian, Gay, Bisexual and Same-Sex-Attracted Mormons and Ex-Mormons." *Mental Health, Religion & Culture* 20, no. 10 (2017): 1028–41. https://doi.org/10.1080/13674676.2018.1435634.

Kahan, Dan M. "Vaccine Risk Perceptions and Ad Hoc Risk Communication: An Empirical Assessment." Yale Law & Economics Research Paper 491, January 27, 2014.

Kahneman, Daniel. *Thinking, Fast and Slow*. New York: Farrar, Straus & Giroux, 2011.

Kant, Immanuel. *The Metaphysics of Morals*, edited by Mary J. Gregor and Rodger J. Sullivan. Cambridge, UK: Cambridge University Press, 1996.

Kaplan, Karen. "Here's What Happened After California Got Rid of Personal Belief Exemptions for Childhood Vaccines." *Los Angeles Times*, October 29, 2018.

Karlamangla, Soumya. "Why Hasn't California Cracked Down on Antivaccination Doctors?" *Los Angeles Times*, November 6, 2017.

Karlamangla, Soumya. "Pushback Against Immunization Laws Leaves Some California Schools Vulnerable to Outbreaks." *Los Angeles Times*, July 13, 2018. https://www.lati mes.com/local/lanow/la-me-ln-sears-vaccines-fight-20180713-story.html.

Kassner, Joshua, and David Lefkowitz. "Conscientious Objection." In *Encyclopedia of Applied Ethics* (2ond ed.), edited by Ruth Chadwick, 594–601. San Diego, CA: Elsevier, 2012.

Kata, Anna. "Anti-Vaccine Activists, Web 2.0, and the Postmodern Paradigm—An Overview of Tactics and Tropes Used Online by the Anti-Vaccination Movement." *Vaccine* 30 (2012): 3778–89.

Katz, Aviva L., and Sally A. Webb. "Informed Consent in Decision-Making in Pediatric Practice." *Pediatrics* 138, no. 2 (2016): e20161485. https://doi.org/10.1542/peds.2016-1485.

Katz, Jay "The Consent Principle of the Nuremberg Code: Its Significance Then and Now." In *The Nazi Doctors and the Nuremberg Code*, edited by George J. Annas and M. A. Grodin, 227–238. New York: Oxford University Press, 1998.

Katznelson, Ira. "What America Taught the Nazis." *The Atlantic*, November 2017, 42–44.

Kaufman, Amanda. "A Supreme Court Case That Originated in Mass. Could Provide a Legal Precedent for President Biden's Vaccine Mandates, Experts Say." *The Boston Globe*, September 10, 2021.

Kaufman, M. "The American Anti-Vaccinationists and Their Arguments." *Bulletin of the History of Medicine* 41, no. 5 (1967): 463–78.

Kaye, James A., Maria del Mar Melero-Montes, and Hershel Jick. "Mumps, Measles, and Rubella Vaccine and the Incidence of Autism Recorded by General Practitioners: A Time Trend Analysis." *BMJ* 322, no. 7284 (2001): 460–63. https://doi.org/10.1136/bmj.322.7284.460.

Khazan, Olga "Wealthy L.A. Schools' Vaccination Rates Are as Low as South Sudan's." *The Atlantic*, September 14, 2014. https://www.theatlantic.com/health/archive/2014/09/wealthy-la-schools-vaccination-rates-are-as-low-as-south-sudans/380252.

Kinsella, Stephan. "Patent and Penicillin." Mises Institute, 2006. https://mises.org/wire/patent-and-penicillin.

Kipling, R. "The White Man's Burden." 1899.

Kirzinger, Ashley, Audrey Kearney, Liz Hamel, and Mollyann Brodie. "KFF COVID-19 Vaccine Monitor: The Increasing Importance of Partisanship in Predicting COVID-19 Vaccination Status." KFF, November 16, 2021. https://www.kff.org/coronavirus-covid-19/poll-finding/importance-of-partisanship-predicting-vaccination-status.

Kiser, Dian, and T. Boschert. "Eliminating Smoking in Bars, Restaurants, and Gaming Clubs in California: Breath, the California Smoke-Free Bar Program." *Journal of Public Health Policy* 22, no. 1 (2001): 81–87. https://doi.org/10.2307/3343554.

Klein, Ezra. *Why We're Polarized.* New York: Avid Reader Press/Simon & Schuster, 2020.

Klosko, George. *The Principle of Fairness and Political Obligation.* Lanham, MD: Rowman & Littlefield, 1992.

Kluender, Raymond, Neale Mahoney, Francis Wong, and Wesley Yin. "Medical Debt in the US, 2009–2020." *JAMA* 326, no. 3 (2021): 250–56. https://doi.org/10.1001/jama.2021.8694.

Kluger, Jeffrey. "'They're Chipping Away': Inside the Grassroots Effort to Fight Mandatory Vaccines." *TIME*, June 13, 2019.

Kluger, Richard. *Simple Justice : The History of Brown v. Board of Education and Black America's Struggle for Equality.* New York: Vintage Books, 1977.

Kopelman, Loretta M. "The Best-Interests Standard as Threshold, Ideal, and Standard of Reasonableness." *Journal of Medicine and Philosophy* 22, no. 3 (1997): 271–89. https://doi.org/10.1093/jmp/22.3.271.

Kornfield, Meryl. "Anti-Vaccine Protesters Temporarily Shut Down Major Coronavirus Vaccine Site at Dodger Stadium in Los Angeles." *The Washington Post*, January 31, 2021.

Korpi, Walter, and Joakim Palme. "New Politics and Class Politics in the Context of Austerity and Globalization: Welfare State Regress in 18 Countries, 1975–95." *American Political Science Review* 97, no. 3 (2003): 425–46. https://doi.org/10.1017/s0003055403000789.

La Corte, Rachel. "Washington House Passes Bill Limiting Vaccine Exemptions." *The Seattle Times*, March 7, 2019. https://www.seattletimes.com/seattle-news/bill-limiting-vaccine-exemptions-passes-washington-state-house.

Labong, Leilani Marie. "Q&A with California Senator Richard Pan." *SacTown Magazine*, 2021. https://www.sactownmag.com/california-senator-richard-pan.

Ladi, Stella. "Austerity Politics and Administrative Reform: The Eurozone Crisis and Its Impact upon Greek Public Administration." *Comparative European Politics* 12, no. 2 (2013): 184–208. https://doi.org/10.1057/cep.2012.46.

Lai, Yun-Kuang, Jessica Nadeau, Louise-Anne McNutt, and Jana Shaw. "Variation in Exemptions to School Immunization Requirements Among New York State Private and Public Schools." *Vaccine* 32, no. 52 (2014): 7070–76.

Laing, R. D. *The Divided Self: A Study of Sanity and Madness*. London: Tavistock, 1960.

Lakshmanan, Rekha, and Jason Sabo. "Lessons from the Front Line: Advocating for Vaccines Policies at the Texas Capitol During Turbulent Times." *Journal of Applied Research on Children* 10, no. 2 (2019): Article 6.

Largent, Mark A. *Vaccine. The Debate in Modern America* Baltimore, MD: Johns Hopkins University Press, 2012.

Larson, Heidi J. *Stuck: How Vaccine Rumors Start and Why They Don't Go Away*. New York: Oxford University Press, 2020.

Larson, Heidi J. "Negotiating Vaccine Acceptance in an Era of Reluctance." *Human Vaccines & Immunotherapeutics* 9, no. 8 (2013): 1779–81. https://doi.org/10.4161/hv.25932.

Lassiter, Matthew D. "How White Americans' Refusal to Accept Busing Has Kept Schools Segregated." *The Washington Post*, April 20, 2021.

Latina, Jodi, Bob Wilson, and Kent Pierce. "Gov. Lamont Signs Bill to Repeal Religious Exemption for Childhood Vaccinations After Large Protest During Senate Vote." WTNH News, April 27, 2021. https://www.wtnh.com/news/connecticut/hartford/state-senate-takes-up-repeal-of-religious-exemption-for-childhood-vaccinations-after-passing-in-house.

Lawlor, Joe. "'No' Vote – to Keep State's New Vaccine Law – Wins by Overwhelming Margin." *Portland Press Herald*, March 4, 2020. https://www.pressherald.com/2020/03/03/no-vote-to-keep-pro-vaccine-law-leading-big-in-referendum/.

Lawrence, Elizabeth M., Richard G. Rogers, and Anna Zajacova. "Educational Attainment and Mortality in the United States: Effects of Degrees, Years of Schooling, and Certification." *Population Research and Policy Review* 35, no. 4 (2016): 501–25. https://doi.org/10.1007/s11113-016-9394-0.

Lawrence, Heidi. *Vaccine Rhetorics*. Columbus: Ohio State University Press, 2020.

Laycock, Douglas. "Regulatory Exemptions of Religious Behavior and the Original Understanding of the Establishment Clause." *Notre Dame Law Review* 81, no. 5 (2006): 1793–842.

Leach, Melissa, and James Fairhead. *Vaccine Anxieties: Global Science, Child Health and Society*, edited by Steve Rayner. London: Earthscan, 2007.

Learn the Risk. "About Us," 2018. https://learntherisk.org/about-us.

Leask, Julie. "Target the Fence-Sitters." *Nature* 473, no. 7348 (2011): 443–5. https://doi.org/10.1038/473443a.

Leask, Julie, and Margie Danchin. "Imposing Penalties for Vaccine Rejection Requires Strong Scrutiny." *Journal of Paediatrics and Child Health* 53, no. 5 (2017): 439–44.

Leask, Julie, Paul Kinnersley, Cath Jackson, Francine Cheater, Helen Bedford, and Greg Rowles. "Communicating with Parents About Vaccination: A Framework for Health Professionals." *BMC Pediatrics* 12, no. 1 (2012): 154. http://www.biomedcentral.com/1471-2431/12/154.

Leask, Julie, Helen E. Quinn, Kristine Macartney, Marianne Trent, Peter Massey, Chris Carr, and John Turahui. "Immunisation Attitudes, Knowledge and Practices of Health Professionals in Regional NSW." *Australian & New Zealand Journal of Public Health* 32, no. 3 (2008): 224–29. https://doi.org/10.1111/j.1753-6405.2008.00220.x.

Lederer, Susan E. *Subjected to Science : Human Experimentation in America Before the Second World War.* Baltimore, MD: Johns Hopkins University Press, 1995.

Lederer, Susan E. "Children as Guinea Pigs: Historical Perspective." *Accountability in Research* 10, no. 1 (2003): 1–16.

Lee, Hye-Yon, and Diana C. Mutz. "Changing Attitudes Toward Same-Sex Marriage: A Three-Wave Panel Study." *Political Behavior* 41, no. 3 (2018): 701–22. https://doi.org/10.1007/s11109-018-9463-7.

Lee, Yueh-Ting, Victor Ottati, and Imtiaz Hussain. "Attitudes Toward 'Illegal' Immigration into the United States: California Proposition 187." *Hispanic Journal of Behavioral Sciences* 23, no. 4 (2001): 430–43. https://doi.org/10.1177/0739986301234005.

Lenthang, Marlene "How School Board Meetings Have Become Emotional Battlegrounds for Debating Mask Mandates." ABC News, August 29, 2021. https://abcnews.go.com/US/school-board-meetings-emotional-battlegrounds-debating-mask-mandates/story?id=79657733.

Lerer, Lisa. "How Republican Vaccine Opposition Got to This Point: News Analysis." *The New York Times*, July 17, 2021. https://www.nytimes.com/2021/07/17/us/politics/coronavirus-vaccines-republicans.html.

Levin, Hillel Y. "Why Some Religious Accommodations for Mandatory Vaccinations Violate the Establishment Clause." *Hastings Law Journal* 68 (2017): 193–1242.

Levin, Hillel Y. "Private Schools' Role and Rights in Setting Vaccination Policy: A Constitutional and Statutory Puzzle." *William and Mary Law Review* 61, no. 6 (2020): 1607.

Levin, Hillel Y., Allan J. Jacobs, and Kavita Shah Arora. "To Accommodate or Not to Accommodate: Should the State Regulate Religion to Protect the Rights of Children and Third Parties?" *Washington and Lee Law Review* 73, no. 2 (2016): 915.

Levin, Hillel Y., Stacie Patrice Kershner, Timothy D. Lytton, Daniel Salmon, and Saad B. Omer. "Stopping the Resurgence of Vaccine-Preventable Childhood Diseases: Policy, Politics, and Law." *University of Illinois Law Review* 2020, no. 1 (2020): 233–72.

Levin, Kelly, Benjamin Cashore, Steven Bernstein, and Graeme Auld. "Overcoming the Tragedy of Super Wicked Problems: Constraining Our Future Selves to Ameliorate Global Climate Change." *Policy Sciences* 45, no. 2 (2012): 123–52. https://doi.org/http://dx.doi.org/10.1007/s11077-012-9151-0.

Lewis, Sinclair. *Arrowsmith.* New York: Harcourt, Brace & World, 1952.

Li, Meng, and Gretchen B. Chapman. "Nudge to Health: Harnessing Decision Research to Promote Health Behaviors." *Social and Personality Psychology Compass* 7, no. 3 (2013): 187–98. https://doi.org/10.1111/spc3.12019.

Lillvis, Denise. "Managing Dissonance and Dissent: Bureaucratic Professionalism and Political Risk in Policy Implementation." *Law & Policy* 41, no. 3 (2019): 310–35. https://doi.org/10.1111/lapo.12131.

Lillvis, Denise F., Anna Kirkland, and Anna Frick. "Power and Persuasion in the Vaccine Debates: An Analysis of Political Efforts and Outcomes in the United States, 1998–2012." *Milbank Quarterly* 92, no. 3 (2014): 475–508. doi:10.1111/1468–0009.12075.

Lillvis, Denise F., Charley Willison, and Katia Noyes. "Normalizing Inconvenience to Promote Childhood Vaccination: A Qualitative Implementation Evaluation of a Novel Michigan Program." *BMC Health Services Research* 20, no. 1 (2020): 683–83. https://doi.org/10.1186/s12913-020-05550-6.

Lim, Megan S. C., Paul A. Agius, Elise R. Carrotte, Alyce M. Vella, and Margaret E. Hellard. "Young Australians' Use of Pornography and Associations with Sexual Risk Behaviours." *Australian and New Zealand Journal of Public Health* 41, no. 4 (2017): 438–43. https://doi.org/10.1111/1753-6405.12678.

Lin II, R.-G., S. Karlamangla, and R. Xia. "California Wants to Pull This Doctor's License. Here's How It's Sparked a New Battle over Child Vaccinations." *Los Angeles Times*, September 12, 2016. https://www.latimes.com/local/lanow/la-me-sears-vaccine-20160909-snap-story.html.

Lin II, R.-G., S. Karlamangla, and R. Xia. "Board Action Renews Vaccine Battle; Medical Panel's Bid to Pull the License of an O.C. Doctor Is Seen as Targeting Those Who Try to Skirt a New Law." *Los Angeles Times*, September 13, 2016.

Liou, Stephanie, Han Yu, Max Kapustin, Monica Bhatt, Cameron Boyd, Christine Cahaney, and Jasmine Thomas. "Impact of the COVID-19 Pandemic on Pediatric Firearm-Related Injuries in the USA." *Pediatrics* 147, no. 3 (2021): 103–05.

Lipka, Michael. "Mainline Protestants Make up Shrinking Number of U.S. Adults." Pew Research Center. 2015. https://www.pewresearch.org/fact-tank/2015/05/18/mainline-protestants-make-up-shrinking-number-of-u-s-adults.

Locke, J. *Second Treatise of Government and a Letter Concerning Toleration.* Oxford, UK: Oxford University Press, 2016.

London, Alex John. *For the Common Good: Philosophical Foundations of Research Ethics.* Oxford, UK: Oxford University Press, 2021.

Loudon, Kyle, and Irvine Loudon. *Medical Care and the General Practitioner, 1750–1850.* Oxford, UK: Oxford University Press, 1986.

Lowenstein, Roger. "A Question of Numbers." *The New York Times*, January 16, 2005.

Luna, Taryn. "Vaccine Bill Protester Threw Blood on California Senators, Investigation Confirms." *Los Angeles Times*, October 2, 2019.

Luyten, Jeroen, and Philippe Beutels. "The Social Value of Vaccination Programs: Beyond Cost-Effectiveness." *Health Affairs* 35, no. 2 (2016): 212–18. https://doi.org/10.1377/hlthaff.2015.1088.

Lyman, George D. "The Beginnings of California's Medical History." *California and Western Medicine* 23, no. 5 (1925): 561–76.

MacDonald, Noni E., and SAGE Working Group on Vaccine Hesitancy. "Vaccine Hesitancy: Definition, Scope and Determinants." *Vaccine* 33, no. 34 (2015): 4161–64. https://doi.org/doi:10.1016/j.vaccine.2015.04.036.

MacKay, Kathryn. "Utility and Justice in Public Health." *Journal of Public Health* 40, no. 3 (2018): e413–e18. https://doi.org/10.1093/pubmed/fdx169.

Macy, Beth. *Dopesick: Dealers, Doctors, and the Drug Company That Addicted America.* Boston: Little, Brown, 2018.

Madigan, Sheri, Dillon Browne, Nicole Racine, Camille Mori, and Suzanne Tough. "Association Between Screen Time and Children's Performance on a Developmental Screening Test." *JAMA Pediatrics* 173, no. 3 (2019): 244–50. https://doi.org/10.1001/jamapediatrics.2018.5056.

Mahdawi, Arwa "Telling Anti-Vaxxers to Get the Jab Should Not Be Controversial—Even Fox News Is Doing It." *The Guardian*, October 27, 2021. https://www.theguardian.com/commentisfree/2021/oct/27/telling-anti-vaxxers-to-get-the-jab-should-not-be-controversial-even-fox-news-is-doing-it.

Mahler, Jonathon. "A Governor on Her Own, with Everything at Stake." *The New York Times*, June 25, 2020.

Mahoney, James, and Kathleen Thelen (Eds.). *Explaining Institutional Change: Ambiguity, Agency, and Power.* Cambridge, UK: Cambridge University Press, 2010.

Mahr, Krista. "'It's a Tsunami': Legal Challenges Threatening Public Health Policy." *Politico*, May 10, 2022. https://www.politico.com/news/2022/05/10/legal-challenges-cdc-public-health-policy-00031253.

Makela, Annamari, J. Pekka Nuorti, and Heikki Peltola. "Neurologic Disorders After Measles–Mumps–Rubella Vaccination." *Pediatrics* 110, no. 5 (2002): 957–63. https://doi.org/10.1542/peds.110.5.957.

Malcolm, Lynne. "Research Says Young People Today Are More Narcissistic Than Ever." ABC Australia, 2014. Accessed April 15, 2023. https://www.abc.net.au/radionational/programs/allinthemind/young-people-today-are-more-narcissistic-than-ever/5457236.

Malik, Amyn A., SarahAnn M. McFadden, Jad Elharake, and Saad B. Omer. "Determinants of COVID-19 Vaccine Acceptance in the US." *EClinicalMedicine* 26 (2020). https://doi.org/10.1016/j.eclinm.2020.100495.

Malm, Heidi. "Military Metaphors and Their Contribution to the Problems of Overdiagnosis and Overtreatment in the 'War' Against Cancer." *American Journal of Bioethics* 16, no. 10 (2016): 19–21. https://doi.org/10.1080/15265161.2016.1214331.

Malm, Heidi, and Mark C. Navin. "Pox Parties for Grannies? Chickenpox, Exogenous Boosting, and Harmful Injustices." *American Journal of Bioethics* 20, no. 9 (2020): 45–57. https://doi.org/10.1080/15265161.2020.1795528.

Malone, K., and A. Hinman. "Vaccination Mandates: The Public Health Imperative and Individual Rights." In *Law in Public Health Practice*, edited by R. Goodman. New York: Oxford University Press, 2007; 262–83.

Margulis, Jennifer. *The Business of Baby: What Doctors Don't Tell You, What Corporations Try to Sell You, and How to Put Your Pregnancy, Childbirth, and Baby Before Their Bottom Line.* New York: Scribner, 2013.

Mariner, Wendy K., George J. Annas, and Leonard H. Glantz. "Jacobson V Massachusetts: It's Not Your Great-Great-Grandfather's Public Health Law." *American Journal of Public Health* 95, no. 4 (2005): 581–90. https://doi.org/10.2105/ajph.2004.055160.

Marsden, George. *The Twilight of the American Enlightenment: The 1950s and the Crisis of Liberal Belief.* New York: Basic Books, 2014.

Mason, Lilliana. *Uncivil Agreement: How Politics Became Our Identity.* Chicago: University of Chicago Press, 2018.

Masters, Nina B., Jon Zelner, Paul L. Delamater, David Hutton, Matthew Kay, Marissa C. Eisenberg, and Matthew L. Boulton. "Evaluating Michigan's Administrative Rule Change on Nonmedical Vaccine Exemptions." *Pediatrics* 148, no. 2 (2021): e2021049942.

Mays, Mackenzie. "Anti-Vaccine Protesters Are Likening Themselves to Civil Rights Activists." *Politico*, September 18, 2019. https://www.politico.com/story/2019/09/18/california-anti-vaccine-civil-rights-1500976.

Mazzei, Patricia. "As G.O.P. Fights Mask and Vaccine Mandates, Florida Takes the Lead." *The New York Times*, November 17, 2021.

McAndrew, Stephen. "Internal Morality of Medicine and Physician Autonomy." *Journal of Medical Ethics* 45, no. 3 (2019): 198–203. https://doi.org/10.1136/medethics-2018-105069.

McCoy, Charles Allan. "Adapting Coercion: How Three Industrialized Nations Manufacture Vaccination Compliance." *Journal of Health Politics, Policy and Law* 44, no. 6 (2019): 823–54. https://doi.org/10.1215/03616878-7785775.

McCrory, Cathal, and Aisling Murray. "The Effect of Breastfeeding on Neuro-Development in Infancy." *Maternal and Child Health Journal* 17 (2013): 1680–88.

McDonald, Pamela, Rupali J. Limaye, Saad B. Omer, Alison M. Buttenheim, Salini Mohanty, Nicola P. Klein, and Daniel A. Salmon. "Exploring California's New Law Eliminating Personal Belief Exemptions to Childhood Vaccines and Vaccine Decision-Making Among Homeschooling Mothers in California." *Vaccine* 37, no. 5 (2019): 742–50. https://doi.org/https://doi.org/10.1016/j.vaccine.2018.12.018.

McIntosh, E. David G., Jan Janda, Jochen H. H. Ehrich, Massimo Pettoello-Mantovani, and Eli Somekh. "Vaccine Hesitancy and Refusal." *Journal of Pediatrics* 175 (2016): 248–49.e1. https://doi.org/10.1016/j.jpeds.2016.06.006.

McPherson, Katie. "How to Ask Loved Ones to Get Vaccinated Before Meeting Your Newborn." *Romper*, November 20, 2020. https://www.romper.com/pregnancy/how-to-ask-family-to-get-the-flu-shot-other-vaccines-before-baby-is-born.

Meldrum, Marcia. "'A Calculated Risk': The Salk Polio Vaccine Field Trials of 1954." *BMJ* 317, no. 7167 (1998): 1233–36. doi:10.1136/bmj.317.7167.1233.

Meleo-Erwin, Zoë, Corey Basch, Sarah A. MacLean, Courtney Scheibner, and Valerie Cadorett. "'To Each His Own': Discussions of Vaccine Decision-Making in Top Parenting Blogs." *Human Vaccines & Immunotherapeutics* 13, no. 8 (2017): 1895–901. https://doi.org/10.1080/21645515.2017.1321182.

Mello, Michelle M., David M. Studdert, and Wendy E. Parmet. "Shifting Vaccination Politics—The End of Personal-Belief Exemptions in California." *New England Journal of Medicine* 373, no. 9 (2015): 785–87. https://search.proquest.com/docview/1707845173?accountid=14681.

Menasce Horowitz, J., R. Igielnik, and R. Kochhar. "Trends in Income and Wealth Inequality." Pew Research Center. 2020. https://www.pewresearch.org/social-trends/2020/01/09/trends-in-income-and-wealth-inequality.

Mendelsohn, Robert. *How to Raise a Healthy Child in Spite of Your Doctor: One of America's Leading Pediatricians Puts Parents Back in Control of Their Children's Health.* New York: Ballantine, 1987.

Menzel, Paul, Marthe R. Gold, Erik Nord, Jose-Louis Pinto-Prades, Jeff Richardson, and Peter Ubel. "Toward a Broader View of Values in Cost-Effectiveness Analysis of Health." *Hastings Center Report* 29, no. 3 (1999): 7–15. https://doi.org/10.2307/3528187.

Metzger, Gillian E. "1930s Redux: The Administrative State Under Siege." *Harvard Law Review* 131, no. 1 (2017): 1.

Miles, Steven H., Laura W. Lane, Janet Bickel, Robert M. Walker, and Christine K. Cassel. "Medical Ethics Education: Coming of Age." *Academic Medicine* 64, no. 12 (1989): 705–14. https://doi.org/10.1097/00001888-198912000-00004.

Milken Institute School of Public Health. "Understanding the Impact of Vaccines: A Conversation with the National Public Health Information Coalition." Milken Institute School of Public Health at The George Washington University, September 18, 2015. Accessed April 15, 2023. https://onlinepublichealth.gwu.edu/resources/niam-2015-nphic/.

Milkman, Katherine L., John Beshears, James J. Choi, David Laibson, and Brigitte C. Madrian. "Using Implementation Intentions Prompts to Enhance Influenza Vaccination Rates." *Proceedings of the National Academy of Sciences of the United States of America* 108, no. 26 (2011): 10415–20.

Mill, John Stuart. *On Liberty* (London: Parker & Son, 1859).

Miller, Courtney. "'Spiritual but Not Religious': Rethinking the Legal Definition of Religion." *Virginia Law Review* 102, no. 3 (2016): 833–94.

Miller, Franklin G. "Liberty and Protection of Society During a Pandemic: Revisiting John Stuart Mill." *Perspectives in Biology and Medicine* 64, no. 2 (2021): 200–10. https://doi.org/10.1353/pbm.2021.0016.

Miller, Patrice M., and Michael L. Commons. "The Benefits of Attachment Parenting for Infants and Children: A Behavioral Developmental View." *Behavioral Development Bulletin* 16, no. 1 (2010): 1–14. https://doi.org/10.1037/h0100514.

Miller, Patrick R., and Pamela Johnston Conover. "Red and Blue States of Mind: Partisan Hostility and Voting in the United States." *Political Research Quarterly* 68, no. 2 (2015): 225–39. https://doi.org/10.1177/1065912915577208.

Millward, G. *Vaccinating Britain: Mass Vaccination and the Public Since the Second World War*. Manchester, UK: Manchester University Press, 2019. https://www.ncbi.nlm.nih.gov/books/NBK545997.

Milosh, Maria, Marcus Painter, Konstantin Sonin, David Van Dijcke, and Austin Wright. "Biden's Plea for Masks Will Fail. Blame Political Polarization." *The Washington Post*, April 13, 2021. https://www.washingtonpost.com/outlook/2021/04/13/masks-mandate-partisanship-politics.

Ministry of Health: Singapore. "COVID-19 Vaccination, Vaccination Statistics." 2022. Accessed April 15, 2023, https://www.moh.gov.sg/covid-19/vaccination.

Mnookin, Seth. *The Panic Virus*. New York: Simon & Schuster, 2011.

Mohanty, Salini, Alison M. Buttenheim, Caroline M. Joyce, Amanda C. Howa, Daniel Salmon, and Saad B. Omer. "California's Senate Bill 277: Local Health Jurisdictions' Experiences with the Elimination of Nonmedical Vaccine Exemptions." *American Journal of Public Health* 109, no. 1 (2019): 96–101.

Mohanty, Salini, Alison M. Buttenheim, Caroline M. Joyce, Amanda C. Howa, Daniel Salmon, and Saad B. Omer. "Experiences with Medical Exemptions After a Change in Vaccine Exemption Policy in California." *Pediatrics* 142, no. 5 (2018): e20181051.

Moisse, Katie. "The Lasting Fallout of Fake Vaccination Programs." ABC News Network, May 21, 2014. https://abcnews.go.com/Health/lasting-fallout-fake-vaccination-progr ams/story?id=23795483;%20.

Mols, Frank, S. Alexander Haslam, Jolanda Jetten, and Niklas K. Steffens. "Why a Nudge Is Not Enough: A Social Identity Critique of Governance by Stealth." *European Journal of Political Research* 54, no. 1 (2015): 81–98. https://doi.org/10.1111/1475-6765.12073.

Moreno, Jonathon. *Undue Risk: Secret State Experiments on Humans.* New York: Routledge, 2001.

Moss, Brian G., and William H. Yeaton. "Early Childhood Healthy and Obese Weight Status: Potentially Protective Benefits of Breastfeeding and Delaying Solid Foods." *Maternal and Child Health Journal* 18 (2014): 1224–32. https://doi.org/10.1007/s10 995-013-1357-z.

Mounk, Yascha. *The Great Experiment: Why Diverse Democracies Fall Apart and How They Can Endure.* London: Penguin, 2022.

Mueller, Benjamin, and Jan Hoffman. "Routine Childhood Vaccinations in the U.S. Slipped During the Pandemic." *The New York Times*, April 21, 2022.

Müller, Jan-Werner. "The American Right Is Pushing 'Freedom over Fear': It Won't Stop the Virus." *The Guardian*, July 16, 2020. https://www.theguardian.com/commentisfree/ 2020/jul/16/coronavirus-american-right-freedom-over-fear.

Mulligan, Karen, and Jeffrey E. Harris. "COVID-19 Vaccination Mandates for School and Work Are Sound Public Policy." USC, Schaeffer Center for Health Policy & Economics. 2021. https://healthpolicy.usc.edu/wp-content/uploads/2022/07/USC_Schaef fer_Covid19-VaccineMandates_WhitePaper.pdf.

Nagel, Thomas. "The Problem of Global Justice." *Philosophy and Public Affairs* 33, no. 2 (2005): 113–47. https://doi.org/https://doi.org/10.1111/j.1088-4963.2005.00027.x.

Najera, Rene F., "An Adult Measles Patient Writes Home to Mom and Dad in the 1880s." *History of Vaccines*, October 5, 2022. Accessed October 22, 2022. https://www.histor yofvaccines.org/content/blog/california-immunization-exemption-legislation.

National Conference of State Legislatures. "State Partisan Composition 2022." 2022. Accessed April 15, 2023, https://www.ncsl.org/about-state-legislatures/state-partisan- composition.

National Conference of State Legislatures. "States with Religious and Philosophical Exemptions from School Immunization Requirements." 2022. Accessed July 06, 2022. https://www.ncsl.org/research/health/school-immunization-exemption-state- laws.aspx.

National Vaccine Information Center. "CA: Oppose AB2109 Restricting Personal Belief Exemptions to Mandatory Vaccination." 2012. http://nvicadvocacy.org/members/ Resources/CAOPPOSEAB2109RestrictingVaccineExemptions.aspx.

Navin, Mark C. *Values and Vaccine Refusal: Hard Questions in Ethics, Epistemology, and Health Care.* New York: Routledge, 2016.

Navin, Mark C. "Prioritizing Religion in Vaccine Exemption Policies." In *Religious Exemptions*, edited by Kevin Vallier and Michael Weber, 184–202. Oxford, UK: Oxford University Press, 2018.

Navin, Mark C. "Privacy and Religious Exemptions." In *Core Concepts and Contemporary Issues in Privacy*, edited by Ann Cudd and Mark C. Navin, 121–33. New York: Springer, 2018.

Navin, Mark C., and Katie Attwell. "Vaccine Mandates, Value Pluralism, and Policy Diversity." *Bioethics* 33, no. 9 (2019): 1042–49. https://doi.org/10.1111/bioe.12645.

Navin, Mark C., Andrea T. Kozak, and Katie Attwell. "School Staff and Immunization Governance: Missed Opportunities for Public Health Promotion." *Vaccine* 40, no. 51 (2022): 7433–39. https://doi.org/https://doi.org/10.1016/j.vaccine.2021.07.061.

Navin, Mark C., Andrea T. Kozak, and Emily C. Clark. "The Evolution of Immunization Waiver Education in Michigan: A Qualitative Study of Vaccine Educators." *Vaccine* 36, no. 13 (2018): 1751–56. https://doi.org/https://doi.org/10.1016/j.vaccine.2018.02.046.

Navin, Mark C., and Mark A. Largent. "Improving Nonmedical Vaccine Exemption Policies: Three Case Studies." *Public Health Ethics* 10, no. 3 (2017): 225–34. http://dx.doi.org/10.1093/phe/phw047.

Navin, Mark C., Mark A. Largent, and Aaron M. McCright. "Efficient Burdens Decrease Nonmedical Exemption Rates: A Cross-County Comparison of Michigan's Vaccination Waiver Education Efforts." *Preventive Medicine Reports* 17 (2020): 101049. https://doi.org/https://doi.org/10.1016/j.pmedr.2020.101049.

Navin, Mark C., Lindsay Margaret-Sander Oberleitner, Victoria C. Lucia, Melissa Ozdych, Nelia Afonso, Richard H. Kennedy, Hans Keil, Lawrence Wu, and Trini A. Mathew. "COVID-19 Vaccine Hesitancy Among Healthcare Personnel Who Generally Accept Vaccines." *Journal of Community Health* 47, no. 3 (2022): 519–29. https://doi.org/10.1007/s10900-022-01080-w.

Navin, Mark C., Aaron M. Scherer, Ethan Bradley, and Katie Attwell. "School Staff as Vaccine Advocates: Perspectives on Vaccine Mandates and the Student Registration Process." *Vaccine*, 5, no. 27 (2022): 1169–75.

Navin, Mark C., and Jason Adam Wasserman. "Reasons to Amplify the Role of Parental Permission in Pediatric Treatment." *American Journal of Bioethics* 17, no. 11 (2017): 6–14. https://doi.org/10.1080/15265161.2017.1378752.

Navin, Mark C., Jason Adam Wasserman, Miriam Ahmad, and Shane Bies. "Vaccine Education, Reasons for Refusal, and Vaccination Behavior." *American Journal of Preventive Medicine* 56, no. 3 (2019): 359–67. https://doi.org/10.1016/j.amepre.2018.10.024.

Neelon, Brian, Fedelis Mutiso, Noel T. Mueller, John L. Pearce, and Sara E. Benjamin-Neelon. "Associations Between Governor Political Affiliation and COVID-19 Cases, Deaths, and Testing in the U.S." *American Journal of Preventive Medicine* 61, no. 1 (2021): 115–19. https://doi.org/10.1016/j.amepre.2021.01.034.

Neuman, Scott. "1 in 4 Americans Thinks the Sun Goes Around the Earth, Survey Says." NPR. February 14, 2014. https://www.npr.org/sections/thetwo-way/2014/02/14/277058739/1-in-4-americans-think-the-sun-goes-around-the-earth-survey-says.

New York State. "Title: Section 69-3.10—Religious Exemption from Immunization." In *New York Codes, Rules and Regulations*. 1991. https://regs.health.ny.gov/content/section-69-310-religious-exemption-immunization.

Nie, Jing-Bao, Adam Gilbertson, Malcolm de Roubaix, Ciara Staunton, Anton van Niekerk, Joseph D. Tucker, and Stuart Rennie. "Healing Without Waging War: Beyond Military Metaphors in Medicine and HIV Cure Research." *American Journal of Bioethics* 16, no. 10 (2016): 3–11. https://doi.org/10.1080/15265161.2016.1214305.

Nietzsche, Friedrich. *Human, All Too Human: A Book for Free Spirits*, translated by R. J. Hollingdale. (1878; reprint). Cambridge, UK: Cambridge University Press, 1996.

Nolen, Stephanie. "Sharp Drop in Global Childhood Vaccinations Imperils Millions of Lives." *The New York Times*, July 14, 2022. https://www.nytimes.com/2022/07/14/health/childhood-vaccination-rates-decline.html.

Novak, William J. *The People's Welfare: Law and Regulation in Nineteenth-Century America*. Chapel Hill: University of North Carolina Press, 1996.

Nozick, Robert. *Anarchy, the State, and Utopia*. New York: Basic Books, 1974.

NPR. "The U.S. Has a Long Precedent for Vaccine Mandates." NPR. August 29, 2021. https://www.npr.org/2021/08/29/1032169566/the-u-s-has-a-long-precedent-for-vaccine-mandates.

NPR. "Battle over CDC's Powers Goes Far Beyond Travel Mask Mandate." April 21, 2022. https://oneill.law.georgetown.edu/press/battle-over-cdcs-powers-goes-far-beyond-travel-mask-mandate.

Nuti, Sabina, Federico Vola, Anna Bonini, and Milena Vainieri. "Making Governance Work in the Health Care Sector: Evidence from a 'Natural Experiment' in Italy." *Health Economics, Policy and Law* 11, no. 1 (2016): 17–38. https://doi.org/10.1017/S1744133115000067.

Nyathi, Sindiso, Hannah C. Karpel, Kristin L. Sainani, Yvonne Maldonado, Peter J. Hotez, Eran Bendavid, and Nathan C. Lo. "The 2016 California Policy to Eliminate Nonmedical Vaccine Exemptions and Changes in Vaccine Coverage: An Empirical Policy Analysis." *PLoS Medicine* 16, no. 12 (2019): e1002994. https://doi.org/10.1371/journal.pmed.1002994.

Nyhan, Brendan. "Why California's Approach to Tightening Vaccine Rules Has Potential to Backfire." *The New York Times*, April 14, 2015. https://www.nytimes.com/2015/04/15/upshot/why-californias-approach-to-tightening-vaccine-rules-could-backfire.html.

Nyhan, Brendan, Jason Reifler, Sean Richey, and Gary L. Freed. "Effective Messages in Vaccine Promotion: A Randomized Trial." *Pediatrics* 133, no. 4 (Apr 2014): e835–42. https://doi.org/10.1542/peds.2013-2365.

O'Brien, Thomas V. *The Politics of Race and Schooling: Public Education in Georgia, 1900–1961*. Lanham, MD: Lexington Books, 1999.

O'Connor, Sandra Day. "They Often Are Half Obscure: The Rights of the Individual and the Legacy of Oliver W. Holmes." *San Diego Law Review* 29, no. 3 (1992): 385.

Office of Governor Gavin Newsom. "California Implements First-in-the-Nation Measure to Encourage Teachers and School Staff to Get Vaccinated." News release. August 11, 2021. https://www.gov.ca.gov/2021/08/11/california-implements-first-in-the-nation-measure-to-encourage-teachers-and-school-staff-to-get-vaccinated.

Office of Governor Gavin Newsom. "California Becomes First State in Nation to Announce COVID-19 Vaccine Requirements for Schools." October 1, 2021. Accessed July 31, 2022. https://www.gov.ca.gov/2021/10/01/california-becomes-first-state-in-nation-to-announce-covid-19-vaccine-requirements-for-schools.

Offit, Paul A. *Deadly Choices: How the Anti-Vaccine Movement Threatens Us All*. New York: Basic Books, 2011.

Offit, Paul A. "Incomplete Financial Disclosure in a Viewpoint on Complementary and Alternative Therapies." *JAMA* 308, no. 5 (2012): 454. https://doi.org/10.1001/jama.2012.7751.

Offit, Paul A. "The Anti-Vaccination Epidemic." *The Wall Street Journal*, September 24, 2014. https://www.wsj.com/articles/paul-a-offit-the-anti-vaccination-epidemic-1411598408.

Offit, Paul A. *Bad Faith: When Religious Belief Undermines Modern Medicine*. New York: Basic Books, 2015.

Offit, Paul A. "Vaccine Exemptions: When Do Individual Rights Trump Societal Good?" *Journal of the Pediatric Infectious Diseases Society* 4, no. 2 (2015): 89–90. https://doi.org/10.1093/jpids/piv018.

O'Leary, Sean T., Mandy A. Allison, Tara Vogt, Laura P. Hurley, Lori A. Crane, Michael Brtnikova, Erin McBurney, et al. "Pediatricians' Experiences with and Perceptions of the Vaccines for Children Program." *Pediatrics* 145, no. 3 (2020): e20191207.

O'Leary, Sean T., Jessica R. Cataldi, Megan C. Lindley, Brenda L. Beaty, Laura P. Hurley, Lori A. Crane, and Allison Kempe. "Policies Among US Pediatricians for Dismissing Patients for Delaying or Refusing Vaccination." *JAMA* 324, no. 11 (2020): 1105–107. https://doi.org/10.1001/jama.2020.10658.

Olive, Jacqueline K., Peter J. Hotez, Ashish Damania, and Melissa S. Nolan. "The State of the Antivaccine Movement in the United States: A Focused Examination of Nonmedical Exemptions in States and Counties." *PLoS Medicine* 15, no. 6 (2018): e1002578. https://doi.org/10.1371/journal.pmed.1002578.

Oliver, Adam. "From Nudging to Budging: Using Behavioural Economics to Inform Public Sector Policy." *Journal of Social Policy* 42, no. 4 (2013): 685–700. https://doi.org/10.1017/s0047279413000299.

Oliver, S. E., P. Moro, and A. E. Blain. "Haemophilus Influenzae Type B." In *Epidemiology and Prevention of Vaccine-Preventable Diseases. The Pink Book: Course Textbook.* Centers for Disease Control and Prevention, 2021. Accessed April 5, 2023, https://www.cdc.gov/vaccines/pubs/pinkbook/hib.html

Olson, Mansur. *The Logic of Collective Action : Public Goods and the Theory of Groups.* Cambridge, Mass.: Harvard University Press, 1965.

Omer, Saad. B., K. Allen, D. H. Chang, L. Beryl Guterman, Robert A. Bednarczyk, Alex Jordan, Alison Buttenheim, et al. "Exemptions from Mandatory Immunization After Legally Mandated Parental Counseling." *Pediatrics* 141, no. 1 (2018): e20172364.

Omer, Saad B., Cornelia Betsch, and Julie Leask. "Mandate Vaccination with Care." *Nature* 571, no. 7766 (2019): 469. https://doi.org/10.1038/d41586-019-02232-0.

Omer, Saad B., Kyle S. Enger, Lawrence H. Moulton, Neal A. Halsey, Shannon Stokley, and Daniel A. Salmon. "Geographic Clustering of Nonmedical Exemptions to School Immunization Requirements and Associations with Geographic Clustering of Pertussis." *American Journal of Epidemiology* 168, no. 12 (2008): 1389–96. https://doi.org/10.1093/aje/kwn263.

Omer, Saad B., Dianne Peterson, Eileen A. Curran, Alan R. Hinman, and Walter A. Orenstein. "Legislative Challenges to School Immunization Mandates, 2009–2012." *JAMA* 311, no. 6 (2014): 620–21. https://doi.org/10.1001/jama.2013.282869.

Omer, Saad B., Jennifer Richards, Michelle A. B. Ward, and Robert A. Bednarczyk. "Vaccination Policies and Rates of Exemption from Immunization, 2005–2011." *New England Journal of Medicine* 367, no. 12 (2012): 1170–71.

O'Neill, Onora. "Accountability, Trust and Informed Consent in Medical Practice and Research." *Clinical Medicine* 4, no. 3 (2004): 269–76. https://doi.org/10.7861/clinmedicine.4-3-269.

Opel, Douglas J., Douglas S. Diekema, and Lainie Friedman Ross. "Should We Mandate a COVID-19 Vaccine for Children?" *JAMA Pediatrics* 175, no. 2 (2021): 125–26. https://doi.org/10.1001/jamapediatrics.2020.3019.

Opel, Douglas J., John Heritage, James A. Taylor, Rita Mangione-Smith, Halle Showalter Salas, Victoria DeVere, Chuan Zhou, and Jeffrey D. Robinson. "The Architecture of Provider–Parent Vaccine Discussions at Health Supervision Visits." *Pediatrics* 132, no. 6 (2013): 1037–46. https://doi.org/10.1542/peds.2013-2037.

Opel, Douglas J., Matthew P. Kronman, Douglas S. Diekema, Edgar K. Marcuse, Jeffrey S. Duchin, and Eric Kodish. "Childhood Vaccine Exemption Policy: The Case for a Less Restrictive Alternative." *Pediatrics* 137, no. 4 (2016): e20154230.

Opel, Douglas J., and Saad B. Omer. "Measles, Mandates, and Making Vaccination the Default Option." *JAMA Pediatrics* 169, no. 4 (2015): 303–304. https://doi.org/10.1001/jamapediatrics.2015.0291.

O'Reilly, Michelle, Nisha Dogra, Natasha Whiteman, Jason Hughes, Seyda Eruyar, and Paul Reilly. "Is Social Media Bad for Mental Health and Wellbeing? Exploring the Perspectives of Adolescents." *Clinical Child Psychology and Psychiatry* 23, no. 4 (2018): 601–13. https://doi.org/10.1177/1359104518775154.

Orenstein, Walter A., and Alan R. Hinman. "The Immunization System in the United States—The Role of School Immunization Laws." *Vaccine* 17 (1999): 19–24.

Orenstein, Walter A., Jerome A. Paulson, Michael T. Brady, Louis Z. Cooper, and Katherine Seib. "Global Vaccination Recommendations and Thimerosal." *Pediatrics* 131, no. 1 (2013): 149–51. https://doi.org/10.1542/peds.2012-1760.

Oshin, Olafimihan. "Trump Takes Credit for Vaccine Rollout: 'One of the Greatest Miracles of the Ages.'" *The Hill*, May 25, 2021. https://thehill.com/homenews/administration/555247-trump-takes-credit-for-vaccine-rollout-one-of-the-greatest-miracles.

Oshinsky, David M. *Polio: An American Story.* Oxford, UK: Oxford University Press, 2005.

Ostroff, S., P. Drotman, and A. M. Levitt. "Control of Infectious Diseases—A 20th Century Public Health Achievement." In *Silent Victories: The History and Practice of Public Health in Twentieth-Century America*, edited by J. Ward and C. Warren, 3–16. Oxford, UK: Oxford University Press, 2006.

Otterman, Sharon. "Why Polio, Once Eliminated, Is Testing N.Y. Health Officials." *The New York Times*, October 3, 2022. https://www.nytimes.com/2022/10/03/nyregion/polio-new-york-eradication.html.

"Our View: Vaccine Mandates Serve the Greater Good in Ending a Never-Ending Pandemic." *USA Today*, October 1, 2021. Accessed December 22, 2021. https://www.lohud.com/story/opinion/editorials/2021/10/01/gannett-new-york-editorial-support-covid-19-vaccination-mandates/5936672001.

Owens, Eric W., Richard J. Behun, Jill C. Manning, and Rory C. Reid. "The Impact of Internet Pornography on Adolescents: A Review of the Research." *Sexual Addiction & Compulsivity* 19, no. 1–2 (2012): 99–122. https://doi.org/10.1080/10720162.2012.660431.

Ozawa, Sachiko, Tatenda T. Yemeke, Daniel R. Evans, Sarah E. Pallas, Aaron S. Wallace, and Bruce Y. Lee. "Defining Hard-to-Reach Populations for Vaccination." *Vaccine* 37, no. 37 (2019): 5525–34. https://doi.org/10.1016/j.vaccine.2019.06.081.

Pack, Robert. *Jerry Brown, the Philosopher-Prince.* New York: Stein & Day, 1978.

Page, Matthew J. L., Kristin M. Lindahl, and Neena M. Malik. "The Role of Religion and Stress in Sexual Identity and Mental Health Among Lesbian, Gay, and Bisexual Youth." *Journal of Research on Adolescence* 23, no. 4 (2013): 665–77. https://doi.org/10.1111/jora.12025.

Pan, Richard. "Senate Bill 277 Introduced to End California's Vaccine Exemption Loophole." February 19, 2015. https://sd06.senate.ca.gov/news/2015-02-19-senate-bill-277-introduced-end-california%E2%80%99s-vaccine-exemption-loophole.

Pan, Richard. "Governor Signs Legislation SB 742 to Protect Our Right to Get Vaccinated." News release. October 9, 2021. https://sd06.senate.ca.gov/news/2021-10-09-governor-signs-legislation-sb-742-protect-our-right-get-vaccinated.

Pan, Richard. "Richard Pan Pediatrician | California State Senator (D-Sacramento)." LinkedIn. 2022. Accessed November 02, 2022. https://www.linkedin.com/in/drrichardpan.

Pan, Richard. "State Senator Dr. Richard Pan Statement on Holding School Vaccination Requirement Legislation." News release. April 14, 2022. https://sd06.senate.ca.gov/news/2022-04-14-state-senator-dr-richard-pan-statement-holding-school-vaccination-requirement.

Pan, Richard, and Dorit Reiss. "Vaccine Medical Exemptions Are a Delegated Public Health Authority." *Pediatrics* 142, no. 5 (2018): e20182009. https://doi.org/10.1542/peds.2018-2009.

Panzeri, Francesca, Simona Di Paola, and Filippo Domaneschi. "Does the COVID-19 War Metaphor Influence Reasoning?" *PLoS One* 16, no. 4 (2021): e0250651.

Papania, Mark J., Gregory S. Wallace, Paul A. Rota, Joseph P. Icenogle, Amy Parker Fiebelkorn, Gregory L. Armstrong, Susan E. Reef, et al. "Elimination of Endemic Measles, Rubella, and Congenital Rubella Syndrome from the Western Hemisphere: The US Experience." *JAMA Pediatrics* 168, no. 2 (2013): 148–55. https://doi.org/10.1001/jamapediatrics.2013.4342.

Parenteau, Patrick. "The Supreme Court Has Curtailed EPA's Power to Regulate Carbon Pollution—and Sent a Warning to Other Regulators." The Conversation. June 30, 2022. https://theconversation.com/the-supreme-court-has-curtailed-epas-power-to-regulate-carbon-pollution-and-sent-a-warning-to-other-regulators-185281.

Parker, Kim, Rich Morin, and Juliana Menasce Horowitz. "Looking to the Future, Public Sees an America in Decline on Many Fronts." Pew Research Center. 2019. https://www.pewresearch.org/social-trends/2019/03/21/public-sees-an-america-in-decline-on-many-fronts.

Parmet, Wendy E. "Informed Consent and Public Health: Are They Compatible When It Comes to Vaccines?" *Journal of Health Care Law & Policy* 8, no. 1 (2005): 71–110.

Patel, Kavin M., SarahAnn M. McFadden, Salini Mohanty, Caroline M. Joyce, Paul L. Delamater, Nicola P. Klein, Daniel A. Salmon, Saad B. Omer, and Alison M. Buttenheim. "Evaluation of Trends in Homeschooling Rates After Elimination of Nonmedical Exemptions to Childhood Immunizations in California, 2012–2020." *JAMA Network Open* 5, no. 2 (2022): e2146467. https://doi.org/10.1001/jamanetworkopen.2021.46467.

Paul, Katharina T., and Kathrin Loer. "Contemporary Vaccination Policy in the European Union: Tensions and Dilemmas." *Journal of Public Health Policy* 40 (2019): 166–179. https://doi.org/10.1057/s41271-019-00163-8.

Pear, Robert. "Clinton's Health Plan: A.M.A. Rebels over Health Plan in Major Challenge to President." *The New York Times*, September 30, 1993. https://www.nytimes.com/1993/09/30/us/clinton-s-health-plan-ama-rebels-over-health-plan-major-challenge-president.html.

Peltola, Heikki, Annamari Patja, Pauli Leinikki, Martti Valle, Irja Davidkin, and Mikko Paunio. "No Evidence for Measles, Mumps, and Rubella Vaccine-Associated Inflammatory Bowel Disease or Autism in a 14-Year Prospective Study." *Lancet* 351, no. 9112 (1998): 1327–28. https://doi.org/10.1016/s0140-6736(98)24018-9.

Perlstein, Rick. *Nixonland: The Rise of a President and the Fracturing of America.* New York: Scribner, 2009.

Perlstein, Rick. *The Invisible Bridge: The Fall of Nixon and the Rise of Reagan.* New York: Simon & Schuster, 2015.

Perlstein, Rick. *Reaganland: America's Right Turn 1976–1980*. New York: Simon & Schuster, 2020.

Pete, Natalie. "Oregon Sees Increase of Nonmedical Vaccine Exemptions." *Statesman Journal*, June 7, 2018. https://www.statesmanjournal.com/story/news/education/2018/06/08/oregon-increase-nonmedical-vaccine-exemptions/653460002.

Petts, Judith, and Simon Niemeyer. "Health Risk Communication and Amplification: Learning from the MMR Vaccination Controversy." *Health, Risk & Society* 6, no. 1 (2004): 7–23. https://doi.org/10.1080/13698570410001678284.

Pettypiece, S. "White House Warns of COVID Treatment, Vaccine Cuts Without Added Funding." NBC News. March 15, 2022.

Pew Research Center. "Current Decade Rates as Worst in 50 Years." December 21, 2009. https://www.pewresearch.org/politics/2009/12/21/current-decade-rates-as-worst-in-50-years.

Pew Research Center. "White Evangelicals See Trump as Fighting for Their Beliefs, Though Many Have Mixed Feelings About His Personal Conduct." March 12, 2020. https://www.pewforum.org/2020/03/12/white-evangelicals-see-trump-as-fighting-for-their-beliefs-though-many-have-mixed-feelings-about-his-personal-conduct.

Pew Research Center. "Public Trust in Government: 1958–2021." June 6, 2021. https://www.pewresearch.org/politics/2021/05/17/public-trust-in-government-1958-2021.

Pickering, Larry K., Evan J. Anderson, Michael A Daugherty, Walter A Orenstein, and Ram Yogev. "Protecting the Community Through Child Vaccination." *Clinical Infectious Diseases* 67, no. 3 (2018): 464–71. https://doi.org/10.1093/cid/ciy142.

Pickering, Larry K., Carol J. Baker, Gary L. Freed, Stanley A. Gall, Stanley E. Grogg, Gregory A. Poland, Lance E. Rodewald, et al. "Immunization Programs for Infants, Children, Adolescents, and Adults: Clinical Practice Guidelines by the Infectious Diseases Society of America." *Clinical Infectious Diseases* 49, no. 6 (2009): 817–40. https://doi.org/10.1086/605430.

Pierik, Roland. "Mandatory Vaccination: An Unqualified Defence." *Journal of Applied Philosophy* 35, no. 2 (2018): 381–98. https://doi.org/doi:10.1111/japp.12215.

Pierik, Roland, and Marcel Verweij. "Inducing Immunity." Unpublished manuscript.

Pipes, Richard. *A Concise History of the Russian Revolution*. New York: Vintage Books, 1996.

Plevin, Rebecca. "Discredited Vaccination Opponent Andrew Wakefield Crusades Against California SB277." KPCC Southern California. May 1, 2015. https://www.scpr.org/news/2015/05/01/51367/discredited-vaccination-opponent-andrew-wakefield.

Poland, Gregory A., and Ray Spier. "Fear, Misinformation, and Innumerates: How the Wakefield Paper, the Press, and Advocacy Groups Damaged the Public Health." *Vaccine* 28, no. 12 (2010): 2361–62. https://doi.org/10.1016/j.vaccine.2010.02.052.

Poltorak, Mike, Melissa Leach, James Fairhead, and Jackie Cassell. "'MMR Talk' and Vaccination Choice: An Ethnographic Study in Brighton." *Social Science and Medicine* 61, no. 3 (2005): 709–19.

Porter, Dorothy. *Health, Civilization and the State: A History of Public Health from Ancient to Modern Times*. New York: Taylor & Francis, 2005.

Powell, Tori "Virginia Parent Charged After She Threatens to 'Bring Every Single Gun Loaded' over School's Mask Dispute." CBS News. 2022. https://www.cbsnews.com/news/virginia-school-board-gun-threat-face-mask-dispute.

Powers, Madison, and Ruth R. Faden. *Social Justice: the Moral Foundations of Public Health and Health Policy*. Oxford, UK: Oxford University Press, 2006.

Price, Cristofer S., William W. Thompson, Barbara Goodson, Eric S. Weintraub, Lisa A. Croen, Virginia L. Hinrichsen, Michael Marcy, et al. "Prenatal and Infant Exposure to Thimerosal from Vaccines and Immunoglobulins and Risk of Autism." *Pediatrics* 126, no. 4 (2010): 656–64. https://doi.org/10.1542/peds.2010-0309.

Priest, Dana, and Michael Weisskopf. "AMA Split on Clinton Health Plan." *The Washington Post*, December 6, 1993. https://www.washingtonpost.com/archive/politics/1993/12/06/ama-split-on-clinton-health-plan/97c9f379-c9c0-4467-b991-b7bc8f8839c5.

Public Religion Research Institute. "Understanding QAnon's Connection to American Politics, Religion, and Media Consumption." May 27, 2021. accessed February 10, 2022. https://www.prri.org/research/qanon-conspiracy-american-politics-report.

Purtle, Jonathan, Neal D. Goldstein, Eli Edson, and Annamarie Hand. "Who Votes for Public Health? U.S. Senator Characteristics Associated with Voting in Concordance with Public Health Policy Recommendations (1998–2013)." *SSM – Population Health* 3 (2017): 136–40. https://doi.org/10.1016/j.ssmph.2016.12.011.

Putnam, Robert D. *Bowling Alone : The Collapse and Revival of American Community.* New York: Simon & Schuster, 2000.

Qian, Zhenchao. "Breaking the Last Taboo: Interracial Marriage in America." *Contexts* 4, no. 4 (2005): 33–37. https://doi.org/10.1525/ctx.2005.4.4.33.

Qian, Zhenchao, and Daniel T. Lichter. "Changing Patterns of Interracial Marriage in a Multiracial Society." *Journal of Marriage and Family* 73, no. 5 (2011): 1065–84. https://doi.org/10.1111/j.1741-3737.2011.00866.x.

Quinn, P. J., M. O'Callaghan, G. M. Williams, J. M. Najman, M. J. Andersen, and W. Bor. "The Effect of Breastfeeding on Child Development at 5 Years: A Cohort Study." *Journal of Paediatrics and Child Health* 37, no. 5 (2001): 465–69.

Rabin, Roni Caryn. "Eager to Limit Exemptions to Vaccination, States Face Staunch Resistance." *The New York Times*, June 14, 2019. https://www.nytimes.com/2019/06/14/health/vaccine-exemption-health.html.

Race Forward. "Historical Timeline of Public Education in the US." 2006. Accessed July 19, 2022. https://www.raceforward.org/research/reports/historical-timeline-public-education-us.

Rainie, Lee, and Andrew Perrin. "Key Findings About Americans' Declining Trust in Government and Each Other." Pew Research Center. July 22, 2019. https://www.pewresearch.org/fact-tank/2019/07/22/key-findings-about-americans-declining-trust-in-government-and-each-other.

Ramsey, Sonya. "The Troubled History of American Education After the Brown Decision." *The American Historian*. 2022. https://www.oah.org/tah/issues/2017/february/the-troubled-history-of-american-education-after-the-brown-decision.

Rawls, John. *Theory of Justice.* Rev. ed. Cambridge, MA: Harvard University Press, 1999.

Razzaghi, Hilda, Masalovich Svetlana, Anup Srivastav, Carla L. Black, Kimberly H. Nguyen, Marie A. de Perio, A. Scott Laney, and James A. Singleton. "COVID-19 Vaccination and Intent Among Healthcare Personnel, US." *American Journal of Preventive Medicine* 62, no. 5 (2022): 705–15.

Reeves, James E. "The Eminent Domain of Sanitary Science, and the Usefulness of State Boards of Health in Guarding the Public Welfare." *JAMA* 1 (1883): 612–18. https://doi.org/10.1001/jama.1883.02390210008001a.

Reich, Jennifer. "Neoliberal Mothering and Vaccine Refusal: Imagined Gated Communities and the Privilege of Choice." *Gender & Society* 28, no. 5 (2014): 679–704. https://doi.org/10.1177/0891243214532711.

Reich, Jennifer. *Calling the Shots: Why Parents Reject Vaccines.* New York: New York University Press, 2016.

Reich, Jennifer. "Of Natural Bodies and Antibodies: Parents' Vaccine Refusal and the Dichotomies of Natural and Artificial." *Social Science & Medicine* 157 (2016): 103–10. https://doi.org/10.1016/j.socscimed.2016.04.001.

Reilly, Katie. "School Masking Mandates Are Going to Court. Here's Why the Issue Is So Complicated." *TIME*, October 1, 2021. https://time.com/6103134/parents-fight-sch ool-mask-mandates.

Reiss, Dorit. "California Court of Appeal Rejects Challenge to Vaccine Law." Bill of Health. July 30, 2018. Accessed March 4, 2021. https://blog.petrieflom.law.harvard.edu/?s=Cal ifornia+Court+of+Appeal+Rejects+Challenge+to+Vaccine+Law&submit=Search.

Reiss, Dorit. "Thou Shalt Not Take the Name of the Lord Thy God in Vain: Use and Abuse of Religious Exemptions from School Immunization Requirements." *Hastings Law Journal* 65, no. 6 (2014): 1551–602.

Reiss, Dorit. "Litigating Alternative Facts: School Vaccine Mandates in the Courts." *University of Pennsylvania Journal of Constitutional Law* 21, no. 1 (2018): 207.

Reiss, Dorit, and Nili Karako-Eyal. "Informed Consent to Vaccination: Theoretical, Legal, and Empirical Insights." *American Journal of Law & Medicine* 45, no. 4 (2019): 357–419. https://doi.org/10.1177/0098858819892745.

ReliefWeb. "Situation Report: UNICEF Somalia Monthly SitRep 9, September 2014." 2014. Accessed December 17, 2021. https://reliefweb.int/report/somalia/unicef-soma lia-monthly-sitrep-9-september-2014.

"Religious Exemptions Under the Free Exercise Clause: A Model of Competing Authorities." *Yale Law Journal* 90, no. 2 (1980): 350–76. https://doi.org/10.2307/795990.

Riccardi, Nicholas. "Vaccine Skeptics Find Unexpected Allies in Conservative GOP." PBS NewsHour. February 6, 2015. https://www.pbs.org/newshour/health/vaccine-skept ics-find-unexpected-allies-conservative-gop.

Riley, James C. "Smallpox and American Indians Revisited." *Journal of the History of Medicine and Allied Sciences* 65, no. 4 (2010): 445–77. https://doi.org/10.1093/jhmas/jrq005.

Rimrod, Fran. "Perth Alternative School Slammed for Refusing to Support to Vaccinate Students." WA Today. August 9, 2017. https://www.watoday.com.au/national/western-australia/perth-alternative-school-slammed-for-refusing-to-support-to-vaccinate-students-20170809-gxsv1h.html.

Rini, Regina. "Fake News and Partisan Epistemology." *Kennedy Institute of Ethics Journal* 27, no. 2 (2017): E-43–E-64. https://doi.org/10.1353/ken.2017.0025.

Rivers, Ian, Cesar Gonzalez, Nuno Nodin, Elizabeth Peel, and Allan Tyler. "LGBT People and Suicidality in Youth: A Qualitative Study of Perceptions of Risk and Protective Circumstances." *Social Science & Medicine* 212 (2018): 1–8. https://doi.org/10.1016/j.socscimed.2018.06.040.

Robert Wood Johnson Foundation and Harvard T.H. Chan School of Public Health. "The Public's Perspective on the United States Public Health System." May 2021. https://cdn1. sph.harvard.edu/wp-content/uploads/sites/94/2021/05/RWJF-Harvard-Report_ FINAL-051321.pdf.

Roche, Darragh "What Is Jacobson V. Massachusetts? How Supreme Court Ruled on Vaccine Mandate in 1905." *Newsweek*, September 10, 2021. https://www.newsweek. com/what-jacobson-v-massachusetts-how-supreme-court-ruled-vaccine-mandate-1905-1627761.

Rogers, Richard G., Robert A. Hummer, and Bethany G. Everett. "Educational Differentials in US Adult Mortality: An Examination of Mediating Factors." *Social Science Research* 42, no. 2 (2013): 465–81. https://doi.org/10.1016/j.ssresearch.2012.09.003.

"Ronald Reagan Speaks out on Socialized Medicine." YouTube. 1961. https://www.yout ube.com/watch?v=AYrlDlrLDSQ.

Rose, Nikolas. *The Politics of Life Itself: Biomedicine, Power, and Subjectivity in the Twenty-First Century.* Princeton, NJ: Princeton University Press, 2007.

Rosen, George. *A History of Public Health.* Baltimore, MD: Johns Hopkins University Press, 1993.

Rosenhall, Laurel. "Parents Lobby California Lawmakers from Both Sides of Vaccine Debate." *The Sacramento Bee*, February 25, 2015. https://www.sacbee.com/news/polit ics-government/capitol-alert/article11174378.html.

Rosenhall, Laurel, and Emily Hoeven. "At California Capitol, Lawmakers Want Info; Protesters Want to End Coronavirus Stay-at-Home." Cal Matters. April 20, 2020. https://calmatters.org/politics/california-legislature/2020/04/california-lawmakers-protesters-gov-newsom-coronavirus-stay-at-home.

Rosenthal, Albert J. "Conditional Federal Spending and the Constitution." *Stanford Law Review* 39, no. 5 (1987): 1103–64. https://doi.org/10.2307/1228790.

Ross, Alex. "The Hitler Vortex: How American Racism Influenced Nazi Thought." *The New Yorker* 94, no. 11 (2018). https://www.newyorker.com/magazine/2018/04/30/how-american-racism-influenced-hitler.

Ross, Catherine E., and Chia-ling Wu. "The Links Between Education and Health." *American Sociological Review* 60, no. 5 (1995): 719–45. https://doi.org/10.2307/2096319.

Ross, Lainie Friedman. *Children, Families, and Health Care Decision Making.* Oxford, UK: Oxford University Press, 1998.

Ross, Lainie Friedman. "Better Than Best (Interest Standard) in Pediatric Decision Making." *Journal of Clinical Ethics* 30, no. 3 (2019): 183–95.

Ross, Lainie Friedman, and Alissa Hurwitz Swota. "The Best Interest Standard: Same Name but Different Roles in Pediatric Bioethics and Child Rights Frameworks." *Perspectives in Biology and Medicine* 60, no. 2 (2017): 186–97. https://doi.org/10.1353/pbm.2017.0027.

Ross, W. D. *The Right and the Good.* 2nd ed. Edited by P. Stratton-Lake. Oxford, UK: Oxford University Press, 2002.

Rossen, Isabel, Mark J. Hurlstone, and Carmen Lawrence. "Going with the Grain of Cognition: Applying Insights from Psychology to Build Support for Childhood Vaccination." *Frontiers in Psychology* 7 (2016): 1483. https://doi.org/10.3389/fpsyg.2016.01483.

Rota, Jennifer S., Daniel A. Salmon, Lance E. Rodewald, Robert T. Chen, Beth F. Hibbs, and Eugene J. Gangarosa. "Processes for Obtaining Nonmedical Exemptions to State Immunization Laws." *American Journal of Public Health* 91, no. 4 (2001): 645–48. https://doi.org/10.2105/ajph.91.4.645.

Rothman, David J. *Strangers at the Bedside: A History of How Law and Bioethics Transformed Medical Decision Making.* New York: Basic Books, 1992.

Rothstein, Mark A. "Are Traditional Public Health Strategies Consistent with Contemporary American Values?" *Temple Law Review* 77, no. 2 (2004): 175–92.

Rothstein, Mark A., Wendy E. Parmet, and Dorit Reiss. "Employer-Mandated Vaccination for COVID-19." *American Journal of Public Health* 111, no. 6 (2021): 1061–64. https://doi.org/10.2105/ajph.2020.306166.

Rothstein, Richard. *The Color of Law: A Forgotten History of How Our Government Segregated America Paperback.* New York: Liveright, 2018.

Rothstein, William G. *American Physicians in the Nineteenth Century: From Sects to Science.* Baltimore, MD: Johns Hopkins University Press, 1992.

Rough, Elizabeth. *UK Vaccination Policy, Research Briefing.* London: House of Commons Library, 2022.

Rousseau, Jean-Jacques. "On the Social Contract." In *The Basic Political Writings*, edited and translated by Donald A. Cress, 141–227. Indianapolis, IN: Hackett, 1987.

Royal Commission on Vaccination. *Final Report of the Royal Commission Appointed to Inquire into the Subject of Vaccination.* London: Printed for Her Majesty's Stationery Office by Eyre and Spottiswoode, 1896.

Rozbroj, Tomas, Anthony Lyons, and Jayne Lucke. "The Mad Leading the Blind: Perceptions of the Vaccine-Refusal Movement Among Australians Who Support Vaccination." *Vaccine* 37, no. 40 (2019): 5986–93. http://www.sciencedirect.com/science/article/pii/S0264410X19310734.

Rozbroj, Tomas, Anthony Lyons, and Jayne Lucke. "Understanding How the Australian Vaccine-Refusal Movement Perceives Itself." *Health & Social Care in the Community* 30, no. 2 (2022): 695–705. https://doi.org/10.1111/hsc.13182.

Rozsa, Lori. "Desantis Brings Back Florida Lawmakers to Crack Down on Pandemic Mandates." *The Washington Post*, November 14, 2021. https://www.washingtonpost.com/nation/2021/11/14/desantis-brings-back-florida-lawmakers-crack-down-pandemic-mandates.

Saks, Mike. *The Professions, State and the Market: Medicine in Britain, the United States and Russia.* London: Routledge, 2015.

Salahieh, Noura, Mary Beth McDade, Gene Kang, and Megan Telles. "Parents Keep Kids Home from School to Protest California COVID Vaccine Mandate." KTLA. October 18, 2021. Accessed September 9, 2022. https://ktla.com/news/local-news/california-parents-to-keep-kids-home-from-school-to-protest-covid-vaccine-mandate.

Salcedo, Andrea. "Retired Doctor's License Suspended After State Found She Mailed Fake Vaccine Exemption Forms: 'Let Freedom Ring!'" *The Washington Post*, September 29, 2021.

Salganicoff, A., U. R. Ranji, and R. Wyn. *Women and Healthcare: A National Profile.* Kaiser Family Foundation. 2005. https://www.kff.org/wp-content/uploads/2013/01/women-and-health-care-a-national-profile-key-findings-from-the-kaiser-women-s-health-survey.pdf.

Salmon, Daniel, A., and W. Siegel Andrew. "Religious and Philosophical Exemptions from Vaccination Requirements and Lessons Learned from Conscientious Objectors from Conscription." *Public Health Reports (1974–)* 116, no. 4 (2001): 289–95.

Salmon, Daniel A., Stephen P. Teret, C. Raina MacIntyre, David Salisbury, Margaret A. Burgess, and Neal A. Halsey. "Compulsory Vaccination and Conscientious or Philosophical Exemptions: Past, Present, and Future." *Lancet* 367, no. 9508 (2006): 436–42. https://doi.org/10.1016/s0140-6736(06)68144-0.

Salter, Erica K. "Deciding for a Child: A Comprehensive Analysis of the Best Interest Standard." *Theoretical Medicine and Bioethics* 33, no. 3 (2012): 179–98. https://doi.org/10.1007/s11017-012-9219-z.

Sanders, Northe. "Emerging Vaccine Legislation and its Impact on Access." Presentation at the World Vaccine Congress, Washington D.C., April 3–6, 2023.

Sanders, Rachel. "The Color of Fat: Racializing Obesity, Recuperating Whiteness, and Reproducing Injustice." *Politics, Groups & Identities* 7, no. 2 (2019): 287–304. https://doi.org/10.1080/21565503.2017.1354039.

Savage, Charlie. "E.P.A. Ruling Is Milestone in Long Pushback to Regulation of Business." *The New York Times*, June 30, 2022. https://www.nytimes.com/2022/06/30/us/supreme-court-epa-administrative-state.html.

Savage, Glenn C. "What Is Policy Assemblage?." *Territory, Politics, Governance* 8, no. 3 (2019): 319–35. https://doi.org/10.1080/21622671.2018.1559760.

Savulescu, Julian, Ingmar Persson, and Dominic Wilkinson. "Utilitarianism and the Pandemic." *Bioethics* 34, no. 6 (2020): 620–32. https://doi.org/10.1111/bioe.12771.

Schechter, Robert, and Judith Grether. "Continuing Increases in Autism Reported to California's Developmental Services System: Mercury in Retrograde." *Archives of General Psychiatry* 65, no. 1 (2008): 19–24. https://pubmed.ncbi.nlm.nih.gov/18180424.

Schickler, Eric. *Disjointed Pluralism: Institutional Innovation and the Development of the U.S. Congress.* Princeton, NJ: Princeton University Press, 2001.

Schmid, D., H. Holzmann, S. Abele, S. Kasper, S. König, S. Meusburger, H. Hrabcik, et al. "An Ongoing Multi-State Outbreak of Measles Linked to Non-Immune Anthroposophic Communities in Austria, Germany, and Norway, March–April 2008." *Euro Surveillance* 13, no. 16 (2008). https://doi.org/10.2807/ese.13.16.18838-en.

Schneider, Rob. "Tweet, 05 September 2019." Twitter. 2019. https://twitter.com/RobSchneider/status/1169399763151679488.

Schneider, Rob, and Tim Donnelly. "Rob Schneider and Tim Donnelly on Medical Freedom." Vimeo. September 28, 2012. https://vimeo.com/60227438.

Schoeppe, Jennie, Allen Cheadle, Mackenzie Melton, Todd Faubion, Creagh Miller, Juno Matthys, and Clarissa Hsu. "The Immunity Community: A Community Engagement Strategy for Reducing Vaccine Hesitancy." *Health Promotion Practice* 18, no. 5 (2017): 654–61.

Schreier, Herbert A. "On the Failure to Eradicate Measles." *New England Journal of Medicine* 290, no. 14 (1974): 803–804. https://doi.org/10.1056/nejm197404042901412.

Schröder-Bäck, Peter, Peter Duncan, William Sherlaw, Caroline Brall, and Katarzyna Czabanowska. "Teaching Seven Principles for Public Health Ethics: Towards a Curriculum for a Short Course on Ethics in Public Health Programmes." *BMC Medical Ethics* 15, no. 1 (2014): 73. https://doi.org/10.1186/1472-6939-15-73.

Schumaker, Erin. "Vaccination Rates Lag in Communities of Color, but It's Not Only Due to Hesitancy, Experts Say. Focusing on Hesitancy, Rather Than Access, Is Looking at the Problem Backward." ABC News. May 8, 2021. https://abcnews.go.com/Health/vaccination-rates-lag-communities-color-due-hesitancy-experts/story?id=77272753.

Schwartz, Jason L. "The First Rotavirus Vaccine and the Politics of Acceptable Risk." *Milbank Quarterly* 90, no. 2 (2012): 278–310. https://doi.org/10.1111/j.1468-0009.2012.00664.x.

Schwartzman, Micah. "What If Religion Is Not Special?" *University of Chicago Law Review* 79, no. 4 (2012): 1351–427.

Scottish Government. "Report of the MMR Expert Group." April 2002.

Sears, Bob. "California Bill AB2109 Threatens Vaccine Freedom of Choice." Huffington Post. March 22, 2012. Updated May 22, 2012. https://www.huffpost.com/entry/california-vaccination-bill_b_1355370.

Sears, Robert. *The Vaccine Book: Making the Right Decision for Your Child.* New York: Little, Brown, 2007.

Sears, William, and Martha Sears. *The Attachment Parenting Book: A Commonsense Guide to Understanding and Nurturing Your Child.* New York: Little, Brown Spark, 2001.

Sears, William, Martha Sears, Bob Sears, and James Sears. *The Baby Sleep Book: The Complete Guide to a Good Night's Rest for the Whole Family.* New York: Little, Brown Spark, 2005.

Seipel, Tracy, and Jessica Calefati. "Controversial Mandatory Vaccine Bill Easily Clears California Assembly Committee." *The Mercury News,* June 9, 2015.

Seipel, Tracy, and Jessica Calefati. "California Vaccine Bill SB 277 Signed into Law by Jerry Brown." *The Mercury News,* June 30, 2015. https://www.mercurynews.com/2015/06/30/california-vaccine-bill-sb-277-signed-into-law-by-jerry-brown.

Seipel, Tracy, and Jessica Calefati. "California's New Vaccine Law: Freshman Senator Wins Plaudits from Colleagues." *The Mercury News,* July 4, 2015. https://www.mercuryn ews.com/2015/07/04/californias-new-vaccine-law-freshman-senator-wins-plaudits-from-colleagues.

Seither, Ranee, Jessica Laury, Agnes Mugerwa-Kasujja, Cynthia L. Knighton, and Carla L. Black. "Vaccination Coverage with Selected Vaccines and Exemption Rates Among Children in Kindergarten—United States, 2020–21 School Year." *MMWR Morbidity and Mortality Weekly Report* 71, no. 16 (2022): 561–68. https://doi.org/10.15585/mmwr.mm7116a1.

Selden, Steven. "Transforming Better Babies into Fitter Families: Archival Resources and the History of the American Eugenics Movement, 1908–1930." *Proceedings of the American Philosophical Society* 149, no. 2 (2005): 199–225.

Semino, Elena. "'Not Soldiers but Fire-Fighters'—Metaphors and COVID-19." *Health Communication* 36, no. 1 (2021): 50–58. https://doi.org/https://doi.org/10.1080/10410236.2020.1844989.

Sexton, Terri A., Steven M. Sheffrin, and Arthur O'Sullivan. "Proposition 13: Unintended Effects and Feasible Reforms." *National Tax Journal* 52, no. 1 (1999): 99–111. https://doi.org/10.1086/ntj41789379.

Shanor, Amanda. "The New Lochner." *Wisconsin Law Review* 2016, no. 1 (2016): 133–208.

Shortt, S. E. D. "Physicians, Science, and Status: Issues in the Professionalization of Anglo-American Medicine in the Nineteenth Century." *Medical History* 27, no. 1 (1983): 51–68. https://doi.org/10.1017/S0025727300042265.

Sifferlin. A. "9 Ways Advertisers Think We Could Convince Parents to Vaccinate." *TIME,* February 6, 2015. https://time.com/3693767/ad-campaigns-promoting-vaccines.

Silverman, Ross D., and Wendy F. Hensel. "Squaring State Child Vaccine Policy with Individual Rights Under the Individuals with Disabilities Education Act: Questions Raised in California." *Public Health Reports* 132, no. 5 (2017): 593–96. https://www.jstor.org/stable/26374172.

Silverstone, Alicia. *The Kind Mama.* New York: Rodale, 2014.

Simko-Bednarski, Evan. "Maine Bars Residents from Opting out of Immunizations for Religious or Philosophical Reasons." CNN. May 27, 2019. https://edition.cnn.com/2019/05/27/health/maine-immunization-exemption-repealed-trnd/index.html.

Sinnott-Armstrong, Walter. "Consequentialism." In *The Stanford Encyclopedia of Philosophy,* edited by Edward N. Zalta. Stanford, CA: Metaphysics Research Lab, Stanford University, 2021.

Siva, Nayanah. "Thiomersal Vaccines Debate Continues Ahead of UN Meeting." *Lancet* 379, no. 9834 (2012): 2328–28. https://doi.org/10.1016/s0140-6736(12)61002-2.

Skocpol, T., and P. Pierson. "Historical Institutionalism in Contemporary Political Science." In *Political Science: State of the Discipline*, edited by I. Katznelson and H. V. Milner, 693–721. New York: Norton, 2002.

Sledge, Daniel. *Health Divided: Public Health and Individual Medicine in the Making of the Modern American State*. Lawrence, KS: University Press of Kansas, 2017.

Sloop, John M., and Kent A. Ono. *Shifting Borders. Rhetoric, Immigration, and California's Proposition 187*. Philadelphia, PA: Temple University Press, 2002.

"Smallpox Epidemic." *New York Times*, December 4, 1900.

Smeeth, Liam, Claire Cook, Eric Fombonne, Lisa Heavey, Laura C. Rodrigues, Peter G. Smith, and Andrew J. Hall. "MMR Vaccination and Pervasive Developmental Disorders: A Case–Control Study." *Lancet* 364, no. 9438 (2004): 963–69. https://doi.org/10.1016/s0140-6736(04)17020-7.

Smith, Jane S. *Patenting the Sun: Polio and the Salk Vaccine*. New York: Morrow, 1990.

Smith, N., and T. Graham. "Mapping the Anti-Vaccination Movement on Facebook." *Information, Communication & Society* 22, no. 9 (2019): 1310–27. https://doi.org/10.1080/1369118X.2017.1418406.

Smith, Philip J., Susan Y. Chu, and Lawrence E Barker. "Children Who Have Received No Vaccines: Who Are They and Where Do They Live?" *Pediatrics* 114, no. 1 (2004): 187–95. http://search.ebscohost.com/login.aspx?direct=true&db=pbh&AN=13466774&site=ehost-live.

Sobo, Elisa J. "Social Cultivation of Vaccine Refusal and Delay Among Waldorf (Steiner) School Parents." *Medical Anthropology Quarterly* 29, no. 3 (2015): 381–99. https://doi.org/10.1111/maq.12214.

Sobo, Elisa J. "Theorizing (Vaccine) Refusal: Through the Looking Glass." *Cultural Anthropology* 31, no. 3 (2016): 342–50. https://doi.org/10.14506/ca31.3.04.

Sobo, Elisa J., Diana Schow, and Stephanie McClure. "US Black and Latino Communities Often Have Low Vaccination Rates—but Blaming Vaccine Hesitancy Misses the Mark." The Conversation. July 7, 2021. https://theconversation.com/us-black-and-latino-communities-often-have-low-vaccination-rates-but-blaming-vaccine-hesitancy-misses-the-mark-163169.

Stalnaker, Robert. "Common Ground." *Linguistics and Philosophy* 25, no. 5–6 (2002): 701–21. https://doi.org/10.1023/a:1020867916902.

Starr, Paul. "Professionalization and Public Health: Historical Legacies, Continuing Dilemmas." *Journal of Public Health Management and Practice* 15, no. 6 (2009): S26–S30.

Starr, Paul. *The Social Transformation of American Medicine: The Rise of a Sovereign Profession and the Making of a Vast Industry*. 2nd ed. New York: Basic Books, 2017.

State of California. "Bill Text—AB-2109 Communicable Disease: Immunization Exemption." California Legislative Information. 2012. https://leginfo.legislature.ca.gov/faces/billNavClient.xhtml?bill_id=201120120AB2109.

State of Maine. "An Act to Protect Maine Children and Students from Preventable Diseases by Repealing Certain Exemptions from the Laws Governing Immunization Requirements. H.P. 586–L.D. 798." 2019.

State of New York, The New York State Senate. "Assembly Bill A2371A, Exemptions from Vaccinations Due to Religious Beliefs." 2019.

Statistica. "Vaccination Rate Against the Coronavirus (COVID-19) in Germany June 2022." 2022. Accessed July 5, 2022. https://de.statista.com/statistik/daten/studie/1196 966/umfrage/impfquote-gegen-das-coronavirus-in-deutschland/#professional.

Stefanoff, Pawel, Svenn-Erik Mamelund, Mary Robinson, Eva Netterlid, Jose Tuells, Marianne A. Riise Bergsaker, Harald Heijbel, and Joanne Yarwood. "Tracking Parental Attitudes on Vaccination Across European Countries: The Vaccine Safety, Attitudes, Training and Communication Project (VACSATC)." *Vaccine* 28, no. 35 (2010): 5731–37. https://doi.org/http://dx.doi.org/10.1016/j.vaccine.2010.06.009.

Stein, Joel. "Millennials: The Me Me Me Generation." *TIME*, May 20, 2013. https://time.com/247/millennials-the-me-me-me-generation.

Stein, Samantha. "Are Today's Youth Even More Self-Absorbed (and Less Caring) Than Generations Before?." *Psychology Today*, June 5, 2010. https://www.psychologytoday.com/us/blog/what-the-wild-things-are/201006/are-today-s-youth-even-more-self-absorbed-and-less-caring.

Stepan, Nancy. *Eradication: Ridding the World of Diseases Forever?* Ithaca, NY: Cornell University Press, 2011.

Stern, Alexandra Minna. *Eugenic Nation: Faults and Frontiers of Better Breeding in Modern America*. Berkeley: University of California Press, 2005.

Stern, Mark J. "A New Lochner Era." *Slate*, June 29, 2018. https://slate.com/news-and-politics/2018/06/the-lochner-era-is-set-for-a-comeback-at-the-supreme-court.html.

Stetler, Brian. "Tucker Carlson's Fox News Colleagues Call out His Dangerous Anti-Vaccination Rhetoric." CNN. May 6, 2021. https://www.cnn.com/2021/05/06/media/tucker-carlson-anti-vaccination-monologue/index.html.

Stevens, Jack. "Topical Review: Behavioral Economics as a Promising Framework for Promoting Treatment Adherence to Pediatric Regimens." *Journal of Pediatric Psychology* 39, no. 10 (2014): 1097–1103. https://doi.org/10.1093/jpepsy/jsu071.

Stirling, Louis G. "Tendencies of the Times, Medical and Otherwise." *New Orleans Medical and Surgical Journal* 72 (1920): 218–20.

Streeck, Wolfgang, and Kathleen Thelen, eds. *Beyond Continuity: Institutional Change in Advanced Political Economies*. Oxford, UK: Oxford University Press, 2005.

Stroebe, Wolfgang, Michelle R. vanDellen, Georgios Abakoumkin, Edward P. Lemay, William M. Schiavone, Maximilian Agostini, Jocelyn J. Bélanger, et al. "Politicization of COVID-19 Health-Protective Behaviors in the United States: Longitudinal and Cross-National Evidence." *PLoS One* 16, no. 10 (2021): e0256740–e40. https://doi.org/10.1371/journal.pone.0256740.

Sulaski Wyckoff, Alyson. "Eliminate Nonmedical Immunization Exemptions for School Entry, Says AAP." American Academy of Pediatrics. August 29, 2016. https://publications.aap.org/aapnews/news/8969.

Swanson, Emily, and Tom Murphy. "High Trust in Doctors, Nurses in US, AP-NORC Poll Finds." AP News. August 10, 2021. https://apnews.com/article/joe-biden-business-health-coronavirus-pandemic-509835fc9b663bffc83f52d248e9ef4a.

Swindell, J. S., Amy L. J. Halpern, and Scott D. McGuire. "Beneficent Persuasion: Techniques and Ethical Guidelines to Improve Patients' Decisions." *Annals of Family Medicine* 8, no. 3 (2010): 260–64. https://doi.org/10.1370/afm.1118.

Szabo, Liz. "Anti-Vaccine Activists Latch onto Coronavirus to Bolster Their Movement." California Healthline, April 24, 2020. https://californiahealthline.org/news/anti-vaccine-activists-latch-onto-coronavirus-to-bolster-their-movement.

Szasz, Thomas. *The Myth of Mental Illness*. New York: Harper & Row, 1974.

Táíwò, Olúfẹ́mi O. *Elite Capture: How the Powerful Took over Identity Politics (and Everything Else)*. Chicago: Haymarket Books, 2022.

Tajfel, Henri. "Social Identity and Intergroup Behaviour." *Social Science Information* 13, no. 2 (1974): 65–93. https://doi.org/10.1177/053901847401300204.

Talic, Stella, Shivangi Shah, Holly Wild, Danijela Gasevic, Ashika Maharaj, Zanfina Ademi, Xue Li, et al. "Effectiveness of Public Health Measures in Reducing the Incidence of COVID-19, SARS-COV-2 Transmission, and COVID-19 Mortality: Systematic Review and Meta-Analysis." *BMJ* 375 (2021): e068302.

Tang, Terry, Ken Moritsugu, and Lisa Marie Pane. "US Virus Outbreaks Stir Clash over Masks, Personal Freedom." AP NEWS. April 20, 2021. https://apnews.com/article/ health-us-news-ap-top-news-international-news-virus-outbreak-54374ff841dfd8432 3a1fb86d1e93180.

Taub, David. "Pan's Bill Would Further Restrict Vaccine Exemptions for Schoolkids." GV Wire. March 26, 2019. https://gvwire.com/2019/03/26/pans-bill-would-further-restr ict-vaccine-exemptions-for-schoolkids.

Taylor, Brent, Elizabeth Miller, CPaddy Farrington, Maria-Christina Petropoulos, Isabelle Favot-Mayaud, Jun Li, and Pauline A. Waight. "Autism and Measles, Mumps, and Rubella Vaccine: No Epidemiological Evidence for a Causal Association." *Lancet* 353, no. 9169 (1999): 2026–29. https://doi.org/10.1016/s0140-6736(99)01239-8.

Thakar, Emily. "God Bless Texas: Should Texas Eliminate the Vaccine Exemption for Reasons of Conscience?" *Journal of Biosecurity, Biosafety, and Biodefense Law* 9, no. 1 (2018): 20180008. https://doi.org/10.1515/jbbbl-2018-0008.

Thaler, Richard H., and Cass R. Sunstein. *Nudge: Improving Decisions About Health, Wealth, and Happiness*. New Haven, CT: Yale University Press, 2008.

The Canary Party. "Listen to the California Health Committee Hearings on AB2109." 2012. Accessed January 28, 2021. https://canaryparty.org/commentary/listen-to-the- california-health-committee-hearings-on-ab2109.

The Canary Party. "Shake, Rattle, and Roll Tuesday in Sacramento." 2012. Accessed December 28, 2021. https://canaryparty.org/commentary/shake-rattle-and-roll-tues day-in-sacramento.

The Canary Party. "Top 10 Reasons to Oppose California AB2109." 2012. Accessed December 28, 2021. https://canaryparty.org/commentary/top-10-reasons-to-oppose- california-ab2109/.

The National Archives. "Victorian Health Reform. How Did the Victorians View Compulsory Vaccination?" 2022. Accessed August 26, 2022. https://www.natio nalarchives.gov.uk/education/resources/victorian-health-reform/#:~:text=In%201 898%2C%20a%20new%20Vaccination,exempting%20their%20children%20from%20 vaccination.

The Physicians Foundation. "The Physicians Foundation 2020 Physician Survey: Part 3." October 22, 2020. Accessed September 9, 2022. https://physiciansfoundation.org/physic ian-and-patient-surveys/the-physicians-foundation-2020-physician-survey- part-3.

"The Report of the Royal Commission on Vaccination." *British Medical Journal* 2 (1896): 453–58.

"The Spread of Small-Pox by Tramps." *Lancet* 163, no. 4198 (1904): 446–47. https://www. thelancet.com/journals/lancet/article/PIIS0140-6736(01)87511-5/fulltext.

The Takeaway. "Could You Patent the Sun?" Podcast audio. WNYC Studios. December 12, 2016. https://www.wnycstudios.org/podcasts/takeaway/segments/retro-report-patent ing-sun.

"The Vaccination Act, 1898." *British Medical Journal* 2, no. 1974 (1898): 1351–54. https://doi.org/10.1136/bmj.2.1974.1351-a.

Thomas, E. G. "The Old Poor Law and Medicine." *Medical History* 24, no. 1 (1980): 1–19. https://doi.org/10.1017/s0025727300039764.

"Thomas Jefferson Encyclopedia." 2014. Accessed July 29, 2022. https://www.monticello.org/research-education/thomas-jefferson-encyclopedia/government-best-which-governs-least-spurious-quotation.

Thomas, Margaret M. C., Jane Waldfogel, and Ovita F. Williams. "Inequities in Child Protective Services Contact Between Black and White Children." *Child Maltreatment* 28, no. 1 (2023): 42–54. https://doi.org/10.1177/10775595211070248.

Thomas, Susan, Katarzyna Bolsewicz, Julie Leask, Katrina Clark, Sonya Ennis, and David N. Durrheim. "Structural and Social Inequities Contribute to Pockets of Low Childhood Immunisation in New South Wales, Australia." *Vaccine: X* 12 (2022): 100200. https://doi.org/10.1016/j.jvacx.2022.100200.

Thomson, Angus, Karis Robinson, and Gaëlle Vallée-Tourangeau. "The 5As: A Practical Taxonomy for the Determinants of Vaccine Uptake." *Vaccine* 34, no. 8 (2016): 1018–24. https://doi.org/https://doi.org/10.1016/j.vaccine.2015.11.065.

Thornton, Courtney, and Jennifer A. Reich. "Black Mothers and Vaccine Refusal: Gendered Racism, Healthcare, and the State." *Gender & Society* 36, no. 4 (2022): 525–51. https://doi.org/10.1177/08912432221102150.

Tolbert, J., S. Artiga, J. Kates, and R. Rudowitz. "Implications of the Lapse in Federal COVID-19 Funding on Access to COVID-19 Testing, Treatment, and Vaccines." KFF. March 28, 2022. https://www.kff.org/coronavirus-covid-19/issue-brief/implications-of-the-lapse-in-federal-covid-19-funding-on-access-to-covid-19-testing-treatment-and-vaccines.

Topuzoğlu, Ahmet, Pinar Ay, Seyhan Hidiroglu, and Yucel Gurbuz. "The Barriers Against Childhood Immunizations: A Qualitative Research Among Socio-economically Disadvantaged Mothers." *European Journal of Public Health* 17, no. 4 (2006): 348–52. https://doi.org/10.1093/eurpub/ckl250.

Treisman, Rachel. "Some States Are Working to Prevent COVID-19 Vaccine Mandates." NPR. August 2, 2021. https://www.npr.org/2021/08/02/1023809875/states-ban-covid-vaccine-mandates.

Troeskin, Werner. *Water, Race, and Disease.* Cambridge, MA: MIT Press, 2004.

Trotter, Griffin. "COVID-19 and the Authority of Science." *HEC Forum* (2021): 1–28. https://doi.org/10.1007/s10730-021-09455-7.

Twenge, Jean M., and W. Keith Campbell. "Associations Between Screen Time and Lower Psychological Well-Being Among Children and Adolescents: Evidence from a Population-Based Study." *Preventive Medicine Reports* 12 (2018): 271–83. https://doi.org/10.1016/j.pmedr.2018.10.003.

UN Committee on Economic, Social and Cultural Rights. *Economic and Social Council Official Records, Report on the Twenty-Second, Twenty-Third and Twenty-Fourth Sessions, Supplement 2.* New York: United Nations, 2001. https://www.refworld.org/docid/45c30b2eo.html.

UNICEF. "COVID-19 Pandemic Fuels Largest Continued Backslide in Vaccinations in Three Decades." July 15, 2022. Accessed August 5, 2022. https://www.unicef.org/cuba/en/press-releases/covid-19-pandemic-fuels-largest-continued-backslide-vaccinations-three-decades.

United Nations. "OHCHR Dashboard." 2022. Accessed March 9, 2022. https://indicators. ohchr.org.

United States Congress. "H.R. 2264—Omnibus Budget Reconciliation Act of 1993." 1993. https://www.congress.gov/bill/103rd-congress/house-bill/2264.

United States Congress. "H.R. 3590—Patient Protection and Affordable Care Act." 2010. https://www.congress.gov/bill/111th-congress/house-bill/3590.

"United States Constitution, Article 1, Section 8." Government of the United States of America. 1787.

United States Studies Centre. "More Than a Quarter of Democrats Believe in Electoral Replacement Theory." May 18, 2022. Accessed July 5, 2022. https://www.ussc.edu.au/ analysis/the-46th-more-than-a-quarter-of-democrats-believe-in-electoral-replacem ent-theory.

Ureta, S. *Assembling Policy: Transantiago, Human Devices, and the Dream of a World Class Society.* Cambridge, MA: MIT Press, 2015.

U.S. Department of Health & Human Services. "Preventive Care Benefits for Children." 2022. Accessed July 29, 2022. https://www.healthcare.gov/preventive-care-children.

U.S. Department of Labor. "Fact Sheet: General Facts on Women and Job Based Health." n.d. https://www.dol.gov/sites/dolgov/files/EBSA/about-ebsa/our-activities/resource-center/fact-sheets/women-and-job-based-health.pdf.

Vaccinate California. "SB 276 Fact Sheet." 2019. Accessed February 23, 2022. https://vacc inatecalifornia.org/sb-276-fact-sheet.

Vaccinate California. "People." 2022. Accessed September 9, 2022. https://vaccinatecalifor nia.org/about/people.

Vallier, Kevin. *Trust in a Polarized Age.* New York: Oxford University Press, 2020.

van den Hoven, Mariëtte. "Why One Should Do One's Bit: Thinking About Free Riding in the Context of Public Health Ethics." *Public Health Ethics* 5, no. 2 (2012): 154–60. https://doi.org/10.1093/phe/phs023.

Vanderslott, Samantha. "Exploring the Meaning of Pro-Vaccine Activism across Two Countries." *Social Science & Medicine* 222 (February 1, 2019): 59–66.

Vaz, Olivia M., Mallory K. Ellingson, Paul Weiss, Samuel M. Jenness, Azucena Bardají, Robert A. Bednarczyk, and Saad B. Omer. "Mandatory Vaccination in Europe." *Pediatrics* 145, no. 2 (2020): e20190620. https://doi.org/10.1542/peds.2019-0620.

Verweij, Marcel, and Angus Dawson. "Ethical Principles for Collective Immunisation Programmes." *Vaccine* 22, no. 23 (2004): 3122–26. https://doi.org/10.1016/j.vacc ine.2004.01.062.

Voice for Choice. "Staunch Pro-Vaccine Dr. Charity Dean Propels Herself up the CA Public Health Department Ranks to Become CA Vaccine Medical Exemptions Czar." October 1, 2019. https://avoiceforchoiceadvocacy.org/wp-content/uploads/2019/10/ AVFCA-Press-Release-Charity-Dean-self-propelled-Medical-Exemption-Czar.pdf.

Voice of San Diego. "San Diego Unified Vaccine Exemptions." 2019. https://docs.google. com/spreadsheets/d/e/2PACX-1vSVXxFMi1pgUGkhLzCzVXdKVZGBLoRN94I 2gG3FN1-t18zZNUdWN8v0bdTt93_0criAClUQslXDgT78/pubhtml#.

Voo, Teck Chuan, Hannah Clapham, and Clarence C. Tam. "Ethical Implementation of Immunity Passports During the COVID-19 Pandemic." *Journal of Infectious Diseases* 222, no. 5 (2020): 715–18. https://doi.org/10.1093/infdis/jiaa352.

Waddington, Ivan. "The Movement Towards the Professionalization of Medicine." *British Medical Journal* 301, no. 6754 (1990): 688–90. https://doi.org/10.1136/ bmj.301.6754.688.

Wagner, Dennis. "The COVID Culture War: At What Point Should Personal Freedom Yield to the Common Good?" *USA Today*, August 2, 2021. Accessed December 21, 2022. https://www.usatoday.com/story/news/nation/2021/08/02/covid-culture-war-masks-vaccine-pits-liberty-against-common-good/5432614001.

Wakefield, A. J., S. H. Murch, A. Anthony, J. Linnell, D. M. Casson, M. Malik, M. Berelowitz, et al. "Retracted: Ileal-Lymphoid-Nodular Hyperplasia, Non-Specific Colitis, and Pervasive Developmental Disorder in Children." *Lancet* 351, no. 9103 (1998): 637–41. https://doi.org/10.1016/S0140-6736(97)11096-0.

Wallack, Lawrence, and Regina Lawrence. "Talking About Public Health: Developing America's 'Second Language.'" *American Journal of Public Health* 95, no. 4 (2005): 567–70. https://doi.org/10.2105/ajph.2004.043844.

Waller, A. "Protesters Disrupt Motorists from Entering Dodger Stadium Vaccination Site." *The New York Times*, January 30, 2021. https://www.nytimes.com/2021/01/30/world/dodger-stadium-covid-vaccine-protest.html.

Ward, Jeremy K., James Colgrove, and Pierre Verger. "Why France Is Making Eight New Vaccines Mandatory." *Vaccine* 36, no. 14 (2018): 1801–03. https://doi.org/https://doi.org/10.1016/j.vaccine.2018.02.095.

Ward, Paul R., K. Attwell, S. B. Meyer, P. R. Rokkas, and J. Leask. "Understanding the Perceived Logic of Care by Vaccine-Hesitant and Vaccine-Refusing Parents: A Qualitative Study in Australia." *PLoS One* 12, no. 10 (2017): e0185955. https://journals.plos.org/plosone/article?id=10.1371/journal.pone.0185955.

Wardle, Jon, Jane Frawley, Amie Steel, and Elizabeth Sullivan. "Complementary Medicine and Childhood Immunisation: A Critical Review." *Vaccine* 34, no. 38 (2016): 4484–500. https://doi.org/10.1016/j.vaccine.2016.07.026.

Warzel, Charlie. "Protesting for the Freedom to Catch the Coronavirus." *The New York Times*, April 19, 2020.

Washington State Department of Health. "MMR Vaccine Exemption Law Change 2019." 2019. Accessed May 20, 2020. https://www.healthygh.org/in-the-news-1/2019/5/30/mmr-vaccine-exemption-law-change-2019.

Washington State Department of Health. "Washington State School Immunization Slide Set, 2013–2014 School Year." Updated April 2015.

Washington State Legislature. "2011 Senate Bill 5005: Certification of Exemption from Immunization." 2011. http://www.washingtonvotes.org/2011-SB-5005.

Weaver, Rachel. *Be Your Child's Pediatrician*. Reinholds, PA: Share-A-Care Publications, 2016.

Welch, H. Gilbert, Steven Woloshin, and Lisa Schwartz. *Overdiagnosed: Making People Sick in the Pursuit of Health*. Boston: Beacon Press, 2018.

Wertheimer, A. *Coercion*. Princeton, NJ: Princeton University Press, 1987.

West, Ellis M. "The Right to Religion-Based Exemptions in Early America: The Case of Conscientious Objectors to Conscription." *Journal of Law and Religion* 10, no. 2 (1993): 367–401. https://doi.org/10.2307/1051141.

Westwood, Sean J., Shanto Iyengar, Stefaan Walgrave, Rafael Leonisio, Luis Miller, and Oliver Strijbis. "The Tie That Divides: Cross-National Evidence of the Primacy of Partyism." *European Journal of Political Research* 57, no. 2 (2018): 333–54. https://doi.org/10.1111/1475-6765.12228.

WFAE 90.7 Charlotte's NPR News Source. "NC Lawmakers Behind Vaccine Bill Announce It's Dead." April 2, 2015.

White, Jeremy. "From Death Threats to Holocaust Warning, California Vaccine Bill an Extraordinary Fight." *The Sacramento Bee*, June 30, 2015. https://www.sacbee.com/news/politics-government/capitol-alert/article25909216.html.

White, Martin, Jean Adams, and Peter Heywood. "How and Why Do Interventions That Increase Health Overall Widen Inequalities Within Populations?" In *Health, Inequality and Public Health*, edited by S. Barbones, 65–82. Bristol, UK: Policy Press, 2009.

Wickline Wallan, Sarah. "AMA: No More Non-Medical Vaccine Exemptions—Vaccine Exemptions Should Only Be for Medical Reasons, AMA Members Say." Medpage Today. June 7, 2015. https://www.medpagetoday.com/meetingcoverage/ama/52000.

Wiebe, Robert Huddleston. *The Search for Order, 1877–1920*. New York: Hill & Wang, 1966.

Wiggins, Samuel. *Higher Education in the South*. Berkeley, CA: McCutchan, 1966.

Wiley, Kerrie E., Julie Leask, Katie Attwell, Catherine Helps, Chris Degeling, Paul Ward, and Stacy M. Carter. "Parenting and the Vaccine Refusal Process: A New Explanation of the Relationship Between Lifestyle and Vaccination Trajectories." *Social Science & Medicine* 263 (2020): 113259. https://doi.org/https://doi.org/10.1016/j.socscimed.2020.113259.

Wilkinson, Dominic, Alberto Giubilini, and Julian Savulescu. "Is This the End of the Road for Vaccine Mandates in Healthcare?" The Conversation. February 4, 2022. https://theconversation.com/is-this-the-end-of-the-road-for-vaccine-mandates-in-healthcare-176310.

Wilkinson, Timothy Martin. "Making People Be Healthy." *Journal of Primary Health Care* 1, no. 3 (2009): 244–46. https://doi.org/10.1071/hc09244.

Williams, Gareth. *Paralysed with Fear: The Story of Polio*. Basingstoke, UK: Palgrave Macmillan, 2013.

Williams, Pete. "Man Who Plotted to Kidnap Michigan Gov. Gretchen Whitmer Sentenced to over 6 Years." NBC News Digital, August 26, 2021. https://www.nbcnews.com/politics/justice-department/man-who-plotted-kidnap-michigan-gov-gretchen-whitmer-sentenced-over-n1277582.

Willrich, Michael. *Pox: An American History*. New York: Penguin, 2011.

Wolfe, Robert M., and Lisa K. Sharp. "Anti-Vaccinationists Past and Present." *BMJ* 325, no. 7361 (2002): 430–32. https://doi.org/10.1136/bmj.325.7361.430.

Wolff, Jonathan. "Political Obligation, Fairness, and Independence." *Ratio* 8, no. 1 (1995): 87–99. https://doi.org/10.1111/j.1467-9329.1995.tb00071.x.

Wong, Julia Carrie. "Masks off: How US School Boards Became 'Perfect Battlegrounds' for Vicious Culture Wars." *The Guardian*, August 24, 2021. https://www.theguardian.com/us-news/2021/aug/24/mask-mandates-covid-school-boards.

World Health Organization. "Commemorating the 40th Anniversary of Smallpox Eradication." May 8, 2020. Accessed August 31, 2022. https://www.who.int/newsroom/events/detail/2020/05/08/default-calendar/commemorating-the-40th-anniversary-of-smallpox-eradication.

YouGov America. "Do You Think Parents Should Be Required to Have Their Children Vaccinated Against COVID-19 If They Are Eligible for the Vaccine?" 2022. Accessed August 5, 2022. https://today.yougov.com/topics/health/survey-results/daily/2022/02/17/1ad4c/3.

YouGov America. "Do You Think Parents Should Be Required to Have Their Children Vaccinated Against Infectious Diseases?" 2022. Accessed August 5, 2022. https://today.yougov.com/topics/health/survey-results/daily/2022/02/17/1ad4c/1.

YouGov America. "Do You Think Parents Should Be Required to Have Their Children Vaccinated Against Measles, Mumps, and Rubella?" 2022. Accessed August 5, 2022. https://today.yougov.com/topics/health/survey-results/daily/2022/02/17/1ad4c/2.

Zadrozny, Brandy, and Ben Collins. "As Vaccine Mandates Spread, Protests Follow— Some Spurred by Nurses." NBC News. July 31, 2021. https://www.nbcnews.com/tech/social-media/vaccine-mandates-spread-protests-follow-spurred-nurses-rcna1654.

Zaki, Jamil. "What, Me Care? Young Are Less Empathetic." Scientific American. January 1, 2011. https://www.scientificamerican.com/article/what-me-care.

Zipprich, Jennifer, Kathleen Winter, Jill Hacker, Dongxiang Xia, James Watt, and Kathleen Harriman. "Measles Outbreak—California, December 2014–February 2015." *MMWR Morbidity and Mortality Weekly Report* 64, no. 6 (2015): 153–54.

Zola, Irving Kenneth. "Medicine as an Institution of Social Control." *Ekistics* 41, no. 245 (1976): 210–14.

Zuckerman, Jake. "Pandemic Brings Protests, and Guns, to Officials' Personal Homes." *Ohio Capital Journal*, January 27, 2021. Accessed April 15, 2023. https://ohiocapitaljournal.com/2021/01/27/pandemic-brings-protests-and-guns-to-officials-personal-homes/

Index